Cheatgrass

CHEATGRASS

Fire and Forage on the Range

James A. Young

Charlie D. Clements

UNIVERSITY OF NEVADA PRESS   RENO & LAS VEGAS

University of Nevada Press, Reno, Nevada 89557 USA
Manufactured in the United States of America

Library of Congress Cataloging-in-Publication Data
Young, James A. (James Albert), 1937–
Cheatgrass : fire and forage on the range / James A. Young and
Charlie D. Clements.
p. cm.
Includes bibliographical references and index.
ISBN 978-0-87417-765-7 (hardcover : alk. paper)
1. Cheatgrass brome—Great Plains. 2. Invasive plants—Great Plains.
3. Forage plants—Great Plains. I. Clements, Charlie D., 1965– II. Title.
QK495.G74Y68   2009
633.2'020978—dc22   2008049064
ISBN 978-1-64779-070-7 (paper : alk. paper)

The paper used in this book is a recycled stock made from
30 percent post-consumer waste materials, certified by FSC,
and meets the requirements of American National Standard for
Information Sciences—Permanence of Paper for Printed Library
Materials, ANSI/NISO Z39.48-1992 (R2002). Binding materials
were selected for strength and durability.

This book has been reproduced as a digital reprint.

To the memory of

A. C. HULL JR. *and* GERALD JACKSON KLOMP,

pioneer cheatgrass scientists

Contents

Illustrations

TABLES

Preface

Cheatgrass and the twentieth-century history of Great Basin rangelands are synonymous. Cheatgrass was accidentally introduced early in the twentieth century and became a wandering waif along unpaved roads snaking through an ocean of silver gray sagebrush. A quarter century of excessive, improperly timed, and continuous grazing of domestic livestock in the late nineteenth century had paved the way for cheatgrass to spread into native plant communities. This spread was made possible by the inherent genetic composition of cheatgrass; most important is its exceedingly plastic growth requirements and a breeding system that allows it to adapt along with a continuously changing environment. Cheatgrass transformed Great Basin environments and then evolved to exploit the changes. Among its many unwelcome contributions to rangeland ecology is the term "firestorm," used to describe the catastrophic wildfires cheatgrass causes in sagebrush plant communities. The abundance of finetextured, early-maturing herbage it produces in certain years greatly increases the chance of ignition and the rate of spread of wildfires. In 1999, more than a million acres of rangelands burned during a ten-day period. The ocean of silver gray sagebrush has begun a transition through black ash to golden brown annual grass as the aspect-dominant characteristic of Great Basin landscapes.

As this process of landscape transformation occurred, cheatgrass became one of the most important forage plants for the Great Basin livestock industry and a dietary staple for many wildlife species. This is especially apparent in Nevada, where rangelands constitute most of the landscape. Cheatgrass and the related species red brome are the driving force on Nevada rangelands.

Its abundance, distribution, and fuel characteristics make cheatgrass a highly significant biological issue for all residents of the Great Basin, not just range managers and livestock producers. As the human population of the region continues to expand, the specter of wildfires hovers just at the edge of suburban sprawl. Cheatgrass-fueled wildfires can kill humans. Water is the limiting factor in the growth of most Great Basin urban areas. Cheatgrass changes the characteristics of watersheds and can induce accelerated erosion even in forested watersheds far removed from the characteristic sagebrush of the Great Basin. Windblown ash from wildfires and dust from postfire erosion create deadly driving hazards in the interstate highways transecting the Great Basin. Big game and upland bird hunting are important to many residents of the Great Basin, and cheatgrass is affecting wildlife populations in numerous ways. Cheatgrass is a highly significant weed in cereal grain production in much of western North America.

Above all else, however, cheatgrass represents a stage in transition toward an environment dominated by exotic weeds growing on eroded landscapes. This is not hyperbole or scare tactics. Science-based projections indicate that what has happened to the temperate desert environments of Asia will happen on the rangelands of the Great Basin. Federal environmental laws brought democracy—in the form of public environmental-impact reviews—to the management of natural resources in the United States. The public now plays an important role in government rangelands policy, and it is our hope that the decision-making public will recognize its duty to use reason and knowledge rather than emotions in making well-informed decisions. Knowledge of the turbulent past, present, and future of cheatgrass in the Great Basin is key to making informed decisions. That is the purpose of this book.

Acknowledgments

The authors greatly appreciate many hours of assistance from Daniel Harmon and the contributions of exceptionally knowledgeable reviewers Jim Jeffress, Ken Gray, Dr. Robert Blank, Dr. Ken Sanders, and John McLain. Together they have lived much of the story we tell here. The detailed annotation of the manuscript by Dr. Lynn F. James, based on his decades of experience as a world-class scientist and Great Basin rancher, are especially valued. We appreciate the support of Margaret Dalrymple, acquisitions editor at the University of Nevada Press, for having confidence in this book. We especially appreciate the support and understanding of our wives during the seven years we spent on this manuscript.

Cheatgrass

CHAPTER 1

The Many Faces of Cheatgrass

Newspapers and television news programs have fostered a widespread association between cheatgrass and wildfires in the Great Basin. Cheatgrass has also come to personify the concept of alien invasion. Only a very small percentage of the population living in the large metropolitan centers of the Great Basin can identify cheatgrass. Residents of smaller towns, however, are far more familiar with cheatgrass-dominated landscapes. They know the annual cycle from springtime green to flushes of pinkish red near maturity to the tawny tan of mature herbage. Anyone who has walked in cheatgrass when the seeds are mature knows the pain of trying to pick the sharp-pointed, barbed seeds from socks or from a dog's coat. Those whose livelihood is directly dependent on Great Basin rangelands for livestock production, along with hunters who enjoy upland bird and big game hunting, may have a different perspective of the nature and environmental consequences of cheatgrass invasion. The beauty or evil of cheatgrass is in the eyes of the beholder. The perception of this exotic species has changed over time.

The worst sound a sagebrush country rancher could hear was the bawling of starving cows. It was a familiar sound in the nineteenth-century Great Basin, where 160-acre homesteads could not produce enough hay to bring cows through the winter. It took awhile, but early in the twentieth century

Congress finally gave in to pressure from the western states and increased the size of homesteads from 160 to 640 acres.[1] Easterners did not understand the problem. They found it inconceivable that a person could farm 160 acres, even if he was lucky enough to raise three or four strapping boys to work with him. In the arid portions of Spanish colonial domains, the basic land grant for extensive livestock production had been 40,000 acres, so even the 640-acre stock-raising homestead was in many cases far from adequate for range livestock production. If a cowboy with the urge to work for himself could find a creek in the Great Basin with sufficient watershed to provide irrigation water, he or she had the makings of a stock-raising homestead.

The hard winter of 1889–1890 had driven home the absolute necessity of setting aside one ton of hay for each wintering brood cow.[2] There was a significant difference between realizing that conserved forage as hay was necessary for wintering livestock, though, and having the required infrastructure to produce that hay in the sagebrush country. Hay production required irrigation, and only mountain ranges with appreciable areas of watershed near or above 10,000 feet in elevation could produce sufficient runoff to support the streams that provided sufficient irrigation water. These constraints limited irrigation to about 5 percent of the total landscape.[3]

Hay meadows produced only three-quarters of a ton to a ton of marginal-quality hay per acre, which meant that a rancher needed an acre of hay ground for every brood cow he planned to overwinter, plus additional hay for replacement heifers and 2-year-old steers. By the time the stock-raising homestead law was enacted, the big ranchers had bought most of the irrigable land in Nevada.[4] Stock-raising homesteaders had to look to the little creeks on lower mountain ranges to locate their prospective ranches. The flow of potential irrigation water in these small creeks was limited and undependable. The ranchers had to laboriously clear small, semicircular fields on the opposite sides of rocky canyons along the creeks that wound down from the uplands. The cleared land was often more creek-washed rock than soil. During the winter, the rancher and his family would follow a mule pulling a sled across the fields in endless efforts to remove enough rock to make mowing hay less hazardous.

When the rancher finished stacking his hay the previous summer he knew he was in trouble.[5] The creek had dried up early in the season, and the hay he had been able to cut made a pretty small stack. Putting up hay was backbreaking toil; the entire family worked from daylight until dark. The oldest boy drove the dump rake and the youngest the derrick horse. The rancher's wife and daughter cooked for the haying crew that he could not afford but had to have. Between the meals, his wife and daughter canned vegetables from the garden they had painstakingly watered from the well. Just as the cows needed hay for the winter, the family had to preserve enough food to get them through. Last summer's haying campaign had been mercifully short, but the size of the resulting haystack had cast a pall over the family's future.

The rancher put off feeding hay until the cows had cleaned the dry field of every possible straw of forage. After the first snow fell and failed to completely melt, he grudgingly began digging into the stack of hay. The cows would stand at the fence bawling with hunger as he drove the team and hay wagon out of the stack yard each morning. Cutting the frozen, snow-encrusted long hay with a hay knife was tough, hard work in the early morning cold. There was never as much hay on the wagon as he knew there should be, but he had to gauge what was left in the stack while trying to figure how long the winter was going to last.

As the summer accumulation of flesh wasted away from the frames of the old cows, they licked up every piece of hay off the feed ground and drifted to the fence to bawl. The sound grated on everyone's nerves. The rancher's wife became more and more silent. He could see the fear in her eyes. She looked out the single window above the tin sink at her flowerbeds watered from the dishwater drainage. Her look seemed to say, "We are going to lose everything we have worked so hard to build together as a family."

Finally, he opened the gate to the field so the starving cows could work the woody vegetation along the creek. They savaged the wild rose and willow thickets and even began to hedge the big sagebrush plants that grew on the gravel bars along the stream. Twice the rancher had to run off a band of wild horses that starvation and deep snow had brought down off the ridges, their fear of humans overcome by the chance for a mouthful of hay. One morning a single thin mule deer doe joined the cows at the fence. The ranch dogs poised ready to jump from the wagon as they approached the stack yard each morning, knowing that a mob of jackrabbits had spent the night eating tunnels into the bottom of the stack. Normally a mature jackrabbit was very difficult prey for the average ranch dog, but as the winter wore on, the dogs made multiple kills every morning.

As the rancher closed the door of his small stone barn and started for the house with a bucket of warm milk, he paused to look down the canyon toward the valley below. All day long, the March snow showers had draped their cold touch across the ridges. The endless north wind brought a whiff of pinyon smoke from the kitchen stove chimney. Normally he loved the pinyon or juniper smoke, but the bawling cows at the hay field fence made enjoyment impossible. "Is this winter never going to end?" he wondered as he walked toward the square of lamplight in the kitchen window.

Something kept tickling at the back of the rancher's mind, bothering him as he harnessed the team to do the morning feeding. The days were getting longer because he was able to harness the team and hitch the hay wagon without using a lantern. He had fed the last of the stack the morning before. All he could do this morning was scrape mostly rotten flakes of hay from the mud where the stack had stood. He would get more rabbit pellets than hay, but it was all he had to offer.

The team seemed especially frisky this morning as they rounded the house and started down to the field. Suddenly it dawned on him: the cows were not bawling at the fence. "My God," he thought, his heart sinking down into his boots, "they all died last night." Reason told him that was impossible; but where had they gone? As the morning sun broke over the ridge top and lighted the brush-covered slope across the creek, he saw the cows widely spread out and working their way up the slope. Their heads were down and they were rustling for feed around the sagebrush plants. As he left the team and ran up the slope, he caught a hint of green under the shrubs. He dropped to his knees and looked closely under the sagebrush canopies. There was not much to see except fine, hairy green grass leaves emerging from a rosette of leaves on the soil surface. It was not much forage, but it was enough to keep the cows moving from shrub to shrub as they climbed the hill.

Cows typically left the feed grounds once there was enough spring growth on the grasses to make grazing worthwhile, but the rancher had been overutilizing the native bunchgrasses close to the ranch headquarters, and each year they had been getting scarcer. He hunted around under the shrubs, found a dried seed head from the previous season, and held it up to the light. "I'll be damned," he thought. "It's that new grass that came in along the railroad." Most of the cowboys called it bronco grass because the sharp seeds injured the eyes and mouths of horses, but other people called it cheatgrass. The foreman of the big ranch nearby had talked about trying to get seed of this cheatgrass, but before he could find any it had appeared spontaneously on the bed grounds of a sheep outfit that moved through the valley in the spring and fall. The rancher recalled that a tramp band of sheep had moved down the creek a couple of summers ago before he got his rifle and ran them off. Apparently, they had left behind a present of cheatgrass. "Cheat be damned," he thought. "It sure as hell cheated death today."

We have heard variations of this story from numerous sources, including James A. Young's grandfather. Cheatgrass spread in the biological near-vacuum left by the destruction of the native perennial grasses. At first the ranchers welcomed it. Some ranges that had been excessively grazed had been nearly bare for more than a decade when the cheatgrass arrived. It was better forage than nothing.[6]

As the farmer pulled his buggy into the wheat field, the big Harris combine stopped. The twenty mules that provided the power to move it needed water, and the maze of chains and sprockets that transformed the traction power generated by the mules into a sophisticated machine that cut, threshed, separated, and cleaned wheat from straw needed oil. The part of the crew not engaged in watering or oiling were resting in the shade of the combine, tipping back water jugs covered with wet burlap. The owner did not begrudge the crew their break. The ones resting did the brutal work of sacking the wheat, sewing sacks, and throwing the heavy bags onto wagons. The farmer always felt a surge of pride when he saw the combine. It was a marvel of engineering with a hardwood frame and body supporting the maze of pulleys and chains. Most of his neighbors in the Big Meadows on the Humboldt were still cutting grain in the field with a binder and threshing it with a stationary steam engine and a separator. The Harris had cost a lot of money, but it eliminated the cost of hauling the bundles of straw, chaff, and grain to the separator. Handling the bundles always resulted in unnecessary loss of grain. The farmer hoped to pay off the balance he owed on the combine with this season's wheat crop.

He had sent all the way to California for his seed grain. When it had arrived at the rail station in Lovelock, Nevada, he could not wait to open a sack and check the grain. It was Pacific bluestem, the wonder wheat that had helped California to become the nation's leading wheat producer. Farmers grew wheat on dry land in the central valleys of California during the rainy winter months, but here in the deserts of Nevada, grains had to be grown under irrigation during the spring and summer. He had hauled manure from his feed yard all winter and supervised the preparation of the seedbed with great care. He was out for bragging rights with a bumper crop. The seed grain had looked to be of excellent quality. He had noticed a little bur clover seed mixed in with the wheat, but that was no problem, and also a few seeds of a grass he did not recognize. He would probably not even have noticed had the sharp point on the grass seed not pricked his finger as he dipped into the sack to get a good sample of the wheat.

With carefully timed irrigation, the bluestem had grown as high as the barbed-wire fence surrounding the field. As the season progressed, the farmer worried about the tall wheat lodging or a stray thunderstorm dropping damaging hail. His worries were all in the past now; the wheat was almost in the sack.

When the threshing crew's foreman walked across the wheat stubble wearing a worried frown, the farmer glanced at the combine to see if an equipment breakdown had occurred. "Boss, you are not going to be happy with the yield of wheat," the foreman said apologetically. Before the stunned farmer could reply, the foreman continued, "Come look at the cleaner." The Harris combine was equipped with a fanning mill re-cleaner that produced exceptionally clean grain. When they reached the back of the combine, the foreman snapped open an access door to a paddle elevator that raised grain for the re-cleaning cycle. He caught a handful of wheat grains as they spilled out and held them up for inspection. The wheat grains were small, and many were shriveled. A closer look revealed the presence of the slender, pointed seeds of the grass the farmer had noticed in the seed grain shipment.

The foreman pointed to the standing wheat and said, "It's thick under the wheat. I saw it before when I worked up in the Palouse in Washington. Farmers up there called it 'cheat' because it robbed you of a crop." The stunned farmer looked from the cheatgrass seeds in his hand to the abundant plants below the canopy of uncut wheat and wondered what he had done to his best field. How would he ever get rid of this new pest?

Human cultures are built on simple things. Primitive farmers cleared away competing vegetation and scratched a slot in the soil surface to accept seeds saved from the previous year's harvest and protected during the long winter from the damp and from hungry mice. It required faith and patience to watch these plants grow while weeds were pulled and water was carried in earthen jars to nourish the crop's growth. The harvest with a manyfold increase in seeds brought joy and the promise of surviving another winter. Humans domesticated certain animals, and in return for a promise to protect them from predators and provide access to adequate forage received living storehouses of nutrition for use in times of want. Before either of these things occurred, humans ventured into the wilderness in search of prey, armed with weapons to help them overcome their prey's advantages of strength, agility, or cunning. Hunting is one of the most basic instincts of humans.

When Peter Skene Ogden brought a brigade of Hudson's Bay Company trappers to the Humboldt River in the 1820s, his hunters found no bison and very few mule deer.[7] In fact, the band of professional hunters had trouble finding enough big game animals to feed the brigade. The only large grazer present in the vast sagebrush-covered valleys was the pronghorn, which the men complained was as tough as goat and tasted like sagebrush.

Another native animal that characterized the sagebrush landscapes and often tasted like sagebrush was the sage grouse. The first naturalist to be startled when a sage grouse exploded from beneath his feet was certain that he had discovered the western version of the turkey. The leaves of sagebrush are essential items in the diet of the sage grouse. Those privileged to stumble on a flock of sage grouse and watch them rise in a magnificent succession of slate gray waves from the silvery sagebrush canopy never forget the sight. In search of mates, sage grouse males strut their stuff on traditional leks each spring. Many a homesteader, cowboy, or prospector baked sage grouse in a cast iron Dutch oven and was grateful for the meal.

During the 1930s, hunting enthusiasts introduced the chukar partridge to the Great Basin.[8] This native of Pakistan and Afghanistan was preadapted to survive in degraded sagebrush-grass rangelands dominated by cheatgrass. The chukar became the number one upland game bird in the Intermountain Area. Smaller than sage grouse, chukars are challenging birds to hunt. They rise from the sagebrush in a whir of wings beating so fast that the birds are a blur as they dive off the high ridges and glide into the distance. Once they land, the coveys scatter and the birds run with remarkable speed. Insanity is not a requirement to hunt chukar, but having the lungs and legs of a Sherpa is a distinct advantage. The first time you hunt chukars it is for the enjoyment of the sport; subsequent hunts are for revenge.

Hunting is a cherished tradition in the West. Each generation learns the skills and responsibilities of game hunting from the one before it. Hunting chukars is often the first hunting experience for young people in the Great Basin.

The boy jumped when his father touched his shoulder and told him to get out of bed; it was opening day of chukar season. He had not been able to sleep the night before because he was so excited. This would be his first "real" hunt. The pickup's seats were cold in the predawn darkness even though it was only early October. Their bird dog eagerly jumped into the camper shell. The pointer would be too sore-footed to walk tomorrow, but today he was an eager participant in the quest for the elusive chukar. The boy had been going with his father on chukar hunts since he was big enough to walk, packing his toy shotgun and trying to remember to "shoot" when the birds broke cover. All the while he had been acquiring the knowledge essential for a hunter. All guns are considered loaded. You always know where your hunting partners are, and you never follow a bird toward another hunter. You eat what you kill. You do not repeat at the dinner table the stories Dad's hunting buddies told.

The pickup's headlights caught the nocturnal animal life of the high deserts. Kangaroo rats made a bipedal race to escape the onrushing vehicle. The stars were fading and the eastern horizon was beginning to show the first hint of dawn. The heater had finally warmed the truck cab. The boy's anticipation had reached fever pitch. It was barely light when they let the bird dog out of the camper at the canyon bottom. Dad allowed one mad dash in a circle around the truck before he sternly brought the dog to heal. The dry big sagebrush released a pungent odor as they brushed against it and started up the steep slope. It was cold, but the steepness of the climb soon generated some warmth. Until the first fall rains increased the number of potential watering places, the birds were restricted to the few available springs, and Dad knew where the Department of Wildlife had installed a guzzler for chukars. He explained to his son that a guzzler consisted of an inclined roof that captured rainfall and channeled the runoff into a large tank buried under the shadow of the roof. An opening at one end of the tank and a ramp provided access to the water.

Just as the boy thought his lungs were going to burst and he could not lift his legs for another step they heard the chukar's call. It started as a low-intensity *chuck, chuck, chuck,* given slowly with definite breaks between each call. As the intensity of the call rose, it changed to *per* CHUCK, *per* CHUCK, and this in turn gave way to CHU*kar,* CHU*kar,* CHU*kar.* The throaty *chuck* reverberated among the canyons and cliffs. Dad explained that the birds were greeting the sun as it touched the rim rocks.

The first time the dog came to point, the father gave his son a nod of reassurance. When the birds burst from the sagebrush, Dad hesitated long enough to give the boy a chance for the first shot. The .410 slammed against his shoulder before he had a chance to hold it tight, and he knew he had shot too soon. His bird continued in flight. Dad fired two quick shots at maximum range, and two birds dropped, much to the delight of the pointer-turned-retriever. "No big deal, Son, just be ready next time," reassured his father.

After a long morning of tough hiking interspersed with brief moments of excitement, they returned to the truck for a late lunch. First, though, they cleaned the birds and put them in the cooler. As always, they examined the birds' crops to see what they had been foraging on. The crops were full of cheatgrass seeds. Dad spread the crop contents on the pickup's tailgate with the tip of his knife. "Look for insect parts in the crops," he told his son. "Chukars mostly eat cheatgrass seeds, especially in the summer and early fall, but they have to have insects for quality protein. When we get fall rains and the cheatgrass germinates, the birds eat the first green leaves of cheatgrass to balance their diets. If you are going to harvest birds by hunting, you have to understand their habitat and food requirements to help ensure sustainable populations."

"Complete suppression of wildfires in western forests and ranges!" was the rallying cry of natural resource conservationists during the early years of the twentieth century. It was universally accepted within the fledgling conservation movement that the burning of forests and rangelands was evil. At the local level, many people questioned the wisdom of this policy, and some even took their protests to the halls of Congress, but the majority of the general public accepted complete suppression of wildfires as a highly desirable national goal.

In the sagebrush-bunchgrass rangelands of the Great Basin this policy was largely a moot point. Grazing was so intense over most of these rangelands by the time the conservation movement was born that there was insufficient herbaceous fuel to carry a fire from shrub to shrub. The lack of continuous vegetation greatly reduced both the chance of ignition and the rate of spread of wildfires. Only under extreme conditions of high temperatures and low humidity could a strong wind bend the flame heads from the burning sagebrush plants far enough to touch the adjoining shrubs and spread the wildfire.

The U.S. Department of Agriculture (USDA) Forest Service was the first federal wildfire suppression agency. Its core of professional foresters led the attack on wildfires with crews that were literally swept off the streets and out of the bars. The foresters had the authority to draft loggers, cowboys, and even the town drunk for duty on fire lines. Cowboys and ranch hands generally used the standard excuse that they had to go home to milk cows in the morning. We do not know if any resourceful forester ever brought a milk cow along on these sweeps to test their milking ability.

Every rural community across the West included keen observers of nature who realized that a policy of total fire exclusion was not ecologically sustainable. They did not use that term, of course. Most voiced their complaints in more earthy comments such as, "The blasted brush and junipers are going to take over the country without fires." One southwestern rancher complained that the Forest Service tacked notes on every tree telling people to prevent forest fires, but there were no notes telling them how to practice good range management. Unfortunately, very few of these people published their observations to provide a record for future generations. Most of their stories are maintained in the local folk traditions. One such folk tradition concerns Winnemucca Mountain in Humboldt County, Nevada.

Winnemucca Mountain towers over Winnemucca, Nevada. The vast bulk of the Sonoma Range southeast of the town presents a massif of much greater elevation, but its very closeness and familiarity have made Winnemucca Mountain almost a part of the town. Like most Humboldt River towns, Winnemucca used to have a fair and rodeo at the end of the haying season. Haying was much like a military campaign. Timing and organization were crucial. You could not start cutting hay until the annual flood on the Humboldt subsided and the meadows had a chance to dry. Then the hay had to be mowed, raked, and in the stack before the grasses got too tough to cut with horse-drawn mowers. It was an annual war, and when the gate was closed on the last stack yard, the community gathered for a rip-roaring drunk. In the heart of the range livestock industry in the western Great Basin, the festivities included a rodeo where the local cowboys could show off.

Winnemucca's rodeo had a special attraction as the final act. Local Indians staged a suicidal race from the top of Winnemucca Mountain down across the Humboldt River to the fairground. After the winner had been declared and the casualties cleared from the race area, the draw the racers had tumbled down was set on fire as the sun set in the west. The fire would run up the slope, burning the dry perennial grasses. As the draft increased and the fire flared upslope, the fair-goers down below were treated to spectacular fireworks. The fire would burn until it reached the rocky summit, where the fuel supply would be exhausted and the fire would blink out. The last time the draw was burned, the fire was especially bright and spread more rapidly than usual up the mountain; but it did not stop at the top. It burned all night until it reached the Bloody Run Range north of Sand Pass. The fire was the major subject of discussion around town for several days. The general consensus was that the new grass called bronco grass or cheatgrass was the cause of the unprecedented spread of the fire.

Cheatgrass continued to spread in fits and starts. The ranges were very heavily grazed. Ranch families camped out on the summer ranges to ensure that their livestock got every blade of grass available and to make sure neighbors did not get more than their fair share.[9] Or even worse, that some of "their" forage would go to a tramp sheep operation. In dry years, cheatgrass would seem to disappear. In exceptionally wet years, though, not even the excessive stocking rates present on the ranges could suppress the exotic annual, and cheatgrass-fueled wildfires began to be common.

In 1936, in response to concern expressed by the cattle industry, the Agricultural Experiment Station of the University of Nevada launched a research project to do something about the hobo grass of many aliases: cheat, bronco grass, June grass, cheatgrass, and cheatgrass brome. Surprisingly, the project description included two scientific names: *Bromus tectorum* and *Bromus rubens.*[10]

The passage of the Taylor Grazing Act in 1934 and President Franklin Roosevelt's order closing vacant public lands to homesteading ended the era of "free" range in the West. During the 1920s, the Great Basin had been virtually unfenced rangeland, and most ranges were grazed in common by more than one rancher; no one wanted to own them in the traditional American concept of property ownership. The period from 1900 to 1934 was a sordid chapter in the history of natural resource management in the United States. At the same time the infant Forest Service was championing conservation of natural resources on national forest lands, the government refused to end the frontier concept that land in the West was free for the taking. No regulations governed the use of ranges that were not part of a national forest, and conflicts among multiple users of community allotments resulted in improper grazing of some ranges. Congress held hearings on the problems of the western range virtually every session but could not reach a consensus on how to manage the degrading natural resources.[11] Various states tried their own plans to divide the public lands within their boundaries, but such attempts were all patently unconstitutional from the start. Nevada considered an ambitious plan to divide the vacant federal lands within the state based on the ownership of stock water rights.[12] Its constitutionality was never tested, however, because when the claims for rangelands based on water rights were tallied, two-thirds of the claims overlapped. Obviously, a great many ranchers considered the same piece of public domain their personal range.

The economic catastrophe and associated social upheaval of the Great Depression helped to break the political logjam. With the passage of the Taylor Grazing Act in 1934, the U.S. Department of Interior (USDI) Grazing Service was formed. The stated purpose of the Taylor Grazing Act, which instituted a type of benevolent socialism, was to stabilize the range livestock industry. Local grazing boards were appointed to advise federal range conservationists how to manage the often highly degraded ranges. Diehard advocates of the open-range policy found the new policy a bitter pill to swallow. At hearings in Reno, Nevada, to establish the western Nevada grazing district, participants were asked to describe the historic limits of the ranges they grazed. One optimist placed the northern boundary of his range in Canada and the southern one in central Mexico.

By the late 1930s, the Grazing Service had a source of manpower for work on rangeland conservation projects in the form of the Civilian Conservation Corps (CCC), one of the social agencies given birth by the New Deal. The CCC was a sort of paramilitary organization for young men who had spent their early adolescence in the depths of the Great Depression and had virtually no prospects for employment. The CCC was designed to give them physical training, discipline, and on-the-job skills that might lead to future employment. Many of these boys were put to work on projects that improved the environment and stopped erosion.[13] The program was administered by the Works Projects Administration (WPA) in cooperation with the Departments of the Interior, Agriculture, and Army. The army clothed, fed, and housed the boys, who lived in CCC camps constructed all over the country. Extensive road, trail, and water development projects were undertaken in the Great Basin.

The CCC boys provided the Forest Service and Grazing Service with something they had previously lacked: a captive source of disciplined fire-fighters. Currently, it costs between $400 and $1,000 to equip a firefighter for work against wildland fires. The list of essential equipment includes steel-toed high boots, fire-resistant turn-out pants and shirt, hard hat, safety goggles, and a safety blanket to use in extreme emergencies. Modern fire-fighters receive a period of intensive safety training before they are allowed in the field. The CCC boys wore government-issue blue cotton work shirts and hats, if they were lucky. Their safety training was as good as the foreman who commanded their crew (fig. 1.1).

Vast deserts and towering mountain ranges dominate western Humboldt County in northwestern Nevada. Downstream from Winnemucca the Humboldt River swings south to the Humboldt Sink and the Carson Desert. The Humboldt River valley was the route pioneers followed on the trail to California, and it became the route of the transcontinental railroad and eventually the federal highway system. Northwestern Humboldt County is far off the beaten track. The Little Humboldt River branches to the north from the main river just above Winnemucca and has to fight its way to the main river through encroaching sand dunes. When the sand closes the river, Gum Boot Lake forms to the north. The watershed for the Little Humboldt comes from the hills that border the Owyhee Desert and Snowstorm Mountains to the east and, more important, the towering Santa Rosa Mountains to the northwest. Santa Rosa Peak is a pyramid of rock more than 9,700 feet tall, and much of the ridgeline from the Oregon border to the abrupt southern

Fig. 1.1. Civilian Conservation Corps fire crew fighting a wildfire.

end of the range above the Paradise Hill Pass remains at high elevations. North and west of the pinyon-juniper woodlands of the central Great Basin the high slopes of the Santa Rosa Mountains are nearly treeless. On north-facing slopes of the higher peaks, scattered groves of five-needle pines form subalpine parklands. At slightly lower elevations, tangled stands of quaking aspens cling to seeps and wisps of soil moisture around springs, waiting for the return of the next ice age so they can spread their autumn glory across vaster landscapes. The east side of the Santa Rosa Mountains is called Paradise Valley, supposedly because some early traveler who had come across the Black Rock and Desert valleys to the west thought he was in paradise when he got there.

The Quinn River valley on the west side of the Santa Rosa Mountains is named after the river that runs through it. Peter Skene Ogden called it the "valley of lakes" because of the series of ponds that have formed along the course of the river. An arm of pluvial Lake Lahontan once extended up the valley almost to the Oregon state line, where it would have crested the rim of the hydrologic basin and drained to the Pacific through the Snake-Columbia River system. Supposedly, the Quinn was named the Queens River, but a spelling error made it the Quinn. The lower mountains between the Quinn and Kings river valleys are strikingly different in structure from the jagged fault-block mass of the Santa Rosa Mountains. The flat tops of successive basalt flows are apparent in the Kings River Range. The joint Kings and Quinn rivers flow to the northern extension of the Black Rock Desert.

The playa of the Black Rock Desert is not visible from the foothills of the Santa Rosa Mountains near the south end of the valley. The saw-toothed structure of the Jackson Mountains, with King Lear and Parrot peaks rising above 8,000 feet, block the view. It is a long, long way across Desert Valley or the Jungo Flats from the mouth of the Quinn River valley to the Jackson Mountains.

The fine silt and clay-sized particles eroded by winds from the playas and salt deserts tend to be deposited to the northeast, and the southwestern slopes of the Santa Rosa Mountains thus receive the silts eroded from the great deserts. The eolian soils that have developed in the Quinn River area are rich and productive. The valley is now largely filled with farms that produce high-quality alfalfa hay and seed. The desert blooms there because groundwater is pumped from wells for irrigation. In the late 1930s, the irrigated area was much more restricted. Meadow hay for wintering cattle was produced in fields irrigated by water diverted from the streams coming down from the high Santa Rosa Mountains. The creeks and ranches were widely spaced, but the ranches were productive. Cattle grazed the big sagebrush/bunchgrass ranges in the foothills in the spring and fall, the high mountains in the summer, and the deserts to the south and west in the winter. Every fall, fat 2-year-old and 3-year-old steers were driven to the railroad stockyard in Winnemucca and there was money in the bank.

The winter of 1938–1939 had produced a good forage year on the sagebrush ranges of the Quinn River–Santa Rosa Mountains area. It was a welcome relief from the years of severe drought that had occurred earlier in the decade. The range produced more feed than the livestock could consume. The new grass called cheatgrass was especially productive beneath the big sagebrush plants where native perennial grasses had previously grown. Only much later was the strong affinity between cheatgrass and the eolian soils noted.[14]

Some wag once suggested that Orvada, Nevada, exists because there is a gradual turn in the highway about 5 miles south of town just above Cane Springs. The highway down the Quinn River valley from McDermitt is almost perfectly straight for more than 40 miles. If not for Orvada, how would anyone ever remember that the turn was coming? In 1939 Kirk Studebaker was the proprietor of the Orvada Mercantile, a dark, cool place with high shelves holding the variety of goods required by ranch families. Down in the dry goods section there was a tall spool cabinet with just about every kind of thread a rancher's wife could possibly want. Horse collars and harness fittings hung on the walls in the back room. A single hand-operated gas pump stood out in front. A cottonwood planted long ago shaded the front of the store. It was a toss-up what desert highway travelers welcomed most, the gasoline, water for their boiling radiators, or the shade of the cottonwood on a hot afternoon.

In late July 1939, ranchers in the Quinn River valley were hard at work making hay. The summer had been dry and hot. In the foothills, the cheatgrass was dead dry. The big sagebrush plants had dropped their large ephemeral leaves and were doing what they are best at, enduring. Some day next fall the rain would come and biological life would resume after the summer drought.

July 23, 1939, was a different kind of day. In the morning, black thunderclouds rolled across the valley and built up against the wall of the Santa Rosa Mountains.[15] Usually, Great Basin thunderstorms build in late afternoon when maximum heating of the atmosphere occurs. Perhaps a low-pressure cell from the Pacific had slipped past the offshore high-pressure ridge, because by midmorning lighting was flashing and thunder rumbled through the valley. Hay hands cursed and raced to get another load to the stack before the rain started. At 10:40 AM Kirk Studebaker cranked the wall telephone and asked the operator in Winnemucca to connect him with the CCC camp in Paradise Valley. He wanted to report a fire. Mr. Studebaker had one of the most spectacular views in northern Nevada from the front windows of his store. A rocky butte stood out perpendicular to the north–south axis of the Santa Rosa Mountains. The steep upper slopes of the butte were nearly free of vegetation except for a few mountain mahogany trees that clung to fractures in the rock surface. The mass of rock forced the drainageway from the high mountains to jog to the south. Rock Creek, with headwaters near Santa Rosa Peak, was the first canyon south of the butte and the spot where the stream emerged from the mountains. It was close to this point that lightning struck and ignited the dry cheatgrass.

Ranger Paul Travis took Mr. Studebaker's call and was quick to dispatch his standby fire crew. Twenty-three young men under the command of Wilbur Timmons grabbed shovels, fire rakes, and grubbing hoes and mounted flatbed trucks with low stake sides. Rushing down the road to fight a wildfire was a lark, a welcome diversion from the monotony of the camp. The trucks actually hit a cooling 45 miles per hour on the flat, but the young men in the truck bed got awfully hot as the truck ground its way up the Paradise Hill grade. The crew comprised young men cheated out of their adolescence by the Great Depression who had found a substitute home in the CCC camps. Many were escaping the poverty of big eastern cities, although the crew included a sprinkling of dust bowl refugees from the Great Plains. The young men were not necessarily popular with the local people, especially the boys their own age. The sparsely settled desert had a limited number of unmarried young women, and the work-hardened cowboys and farmhands who came to town for Saturday night dances viewed these funny-talking foreigners as competition.

The radiators on the trucks were boiling when the fire crew reached the fire at 1:30 PM. Timmons split the crew, sending half to the upper side of the fire under the direction of Earnest R. Tippin. Sending the men upslope was a critical mistake that violated well-recognized standards of wildland fire-fighting. As the crew started to work on fire lines, the fire literally blew up. The official report described what happened next as the result of an unfortunate change in wind direction and velocity. Perhaps the fire had reached a stage where it had sufficient energy to generate its own winds. In any case, heat from the burning big sagebrush distilled gases from shrubs at the leading edge of the fire front, and these ignited in incandescent sheets of flame that sent the fire jumping ahead at an extremely rapid rate.

Both crews were trapped within a horseshoe-shaped ring of fire. Their only chance was to run for the rocks upslope before the flame front reached them or all the oxygen was exhausted. Discipline, training, and comradeship were forgotten; it was a run motivated by sheer terror. Crew chief Wilbur Timmons and his half of the crew barely made it to safety. The members of Tippin's crew became disoriented in the smoke, and five members, including Tippin, died.

As the fire blew up, it sent a tower of smoke not unlike a nuclear explosion in appearance and energy content roiling into the stratosphere. Everyone from McDermitt to the Jackson Mountains could see that a big fire was running wild. A CCC boy rushed into the Orvada Mercantile with streaks of sweat and tears on his smoke-blackened face. He implored Kirk Studebaker to call the CCC camps at Quinn River, McDermitt, and Golconda and tell them to send everyone; the fire had blown up and crews were burning on the mountainside. A frontier early-warning system went into effect. When Studebaker cranked the telephone to get the operator in Winnemucca, every rancher's wife on the party line picked up her receiver to hear where the fire was located. The word that the CCC boys were dying in the flames spread like another wildfire through the sparsely populated landscape. Every ranch had a bell, triangle, or length of railroad rail suspended by a chain. The cook rang the bell for breakfast, dinner, and supper; otherwise it rang only for emergencies. The clanging brought crews hurrying in from the hay fields. All heard the same thing: "The CCC firefighting boys are dying on the mountain. Put up the horses and get in the truck—we're going to help." Even the staid general manager of the large Utah Construction Ranch at Orvada told his men to get shovels and grubbing hoes at the shop.

Chaos reigned in the alerted CCC camps. Fire crews and new enlistees were loaded on trucks. Camp officers distributed firefighting tools and loaded Cletrac tractors equipped with tiny bulldozer blades on trailers. The ambulance led the convoy, siren wailing. The old trucks hit 60 miles per hour on the flats as the grim-faced crews squatted down out of the wind. This was no lark. Their friends were dying on the mountainside. Hay hands and cowboys came in bib overalls and blue jeans. They had no love for the CCC boys, and less for the government authority they represented, but no one deserves to die in the inferno of a wildfire.

Wilbur Timmons knew he had lost firefighters from the Tippin crew. He organized a search for the bodies. At 9:00 PM, as the sun was dipping beneath the horizon far across the desert, they found the first body. The sight was particularly upsetting to the volunteer fire rescuers. The body of George J. Kennedy of New York City was curled in a ball and burned almost beyond recognition. In the early days of settlement, someone had tried to divert the waters of Rock Creek down the eolian soil foothill slopes to irrigate a field in the valley. The fine silt soils had rapidly eroded, forming a 15-foot-deep cut on the hillside. The irrigation project had been abandoned long ago, but the cut remained and formed a barrier the fleeing firefighters had to cross to escape the oncoming flames. When Kennedy tried to jump the cut, he shattered his right ankle. The party that discovered the body speculated that George might have lain on the bank for several minutes after his injury, watching the wall of flames advance. The story rapidly circulated among the growing army of firefighters.

The search resumed the next morning. About a quarter of a mile past Kennedy's body the searchers found the bodies of Tippin, Frank W. Barker, and Walter James. Tippin was from Oswego, Kansas; the others were from New York City. It was not until 10:30 that morning that the last body, that of Frank J. Vitale from Brooklyn, New York, was found on the top of a high knoll. The rescue party speculated that Vitale might have thought he was running up the great rocky butte and could scramble up the cliffs to safety. In the swirling smoke he reached the apex of the knoll and found no dubious rocky sanctuary.

The fact that his comrades had left George Kennedy to burn particularly angered the rescue party. Only much later, when representatives from the Department of the Interior interviewed the entire group of survivors, was it determined that Tippin and Walter James had run back into the flames to try and rescue Kennedy. In the meantime, things turned very ugly at the fire camp. The locals blamed Wilbur Timmons for the firefighters' deaths; certainly, he had shown a gross lack of judgment. A cowboy went to his pickup and got a rope. Another said the cottonwood tree at the store had a suitable limb. The Humboldt County sheriff and the combined officers from the CCC camps restored calm and prevented a compounded tragedy, but the hard feelings remained for years.

At midday on July 30, the black thunderclouds boiled up again, but this time the clouds opened and an epic rainstorm drowned the fire. The fire died as quickly as it had started. The county coroner and district attorney convened a coroner's jury in Winnemucca to review the deaths of the firefighters. It was obvious to all that Foreman Timmons was on the hot seat. After hearing lengthy testimony from CCC, Forest Service, and Grazing Service officials, the jury ruled that the deaths were due to an act of God—the sudden shifting of the wind.

The Paradise CCC Camp men worked 9 months to construct a memorial to their fallen comrades. They gathered native stones and built a monument 14 feet square at the base and 10 feet tall. The monument was dedicated on April 28, 1940. Attendance was remarkable considering the distances involved and the sparse population; three hundred people attended the dedication. A cold rain fell steadily throughout the ceremony. Attendees could turn and look upon the mountainside at the place where the CCC boys died. The black scar of the wildfire was turning green with sprouting cheat-grass. Cows and calves were searching for cheatgrass leaves rich in nutrients released by the burning of the vegetation and litter. The phoenix was rising from the ashes to rule the environment again.

By chance, we have a research plot located almost exactly on the spot where George Kennedy died. We say "by chance" because the U.S. Department of Interior Bureau of Land Management selected the location from among several hundred thousand acres of cheatgrass in the Santa Rosa Mountain foothills. If you walk around the plot area in the tall, dry cheatgrass you very occasionally find slabs of dense, jet black slate that must have washed down from the high Santa Rosa Mountains during the Pleistocene. Just north of the plot there is an especially large slab at about the place were crew leader Tippin must have tried to rush back into the fire. Perhaps it is nature's tombstone on ground hallowed by blood, sweat, and tears.

In this chapter we examined some of the many faces of the plant we call cheatgrass. It is both the basic forage plant of the Great Basin and a devastating competitor in cereal grain fields. It plays an important part in the diet of chukars and a host of native animals. It fuels deadly wildfires. Obviously, the ecological and economic consequences of cheatgrass invasion have many facets.

CHAPTER 2

Developing a Perspective
of the Environment

Cheatgrass is a highly variable species with the inherent capacity to adapt to a vast range of soils and plant communities and the potential for genotypic as well as phenotypic adaptation. This annual grass occasionally matures at barely an inch in height and sometimes reaches more than 2 feet. Scientists call such flexibility "phenotypic variability." Cheatgrass can germinate in the fall and complete its life cycle as a winter annual or wait until early spring to start growth and complete its life cycle a couple of months later as a very short-lived annual. In any given season, it can exhibit one or both of these growth habits at the same location. Although it rarely happens, cheatgrass can germinate at the same site in both fall and spring, and if late spring rains occur, can have even a third period of germination and emergence. It is common for cheatgrass to produce virtually no herbage in dry years or during a dry spring after a moderately wet winter. During extreme droughts cheatgrass seeds can remain in the seed bank (i.e., the soil and surface litter) for several years with virtually no germination. If the periodicity and amount of rainfall are just right, though, cheatgrass can produce tremendous amounts of herbage that local herbivores cannot consume. After maturity, this fine-textured, superabundant herbage dries and provides fuel for wildfire storms.

Because it adapts to the activities of humans and their domestic livestock, cheatgrass is said to be a "social weed." It also adapts to modern disturbances associated with suburban sprawl and off-highway recreational vehicle use. Cheatgrass is a major weed in certain agronomic crops such as winter wheat. Native animals, both insects and rodents, have adapted to use the seeds of cheatgrass as a source of food. Currently, cheatgrass germinates and completes its life cycle at elevations ranging from the lowest sites in the central and northern Great Basin on soils with extreme accumulations of salts to the 9,000-foot plus elevations of the Shell Creek Range in eastern Nevada.

How can we come to grips with the ecological and economic consequences of this invasive alien species that can adapt to such a vast range of environmental conditions? Cheatgrass occurs in all of the United States and grows in relative abundance in all of the western states, but it reaches its greatest expression as a symbol of environmental degradation in the Intermountain Area between the Sierra-Cascade Mountains on the west and the Rocky Mountains on the east. Within the Intermountain Area, the rangelands of the Great Basin have become a symbol of the hazards associated with the spread of cheatgrass. The Great Basin includes most of Nevada, the eastern and southeastern portions of California, western Utah, southeastern Oregon, and even a small portion of Wyoming. The hydrologic Great Basin actually extends into Baja California, Mexico. No water from this vast area runs into the ocean.

The Great Basin provides a representative sample for the entire Intermountain Area of the ecological and economic importance of cheatgrass. To appreciate the magnitude of the spread of cheatgrass and the amplitude of its ecological disruptiveness, it is necessary first to understand the assemblages of soil, plant, and animal communities that constituted the Great Basin environment before the weed was introduced, as well as how European and Native American societies interacted with this environment.

The Great Basin of western North America is a field geologist's dream. Fault-block mountains without sufficient vegetation to hide their naked rocks tower out of arid basins (fig. 2.1). A trained observer can read the exposed stratigraphic sequences on the mountains like a novel, although admittedly one with a complicated plot. The plot was obscured in some sites as the edge of North America advanced to the west and the underlying tectonic plates were subducted beneath the continent or thrust out over the existing land surface. It is a hard land of extremes where Native Americans struggled for

Fig. 2.1. The Great Basin is characterized by a series of fault-block mountain ranges generally running north and south. Many of the basins between the mountain ranges contained lakes during the glacial periods of the Pleistocene.

centuries to eke out the barest of survivals. Without habitat modification, the region is usually too hot or too cold for the taste of most humans. The legendary mountain man and trapper Jedediah Smith almost died trying to be the first European American to cross the Great Basin.[1] Peter Skene Odgen, leading a Hudson's Bay Company brigade to the Mary's River (modern Humboldt), had little trouble finding forage for his brigade's three hundred horses but had to search hard for potable water. His professional hunters often searched in vain for big game to feed the party. Members of modern societies who are willing to pay the energy cost for heating and air conditioning and providing potable water find it a highly desirable place to live.

Among the many paradoxes of the Nevada portion of the Great Basin is the fact that it had the least dense human population of the adjacent forty-eight states for most of the twentieth century and was among the states most rapidly growing in population by the century's end. Even the common name for the Great Basin's environment, "cold desert," suggests confusion over the region's climate. Vehicles traveling on the unpaved roads of the cold deserts either create long plumes of dust or are in danger of becoming stuck in the mud. In midwinter, it is common to drive across a desert basin on a road covered with a skiff of snow and watch a plume of dust in the rear-view mirror. When it rains in the Great Basin, children are more prone to run outside to experience the phenomenon of rain than they are to rush indoors to escape the wet and cold. Precipitation is almost completely out of phase with temperatures that permit the active growth of plants. Most moisture falls during the winter months in the form of snow; the summer monsoons of the southwestern United States reach the northern Great Basin irregularly and are rare in the northwestern portion. The most constant aspect of the weather is its extreme variability.

Most of the residents of Las Vegas, Reno, and Carson City are newcomers. Thanks to the extreme variability of the weather patterns, though, they are just as expert at amateur weather forecasting as old-timers. Most years are drier than average. Occasionally, the region goes for months without precipitation. During the historic drought of the late 1980s, no rainfall was recorded at the Reno airport for months on end. In late September, television weather forecasters announced that one more twenty-four-hour period without rain in Reno would establish a new record for consecutive days without precipitation. That night an unpredicted brief thunderstorm dashed the chance to break the record. Ironically, the entire community, sick of water rationing and dust blowing in from the desert, was crushed by the occurrence of the event for which they had been collectively praying. The next year, the record for consecutive days without measurable precipitation was broken by more than thirty days.

When wet winters do occur, precipitation is often well over 100 percent above average. Very rarely, winter precipitation is above average and the periodicity of precipitation the following spring is very favorable for the growth of herbaceous vegetation. Cowboys accommodate this rare phenomenon in the statement, "All this range needs is a good warm spring rain." Unfortunately, the infrequent good years for plant growth have often been used as an excuse to justify excessive grazing of Great Basin ranges. Even modern grazing management plans are sometimes based on the "average" forage production. The mathematical average is real, but it never occurs in nature. How appropriate for Nevada that the probability of predicting the long-term weather is often a crapshoot. Lynn F. James, an experienced and keen observer of the Great Basin environment, characterized the climate of the Great Basin quite simply: "It's a desert, damn it."

Although the wealth of the Comstock Lode gave Nevada the nickname "Silver State," the name could just as well have come from the region's dominant vegetation. Much of Nevada's perennial plant cover is silver gray. In an environment where moisture is at a premium, many plants have leaves covered by hairs that reflect the maximum amount of incoming radiation, reducing heat loads and creating moisture-saving turbulence around leaf stomata. Plants lose moisture through the stomatal openings in the leaves, but the stomata are necessary for gas exchange for photosynthesis. These are adaptations to survive in an environment where moisture for plant growth is usually a limiting factor. Major exceptions to the silver gray color are the narrow strands of gallery forest of Fremont cottonwoods that follow rivers to their desert sinks. On a warm summer day, the crowns of the cottonwoods appear almost black because the leaves are absorbing so much incoming radiation.

The lake plains are the beds of a huge system of pluvial lakes that formed and desiccated repeatedly during the Pleistocene.[2] The geological term "pluvial" refers to a former, wetter time. In unison with glacial advances on the mountains that surrounded the Great Basin and on the highest ranges within the basin, the lakes rose in response to decreased evaporation and slightly increased precipitation. Lakes in the higher basins sometimes rose high enough to spill through passes and contribute to lower basin lakes, eventually forming vast, deep lakes in the northwestern (Lake Lahontan) and northeastern (Lake Bonneville) portions of the Great Basin. Lake Lahontan never rose high enough to breach the mountain rim and flow to the ocean. Lake Bonneville did partially drain through the Snake-Columbia River system.[3] The evaporating waters of the pluvial lakes left behind abundant deposits of soluble salts and deltas of sand from the glacier-fed rivers that flowed down from the mountain ranges. These drying lakes left vast plains of silt-textured sediments that were exposed to deflation by wind erosion. The deposition of these windblown sediments dramatically influences soil development in the northern Great Basin.

The soil environments where exotic invasive weeds such as cheatgrass initially proved to be well preadapted to grow were a direct product of the erosion and deposition of sediments from these lake plains. The abundant salts trapped in the soils of the old lake plains are often readily apparent in crusts that form on the surface from effervescence. Anyone who walks on one of these crusts on a hot, still day will instantaneously taste the salts. In an environment where strong winds are common, the salts erode and are deposited, becoming significant factors in soil formation.

The bitter-tasting, eye-irritating salts of the basin naturally led to the name "salt deserts" for the lake plain and adjacent environments. In his early studies of the plant communities of the Carson Desert, W. D. Billings pointed out that salts in the surface soils are not a marked factor in controlling plant distribution in much of the salt deserts.[4] Water is far more important. The native salt desert vegetation exists over much of the area because the plants are able to survive and grow with very little moisture (fig. 2.2).

In contrast to the silver gray upland vegetation, the black greasewood growing on the lake plains is vividly green. These plants have roots that reach down to the salty groundwater. Plants adapted to use groundwater so salty that it would be toxic to normal species can afford the luxury of green leaves in a gray desert. To utilize such water, the plants must have some mechanism to shunt the salts aside once they enter the plant cells and eventually to dispose of the accumulations. The plants get rid of excess salts in the secretions of specialized glands or in deciduous leaves or fruits. Such mechanisms lead to the biological accumulations of salts on the soil surface. Eventually wind action erodes and redeposits these salts. For this entire adaptation to work, the groundwater level must contact the depth of soil wetting at some time. Roots do not grow through dry soils.

The basins of northwestern Nevada lie in the rain shadow of the lofty Sierra-Cascade Mountains. These ranges, which began to rise in the early Pleistocene, lift moisture-laden clouds from the Pacific Ocean high into the atmosphere, where they cool and lose their moisture on the western slopes of the mountains. An old cowboy once said that "the clouds over the Nevada deserts are just empties coming back from Utah." The mountains of the Great Basin tend to have a north–south orientation. Each subsequent range inland after the Sierra-Cascades tends to cast its own rain shadow. In sub-humid to humid areas, the rainfall percolates into the soil and continues down to the groundwater table or bedrock until it moves by gravity to emerge in seeps and springs to continue in the hydrologic cycle. In most of the Great Basin, the systems are much simpler. In the 10–14-inch precipitation zone, the moisture penetrates only about a yard into the soil. In the

Fig. 2.2. Shadscale/Bailey greasewood plant community in the Carson Desert of western Nevada. A desert pavement of black basalt covers the surface of the interspaces between shrubs.

drier portions of the salt deserts, the depth of wetting in the soil may only be 6–8 inches. The depth of wetting encompasses the biologically active zone. Many of the native shrubs enhance their own survival potential by building mounds beneath their canopies. The mounds consist of trapped soil particles, subaerially deposited silts, and sand particles bounced along the soil surface by a process called saltation. The accumulation occurs because the woody shrub canopy changes the aerodynamics of its microenvironment.

The nineteenth-century geologist Israel Cook Russell described Nevada as an area where a person would ride a horse for 100 miles in order to sit in the shade of a tree.[5] In a land so bare of trees, a 2-foot-tall shrub canopy is a major windbreak. Leaf and fruit fall contribute to the microenvironment of shrub mounds. The accumulated organic matter plus the microclimate altered by the shrub canopy create an environment for the growth of microphytic crusts of microscopic blue-green algae and mosses. The roots of the shrubs spread far beyond the edge of their canopies, allowing the shrubs to mine the interspaces between shrubs for moisture and nutrients. All of these factors combine to restrict nutrient cycling largely to the shrub mounds. Great Basin soils are characteristically rich in bases

Fig. 2.3. Playa surface on the lake plain of a desiccated pluvial lake basin. The surface is composed of very fine silts and clay-sized particles that were precipitated as deep-water sediments in the lakes. Many of these playas flood following snowmelt in years with above-average precipitation, and the resulting shallow lakes may persist for several years. After the water evaporates, salts accumulate on the surface from capillary rise. In dry years, wind erosion removes the salts and can deflate the playa surfaces.

and poor in nitrogen. The limited cycling of nitrogen beneath the shrub canopies is of monumental significance in the nutrient dynamics of desert plant communities.

The predominance of mountains and basins within the Great Basin presents another paradox. Alan Brunner, who spent a career literally walking the Nevada desert, explained, "From the top of any of the taller mountain ranges it is obvious the Great Basin is all mountains. From the playa in the center of any of the large basins it is obvious the Great Basin is all valleys." A person standing on the mountain peaks can turn 360 degrees and see apparently endless ranges of mountains stretching to the farthest horizon. Reverse the perspective, though, with the curvature of the earth showing on the flat playa surface, and the mountains appear to be spectral ranges floating in space with their bases hidden in the heat mirage (fig. 2.3).

Closer examination reveals vast alluvial fans partially burying the bases of the mountain escarpments. These fans, composed of material eroded from the mountain slopes, speak of the aridity of the environment, which lacked sufficient energy in the form of waterpower to transport the material completely out into the basins once the steep gradient of the mountain slopes was passed. The presence of a single boulder or an accumulation of very large boulders testifies to the sudden burst of energy associated with some catastrophic local or perhaps regional flood. In the larger basins, the bases of the fans were truncated by the wave action of Pleistocene lakes, proof of the antiquity of the collections of eroded material.

These mountains rise like islands from a sea of aridity. The higher elevations capture moisture, increasing the potential of the environment to support plant and animal life (fig. 2.4). This increase in potential continues with increasing elevation until temperatures become so cool that plant growth is limited. As the elevation rises, the hydrologic cycle gradually changes from a closed system in which the soil is wet to a depth of only a few inches or feet to

Fig. 2.4. The alluvial fan in the foreground supports a plant community dominated by big sagebrush. The lower slopes of the Pine Nut Mountains in western Nevada support single-leaf pinyon–Utah juniper woodlands with big sagebrush in the interspaces. The snow-covered upper slopes support mountain brush communities where big sagebrush is an important species.

an open system in which water percolates down to the bedrock and then out to rills and ravines and springs to produce runoff and seasonal streams. Mountains above 10,000 feet produce permanent streams whose runoff makes irrigated agriculture possible in the desert valleys. Smaller mountain watersheds may produce seasonal streams that run only in the spring and early summer. Often these streams leave the steep mountain escarpments and disappear into the permeable alluvium at the top of the fans that bury the mountain bases.

The creeks flowing out from the mountains determined the location of the early livestock ranches. A permanent source of drinking water for people and stock was an obvious necessity, but equally important for ranching in this environment was irrigation water to produce hay. Why was hay so important? Hay is conserved forage cut and gathered during the peak of the summer growing season and protected in barns or stacks for livestock forage during the winter. In Nevada, the hay was usually stacked because the pioneer ranchers in this nearly treeless environment hardly had sufficient lumber to shelter themselves, much less build barns.

The earliest ranchers were able to operate on the range year-round without feeding hay during the winter. Seasonal movement of the animals from the desert valleys in the winter months, to the foothills in the spring months, to the higher mountains in the summer months, and back to the foothills in the fall provided a yearlong supply of forage. During the late nineteenth century the situation began to change; a series of severe winters caused losses of cattle on the range, and these losses became more severe as the ranges were overgrazed. Finally, the winter of 1889–1890 killed nearly 90 percent of the cattle in the higher-elevation valleys.[6] Ranchers learned the hard way that one ton of hay had to be conserved for every brood cow that was to be kept through the winter.

Nevada is a leader in the production of high-quality alfalfa hay. Most alfalfa is grown in irrigation districts or by pumping groundwater, but traces still remain of the original pattern of establishing ranches where creeks supplied seasonal irrigation. On the road from Lovelock to Winnemucca, Nevada, signs of the old ranches are still visible at the base of the West Humboldt Mountain range where telltale canyons emerge from the mountain escarpment to the cones of the alluvial fans. The cottonwood trees planted around such sites help to identify them. The sparse distribution of the old ranches testifies to the harsh limits the environment set on the husbandry of livestock.

The bulk of the annual forage requirement for the old range livestock operations came from the sagebrush/bunchgrass plant communities. The winterfat communities on the desert winter ranges and the rich meadows and aspen parklands in the high summer ranges were a timely and critical source of forage, but Nevada's 19 million acres of sagebrush/bunchgrass provided the bulk of the herbage for the range cattle industry. The sagebrush/bunchgrass community married a woody shrub overstory with perennial grasses that grew in widely spaced bunches rather than in a continuous sward like a lawn. Only the grasses provided the forage. Cattle did not prefer the browse of most species of sagebrush, and would die if hunger forced them to eat this material in large quantities. Sagebrush leaves contain chemicals that inhibit the microorganisms that live in the ruminant rumens and are critical for forage digestion.

The woody sagebrushes are very interesting plants for a variety of reasons. These evergreen (or in their case ever gray) shrubs produce two forms of leaves. One type grows annually during the spring period when moisture is available. The other leaf type, smaller and more moisture efficient, remains on the plant for several seasons. Sagebrush blooms in the fall at the end of the summer drought, when the flowering stalks are extremely efficient at photosynthesis. Seedlings can produce seeds in their second year and continue to flower when they are more than a century old, when most of their cambium has been destroyed and their stems recline on the ground as twisted wrecks of weathered wood. Big sagebrush plants produce many more seeds each year than are necessary to maintain a stand at the density a particular site is capable of supporting. Neil West, a noted professor of range ecology at Utah State University, suggested that the superabundant seedlings, which have little or no chance of establishing and becoming perennial shrubs, are essentially acting as annuals. This may help to explain why competitive annual species are so rare in these environments and why introduced annuals such as cheatgrass have been so successful.

Cattle will eat big sagebrush only in starvation situations. They relish the herbage of the perennial grasses, however, especially in the early spring when the grasses are rapidly growing. If cattle are allowed to graze the native perennial grasses in the early spring excessively year after year, and the grasses are grazed continuously throughout the season with no chance to restore their carbohydrate reserves, the grasses will disappear from the community, leaving behind only the shrubs. In such situations, the stand renewal process is changed. (The stand renewal process is the way a given generation of a plant community is destroyed; see chapter 15 for a more comprehensive discussion.) In some plant communities the stand renewal process may be as simple as old age or plant senescence; in others it may be herbivory. All indications are that before Europeans arrived in the region, the sagebrush/bunchgrass plant communities of the Great Basin were renewed catastrophically in wildfires. The perennial grasses in the community were not unduly injured by the fires and actually benefited from the soil moisture and nutrients the fires released, especially nitrogen. The sagebrush plants did not sprout after the aerial portion of the plant was burned, however; nor did they produce persistent seed banks that would have provided a source of new plants. Although the initial effect of overgrazing the perennial grasses favored the shrubs by reducing competition for nutrients and moisture, it also reduced wildfires because the fine herbaceous fuel necessary to carry fire among shrubs was no longer present. Big sagebrush communities that lacked an herbage understory were susceptible to plant invasions. This became a crucial ecological factor once cheatgrass became ubiquitous.

Big sagebrush is the characteristic plant species of the alluvial fans that surround the mountains. The subspecies may differ from one site to the next, but big sagebrush continues to be an ecologically significant species at both higher and lower elevations. The northern edge of the pinyon-juniper woodlands runs diagonally across the western Great Basin roughly parallel to the route of Interstate 80. Southeast of this boundary the middle elevations of the mountain slopes support dwarf conifer woodlands of pinyon pine and juniper. Essentially the trees are superimposed on the dominant big sagebrush/bunchgrass communities.

The pinyon-juniper woodlands reached their northwestern limit in very recent times. Analyses of plant remains in fossil packrat middens by Robin Tausch, a scientist with the Rocky Mountain Experiment Station of the USDA Forest Service, indicate that the pinyon-juniper woodlands reached the Truckee River within the last one thousand years. This fact illustrates another aspect of the environmental conditions in the Great Basin that control plant dominance: they are constantly changing. We tend to refer to the precontact condition of the environment as "pristine" and consider it the ideal situation. In fact, Europeans reached the Great Basin at one point in an always-changing panorama of environmental potential. Cheatgrass likewise arrived at one point in the panorama and will eventually yield its dominance to other species.

Just as plant communities change on various time scales, so do the herbivores that eat the plants. The American bison had already withdrawn from the western Great Basin at contact time. The mule deer population expanded exponentially following overgrazing of the perennial grasses and dramatic increases in the density of woody browse species. After the terrible winter of 1889–1890, the range sheep industry underwent a great expansion that lasted until after World War II. Federal legislation protected free-roaming horses and burros and brought concurrent increases in year-round grazing by these animals on certain Great Basin rangelands. The spread of cheatgrass to certain salt desert ranges has greatly increased available herbage for harvest by herbivores. Although we emphasize herbivory here, cheatgrass produces so much seed annually that it has a major impact on granivores as well. In fact, the granivore interaction is a major part of the cheatgrass story.

The environmental template into which the alien cheatgrass was accidentally placed is a semiarid to arid environment born in the wild climatic fluctuations of the Pleistocene. The conglomeration of plant communities that existed in the area was inherently fragile, composed of species of widely varying origins that had very recently (in geologic time) occupied a mid-latitude temperate desert, itself a concept novel in geologic history.

CHAPTER 3

Preadaptation of Cheatgrass for the Great Basin

Cheatgrass has been a successful invader in the Great Basin largely because of a series of preexisting conditions. These include: (1) evolution of sagebrush ecosystems in North America without highly competitive annuals, (2) evolution of cheatgrass in a similar potential environment on a different continent where agriculture had existed for millennia, (3) introduction of cattle into a semiarid to arid environment in North America by ranchers who had no experience with temperate desert environments, (4) mechanization and subsequent expansion of cereal grain production in western North America, and (5) deep-seated inertia resisting the end of open public rangelands in the American West. Even so, these preconditions had to be aligned just so in space and time for cheatgrass to be successful. Was it a fantastic accidental combination of events, or were humans in some sense responsible? It was humans, after all, who discovered that they could increase the quantity and quality of their food supply by domesticating certain herbivores. The animals were ruminants with internal symbiotic fermentation systems adapted to digesting forage so high in cellulose that humans themselves could not use it for food. This system had advantages for both humans and animals. The herbivores used a food not available to human consumers and turned it into high-quality digestible protein, often the limiting nutrient in human diets. Further, these animals were mobile and could cover long distances in search of forage. Nomadic herding societies, a lifestyle known as transhumance, evolved.

Cheatgrass evolved in middle Asia, the same area where cattle, sheep, horses, and goats are thought to have been domesticated. Working at a very early agricultural site in the Zagros Mountains of Iran, archaeologist Kent Flannery enlisted the assistance of the Danish botanist Hans Helbaek to help identify carbonized seeds recovered from fossil hearths.[1] Flannery and Helbaek were able to follow the evolution of small grain crops—especially wheat and barley—from seeds that hunter-gatherers collected from wild grasses through the evolutionary process that produced modern cereal species. It should be no surprise that seeds of species that are considered weeds occurred together with the seeds of cultivated crops. Plants that could mimic the cultural requirements of the cereal grains were preadapted to be weeds in the fields of the earliest agriculturalists. They germinated at the same time as the cereal grains that were intentionally sown, completed their life cycle in the same amount of time, and produced seeds that were difficult to separate from the crop seeds. Wheat started as wild einkorn types and evolved through hybridization and polyploidy to the emmer wheats and then the modern bread wheats. Agriculturists incorporated genes from the various annual grasses related to wheats as they selected for highly valued characteristics such as seed size and heat tolerance. The fact that cheatgrass has persisted as a successful weed in cereal grain fields as farming has become more sophisticated strongly implies that its reproductive system is inherently adaptable. Simply put, cheatgrass is a survivor when it grows in contact with agricultural crops.

The archaeologists working with Helbaek also identified the period during which the primitive residents of the site became farmers rather than hunter-gatherers by examining the morphological changes in the bones of sheep and goats that accompanied domestication and selective breeding and were found in the same hearths that preserved carbonized seeds of evolving wheats and weeds. At the point when the preponderance of bones in the midden switched from wild sheep and goats to domestic animals, carbonized seeds of annual weeds such as Russian thistle, medusahead, goat grass, and annual brome grasses showed up in the fossil hearth samples. Apparently, the exotic annual weeds that are so much a part of the ecology of the sagebrush/bunchgrass ranges of the western United States have shadowed livestock since the dawn of domestication.

It is also not surprising that these annual weed species would be associated with disturbances caused by grazing animals. Their genetic makeup predisposes them to be colonizers. What is surprising is that they were already associated with large herbivores at the time the latter were domesticated. The early Holocene assemblages of plants were descendants of those that remained after the wild climatic fluctuations of the Pleistocene. The five-needle pines that now grow near timberline on the highest mountains in the Great Basin, for example, grew during the Pleistocene glacial maxima on the beaches of the pluvial lakes. The vast and rapid plant migrations forced by the fluctuations in climate during and after the close of the Pleistocene opened plant communities to invasion by other species. Perhaps the annual colonizing weeds associated with cheatgrass had already evolved by the time livestock were domesticated.

Cheatgrass is not the only weedy annual to come to North America's rangelands from Asia. Many of the exotic weeds currently thriving in the Great Basin originated in Central Asia and the Irano-Turanian floristic region.[2] Central Asia comprises the semiarid regions from northern China on the east to the Ob River on the west. It includes the northwestern provinces of China, Mongolia, and southern Siberia between the Ob River and Lake Baikal. This region is considered the center of origin for many groups of plants and is home to abundant endemic species. The Irano-Turanian floristic region begins on the east at the Ob River and extends west to the Caspian Sea. Its northern boundary is the steppe zone of Russia, its southern boundary the Persian Gulf.

The Central Asian and Irano-Turanian floristic regions are nearly synonymous with the Intermountain Area in terms of the range of environments they encompass. Portions of both areas are temperate deserts with areas of salt-influenced soils. Both feature vast landscapes dominated by semiwoody species of sagebrush and, at lower elevations, chenopod shrubs. Sagebrushes, saltbushes, kochia, and winterfat occur in both regions, although the species present vary. Greasewood is endemic to North America, and the woody chenopod haloxylon forms extensive stands in Asia.

The most striking difference between the rangelands of Central Asia and the Great Basin is in the tenure of livestock husbandry. The Great Basin first saw concentrations of domestic large herbivores a little more than a century ago. In portions of Asia, in contrast, nomadic livestock husbandry has been in existence for perhaps 10,000 years. Academician A. G. Babaev of the Institute of Deserts in Ashabad, Turkmenistan, wrote a brilliant account of the livestock industry in Central Asia.[3] In it, he suggested that the development of irrigated agriculture permitted tremendous increases in human population density in very limited areas. The proportion of irrigable land in Central Asia is about the same as that in the Great Basin—about 5 percent of the total landscape. Along with abundant food in irrigated areas came the development of more complex local societies. Eventually population growth outstripped the potential of the sites to produce food while at the same time salinization of irrigated areas and overgrazing of adjacent grazing lands reduced productivity. Civilization brought rulers, and overutilization of resources brought poverty to the basic farmers along with wars for water and grazing rights. Certain people who were more inclined to be herdsmen than farmers struck out into the deserts to find forage for their flocks and freedom from the constraints of civilization. A covenant existed between livestock and their herders; the herder was responsible for his animals. The biological constraints of the desert steppes forced herders to move continually in search of available forage. When a village chief found it necessary to fight the neighboring tribe over irrigation water, he went out to the desert and made a deal with the herdsmen to serve as mercenary soldiers. Is this only a plausible romantic tale, or is there in the human genome a block of "herder" genes that became fixed by natural selection when the first domestic herds wandered the desert ranges of Asia?

Ranchers in western North America were quick to adopt transhumance forms of livestock production; the land gave them no choice. Overstocking and season-long grazing soon took the bloom off previously pristine rangelands. Exceptionally hard winters quickly made it apparent that ranchers either had to conserve forage as hay for feeding during the winter or have access to desert ranges for the winter months.[4] Trappers and miners who visited the high mountain meadows brought back accounts of a summer grazing paradise in the highlands. Range sheep operations in the Intermountain Area became the most extreme wanderers in their annual cycle in search of forage; winter and summer ranges were frequently several hundred miles apart. Twentieth-century transhumance livestock production combined thrills and hard work.

Taking the cattle to the mountains in June was a major undertaking. Young cowboys tried to stay on dancing horses spooked by cows that wanted to run in the dark as the gate to the Forest Service corral was opened at the Forks of Etna Creek, sparking thoughts of pulp westerns and the trail to Dodge City. Even the older ranch hands got caught up in the euphoria of going to the mountains with the cattle. Perhaps it was an expression of the block of genes that defined herdsmen so long ago in Central Asia. The journey meant new country, abundant green forage, and escape from the drudgery of farming. The romance of the cowboy life had worn thin by the next evening when the herd reached the trailhead on the Salmon River. Seeing the moon twice in one working day tended to dampen youthful enthusiasm. The reverse was true in the fall. There was an urgency associated with the fall roundup. The cattle had to be out of the mountains before snow closed the passes.

Bromus is an ancient Greek name for a type of grass and is today the name of the genus in which a particular group of grasses is classified. The genus contains about 150 species distributed worldwide in temperate areas. Several species of *Bromus* are native to western North America; most of these are perennials. In all of the pre-Columbian Americas, only two cereal grasses other than maize (corn) were domesticated: the biannual *Bromus mango,* which was domesticated in Chile, and smooth brome grass, an important hay and pasture grass in the United States. Most of the annual species of brome grasses are self-invasive weeds; that is, they can invade other plant communities without the conscious efforts of humans to introduce them.

Carolus Linnaeus, the founder of the system of binomial nomenclature used to classify plants and animals, first described *Bromus tectorum* in 1753. Its recognition by Linnaeus indicates that cheatgrass was a widely distributed and recognizable type of grass. The specific name, *tectorum,* is apparently a reference to a plant that grows on roofs. In an article on the distribution of cheatgrass in Europe, the Russian-Israeli scientist Vladimir Kostivkovsky mentioned that cheatgrass was found growing on thatched roofs in Russia.[5] Perhaps the Swedish Linnaeus collected his specimen of *Bromus tectorum* from a thatched roof. There are very few thatched roofs in the deserts of the Great Basin, but this did not hinder cheatgrass from finding a home there.

A number of other annual species of brome grass besides cheatgrass occur as dominant plants in specific environments on the western range or as occasional species; they include the following:

Bromus arenarius Labill, no common name; native to Australia

Bromus briziformis Fisher & C. Meyer, rattlesnake chess; native to Eurasia (species name is sometimes spelled *brizaeformis*)

Bromus catharticus Vahl, rescue grass; native to South America

Bromus commutatus Schrader, hairy chess

Bromus japonicus Murr., Japanese chess (brome); native to Eurasia

Bromus madritensis L., foxtail chess; native to Eurasia

B. madritensis ssp. *madritensis* L., foxtail brome; native to Eurasia

B. madritensis ssp. *rubens* (L.) Husnot, red brome (synonym for red brome above)

Bromus mollis = *hordeaceus* L., soft chess; native to Eurasia

Bromus rigidus = *diandrus* Roth, ripgut; native to Eurasia

Bromus rubens L., red brome; native to Eurasia

Bromus secalinus L., cheat; native to Eurasia

Bromus stamineus Desv., no common name; native to Chile

Bromus sterilis L., poverty brome; native to Eurasia

Bromus trinii Desvaux, Chilean chess; native to South America

Cheatgrass has many common names; among the more common ones are bronco grass, cheat, downy chess, downy brome, June grass, six weeks grass, and nodding brome (Canada). Downy brome—a reference to the fine hairs that occur on the leaves and stems of some species of brome grass—is the official common name assigned to *Bromus tectorum* by the Weed Science Society of America, but cheatgrass is the nearly universally accepted common name in western North America. The nomenclature committee of the Weed Science Society was apparently composed of easterners who did not want to risk confusion with the grass *Bromus secalinus,* whose common name is cheat. That name apparently has no botanical significance, but rather refers to depriving someone of something valuable through deceit or fraud. Many of the annual brome grasses turn from green to a rather bright red as they mature and then fade to a light straw color after seed maturity. The species that have a pronounced red period are often referred to as red brome grasses. The common name foxtail refers to the plant's awned and often barbed seeds, or caryopses, which stick in hair, wool, and socks when they are mature. These seeds often injure the eyes and ears of dogs, cattle, horses, and sheep. Bronco grass is a cowboy term.

Cheatgrass is a very widely distributed species. The western edge of its natural range is the Balkan Peninsula, with perhaps adventive populations occurring as far west as Spain. In southwestern Spain on acidic soils, cheatgrass is restricted to old sheep bedding grounds where there is enough nitrogen to allow it to grow. Cheatgrass is common in Israel, Sinai, Jordan, Syria, and the Arabian Peninsula. In the northern part of its distribution, cheatgrass penetrates into the forested zone and is even found in Moscow. It could have arrived in America as a contaminant of crop seeds from the lower Volga and northern Caucasus regions, the Ukraine, Romania, and the Balkan countries. It could also have been a wool contaminant on sheep from Middle Asia.

Cheatgrass thrives in dry conditions. In the forest zone it grows on exposed rocks, in sand pits and stone quarries, and on the roofs of old buildings. On the semiarid and arid plains it is found in the drier environments and not in meadows or irrigated oases. Cheatgrass begins growing after the first spring rains and provides valuable forage until seed maturity occurs in May. During moist autumns, the seeds may germinate and grow before the onset of winter.

All of the native wild large herbivores of Eurasia, including saiga antelopes, onagers, gazelles, and Nubian goats, are typical consumers of cheatgrass herbage on desert ranges. Camels also eat cheatgrass herbage but prefer to browse on woody plants or to graze perennial grasses. These native herbivores may have contributed to the widespread distribution of cheatgrass before the advent of livestock domestication. Monocultures of cheatgrass and the wildfires associated with accumulations of cheatgrass herbage are seldom if ever a feature of the species in Central Asian environments. Natural insect and disease pests, grazing intensity, and nitrogen levels in the soils probably all contribute to cheatgrass's relative lack of dominance in its native habitat.

A. S. Hitchcock's *Manual of the Grasses of the United States* lists cheatgrass as occurring in all states except Alabama, Georgia, South Carolina, and Florida.[6] It was probably introduced in North America early in the nineteenth century. The 1861 edition of Alphonso Wood's *Classbook of Botany* indicates that cheatgrass had already been collected in Pennsylvania.[7] Samuel Betsford Buckley, who apparently collected cheatgrass in Texas, described it in the 1863 *Proceedings of the National Academy of Science* (Philadelphia) as *Bro-*

mus tectorum.[8] Although we commonly view the spread of exotic grasses as being from east to west across North America, following the flow of European civilization, that is not necessarily the case. Spanish settlers reached the Southwest very early and were in California by the eighteenth century. Among the exotic plants they introduced was the forage crop alfalfa, which spread east to the margin of humid areas and north to its winter hardiness limit near the Platte River valley of Nebraska. The origins of alfalfa trace back to the earliest agricultural settlements in Asia; hunter-gatherers collected it as a high-protein seed.[9] When the pioneer Nevada botanist F. H. Hillman conducted a study of weed seeds contaminating samples of legume seeds in the late 1890s, he found abundant soft chess seeds but apparently no cheatgrass.[10] In Idaho in the 1930s, R. L. Piemeisel found cheatgrass in virtually every sample of alfalfa seed he examined.[11] It is probably safe to assume that this contamination had existed since the earliest introductions of alfalfa to a given area.

In 1994, a symposium was held in Medicine Hat, Alberta, Canada, to discuss the consequences of a linear infestation of cheatgrass across the southeastern Alberta prairies and possible control measures. The infestation occurred when intermediate wheatgrass grown and certified in Idaho was seeded on a pipeline right-of-way. The seed bags were correctly labeled to show cheatgrass as a nonnoxious weed contaminant in numbers below the prescribed level for other weed seeds.

Perhaps the most striking case of exotic annual plant invasion in western North America is that of filaree, which apparently was abundant in California before European colonization in 1769. Adobe bricks from the earliest mission buildings in southern California contained large amounts of filaree in the plant material added to strengthen the adobe.[12] When John Frémont visited the San Joaquin Valley in 1844, filaree in full bloom covered the entire valley floor.[13] Filaree's long tenure in the Great Basin is evidenced by the fact that it was the only exotic plant that the Indian tribes of the western Great Basin had named when an ethnologist first studied their cultures.

In *Alien Plants Growing without Cultivation in California* the noted botanist W. W. Robbins suggested that cheatgrass was introduced into California before 1900.[14] Cheatgrass is not a common component of the annual grass–dominated ranges west of the Sierra Nevada that have a true Mediterranean climate. It did, however, find a home in northeastern California, where the climate is similar to that of the Great Basin, and in the trans-Sierra and southern deserts of California that are within the Great Basin. Surprisingly, the first recorded collection of cheatgrass in California was in Yosemite Valley (1900); collections in Klamath River (1901), Sisson (1902), Yreka (1904), Santa Barbara (1910), and Upland (1918) followed.[15] Lt. E. G. Beckwith and his plant collector, James A. Snyder, were the first to describe the plants of the dry regions in the rain shadow of the Sierra Nevada. While exploring a route for the Pacific railroad in the 1850s Beckwith's party encountered abundant bunchgrasses of fine quality for the horses in the sagebrush vegetation of north-central and northwestern Nevada along the forty-first parallel.[16] A decade later, Sereno Watson was the botanist for Clarence King's fortieth parallel expedition. Watson collected and described many new plant species, but none were abundant, highly competitive annual grasses.[17]

The range livestock industry got started in northern Nevada during the 1870s, largely after the completed transcontinental railroad provided a means of marketing beef.[18] In a remarkably short time, the ranges were stocked with cattle driven from Texas. In the 1880s, I. C. Russell traveled widely in northwestern Nevada during his exploration of pluvial Lake Lahontan. His descriptions of the rangelands mention abundant nutritious perennial grasses.[19] It quickly became obvious that the ranges could not support the number of cattle being stocked on them. Observations that the ranges were being overgrazed were not restricted to scientists. The editor of the *Carson Morning Appeal* in Carson City, Nevada, wrote on December 4, 1886, "There should be a call for appropriation of state funds for research to find ways to seed and restore the range."[20] The rapid decline of the native bunchgrasses is often cited as a prime example of why Great Basin rangelands should not be grazed by domestic livestock. All this decline shows, however, is that if excessive numbers of cattle feed on ranges year-round without sufficient supplemental feeding during the winter, the perennial grasses will disappear. Grazing in the early spring every year for several years in a row is particularly detrimental because the perennial grasses never get the opportunity to renew their carbohydrate reserves and produce seed before the summer drought. Grazing of domestic livestock in the temperate deserts is not inherently damaging to the environment if properly designed stocking rates and grazing management systems are used. The first photographs of the Great Basin taken specifically to show range conditions may have been those published by David Griffiths, who bore the impressive title of expert in charge of field management for the Grass and Forage Plant Investigations Unit of the USDA Bureau of Plant Industry.[21] From July 17 through August 30, 1901, Griffiths traveled more than 700 miles by buckboard from Winnemucca, Nevada, to Ontario, Oregon. In the preface to Dr. Griffiths' report of this tour, F. Lamson-Scribner, leader of the Grass and Forage Plant Investigations Unit, stated the reasons for the trip:

> Comparatively little was previously known as to the existing conditions in this region, and the present report shows the pressing need of *reform in range management*—a matter which applies not only to this section, but to all the open ranges of the West. Throughout the entire West, as the better land had been taken up by settlers, the cattle and sheep ranges had become more restricted, and stock were now forced back from fertile river bottoms and other lands so situated as to make irrigation possible, the inevitable result being overstocking of these restricted and poorer ranges, with all the attendant evils.[22]

Lamson-Scribner's analysis is relevant to the situation in the Great Basin in the late nineteenth and early twentieth centuries, but the relative proportion of irrigated lands to rangelands works against it being the primary reason for the decline in range productivity. Less than 5 percent of the landscape of the northern Great Basin was irrigable (without pumping groundwater) because of the limited mountain watersheds that produced runoff, and it does not seem plausible that withdrawal of that land from open grazing caused overgrazing of the rangelands. In any case, the irrigated lands were used almost exclusively to produce hay for wintering cattle. Irrigation increased the forage base rather than diminishing the amount of forage available.

Dr. Griffiths found the ranges of the northern Great Basin heavily overgrazed in certain places (fig. 3.1). Although he visited the Pine Forest, Bartlett Peak, Steins, White Horse, Blue, and Bendire mountains as well as the Humboldt, Quinn, Silvies, and Malheur rivers and the basins of the Alvord Desert and Malheur Lake, Griffiths did not have an unbiased sample of the ranges. His transportation, driver, guides, and cook were furnished by the superintendents of two of the largest cattle-ranching companies in the West: J. P. Irish of the Miller and Lux ranches, and John Gilcrest of the Pacific Livestock Company. It is thus not surprising that most of the examples of excessive

Fig. 3.1. The Steins Mountains of southern Oregon after being grazed by as many as 200,000 sheep during the summer of 1901.

grazing in the Great Basin were the result of *sheep* grazing. Griffiths found three exotic brome grasses along Bartlett Creek in northern Nevada: *Bromus rubens, B. hordeaceus,* and *B. secalinus.* The last species he called "cheat," and many people interested in Great Basin ranges have suggested that he meant *Bromus tectorum.* Lamson-Scribner probably identified the specimens, though, and almost certainly correctly. Griffiths attributed the presence of the annual brome grasses to the importation of an unnamed feedstock from California.

The very large range sheep industry that became established in the Great Basin after the hard winter of 1889–1890 killed up to 90 percent of the cattle in some areas had a substantial negative effect on the cattle industry.[23] Sheep could winter on desert ranges that lacked enough water to support cattle because they could get by with snow as a source of water. The sheep could wander through the sagebrush foothills when spring came and move into the high mountains for the summer. Cattle ranchers were already using these ranges, however, and viewed the land as "their" range even though technically it was public domain. Ranchers petitioned the USDA Forest Service to establish national forests on the mountain ranges of the Great Basin, hoping this would give them exclusive rights to summer grazing based on prior use and ownership of commensurate property and exclude the landless tramp sheep operations.[24]

Winter death losses forced cattle ranchers to develop hay fields for winter feeding. Hay could be produced only on irrigated land, however, and that was a rare commodity. The establishment of hay fields had unintended negative results on the range. The haying process was powered by horses. After the haying, the excess horses were turned loose on the range, and their presence further accentuated the excessive grazing problems. Even the crude level of farming associated with native hay production brought imported seeds and disturbances associated with irrigation structures and limited tillage. These factors coupled with importation of grain (oats) for workhorses presented the means for the introduction and establishment of exotic weeds.

R. N. Mack reported that cheatgrass was first collected in the Intermountain Area at Spence's Bridge, British Columbia, in 1889; Ritzville, Washington, in 1893; and Provo, Utah, in 1894.[25] All of these sites were experiencing rapid expansion in wheat production. The first collections in Nevada were from the northeastern part of the state in Elko County at Tuscarora and Skelton between 1905 and 1910.[26] A. A. Heller was the first to collect cheatgrass in western Nevada. His specimen, collected in Reno in 1912, is in the herbarium of the University of Nevada at Reno. W. D. Billings reported that P. B. Kennedy collected cheatgrass as well as red brome and other annual brome grasses along the railroad tracks in Reno as early as 1906. By 1904, cheatgrass had been collected throughout the dry regions of interior Washington, Oregon, and Utah. Within the next decade the weed appeared in eastern Washington, Oregon, southern Idaho, and northern Nevada and Utah. Within another 15 years it was abundant throughout the entire sagebrush steppe biome.[27]

Why was this spread so rapid? The inherent nature of cheatgrass and chance each played a role. One of the coincidental factors was the mechanization of large-scale cereal grain production in the far West. Large stationary threshing machines powered by steam engines made it possible to greatly enlarge the size of farms producing cereal grains. Separators and steam engines were too expensive for most individual farmers to purchase. Custom operators moved from farm to farm and from farming district to farming district threshing grain, often failing to clean the machines between jobs. Because farmers usually kept part of their crop for seed grain, the contaminated machines made sure the farmers' fields would be infested the next year. Farmers were relatively new to the West, and many were emigrants from Eastern Europe who brought along seed for planting in the new country or had it sent later. This is how Russian thistle came to the region.[28] In Scott Valley in northern California, a rancher with family ties in Ireland wrote home for alfalfa seed.[29] When it arrived, it contained seeds of dyer's woad, a plant cultivated in Europe since medieval times as a source of blue dye stock. The weed naturalized and became a serious pest on rangelands. The potential of multiple introductions of exotic invasive weeds is important in terms of the width of the genetic base of the introduced species.

D. L. Yensen did a very good job of characterizing the spread of cheatgrass into southern Idaho.[30] This area is important because two contrasting types of agriculture are in use there, dryland farming and cooperative irrigation districts. After the Newlands project was initiated at Fallon, Nevada, in 1903, a series of federally sponsored irrigation projects were established across southern Idaho. The water distribution laterals were major horse-powered construction projects that created miles of linear disturbance in native vegetation and bare soil, conditions well suited for the establishment of invasive weeds. The contractors for such projects moved from project to project in search of work, taking their horses and feedstocks with them. There is a documented case of such movement resulting in the introduction of Russian thistle from the northern Great Plains into Washington State.[31]

Widespread promiscuous burning was another major factor in the spread of cheatgrass in former sagebrush/bunchgrass plant communities. A basic tenet of plant ecology says that an existing plant community is substantially the product of the stand renewal process; that is, how the previous plant community existing on the site was destroyed. Burning in sagebrush/perennial grass plant communities destroyed accumulations of woody biomass and subcanopy sagebrush litter. The sagebrush, which does not sprout after fire, was killed, leaving a seedbed rich in bases along with conditions highly favorable for nitrification. Before the introduction of cheatgrass, such a stand renewal process provided ideal growing conditions for the native perennial grasses along with the subdominant shrubs that did sprout from crown or root buds.

Yensen went to considerable lengths to find information on promiscuous burning during the early period of grazing in the Intermountain Area.[32] The first mention of promiscuous burning in the sagebrush is in a figure caption published by David Griffiths in 1902. The figure shows a day's roundup of beef cattle in the Silvies Valley of eastern Oregon, and the caption mentions that smoke from forest fires is visible in the background.[33] The accompanying text notes that the fires were set by sheepherders as they drove their flocks down from the mountains. The photograph, perhaps posed, shows an excellent stand of perennial grasses. The unstated message may have been the contrast between how well the cattlemen were taking care of the range while the sheepherders were setting everything on fire. The large multistate cattle ranches were paying for and providing the guides for Griffiths' trip, after all.

Sheepherders were not the only ones to use wildfires for range management. James A. Young's father told him that cowboys bringing the cows down from the Salmon Mountains for the winter burned everything behind them. It was a common practice in those days. Grass was the primary grazing resource for both sheep and cattle, and promiscuous burning favored grasses over woody species. It also aided in livestock movement. Yensen mentioned that the ranchers believed the burned areas greened up sooner in the spring, and there is an element of truth to that. The black body effect does cause the burned area to warm up more quickly, and the increased availability of nitrogen speeds early spring growth of grasses. The big sagebrush overstory and the perennial grass understory existed in a delicate balance. Neil West suggested that this balance is relatively stable in the northern portions of the sagebrush/bunchgrass type of vegetation, which he called sagebrush steppe. In the Great Basin and on the Colorado Plateau, however, the balance between shrubs and grasses is much more precarious and easily tipped in favor of the woody species.[34] Forest Sneva compared the influence of increasing sagebrush cover on the productivity of understory grasses and found that at a sagebrush overstory cover of 10–20 percent, the productivity of the perennial grass understory decreased by 10 percent for every 1 percent increase in the sagebrush. Obviously, at 20 percent overstory coverage the understory perennial grasses would nearly be gone.[35]

The early herdsmen in the sagebrush-bunchgrass rangelands realized that grazing excessive numbers of domestic livestock was dramatically changing the balance of shrubs versus grasses in the plant communities and saw promiscuous burning as a solution. Where did this remedy originate? Perhaps it is another element of the block of "herder" genes. In fact, the proper application of prescribed burning followed by rest and then grazing management would have restored the balance between the shrub overstory and the perennial grass understory.

The introduction of cheatgrass changed the stand renewal dynamics between perennial grasses and sagebrush. Writing in 1914, the pioneer western ecologist H. T. Shantz described the influence of cheatgrass: "The fire consumes the dry herbaceous growth and the sagebrush plants are normally burned to the ground. They do not sprout up from the old stumps, and the result is usually the complete removal of *Artemisia*. In the following year, a mat of herbaceous vegetation composed chiefly of cheatgrass and redstem filaree covers the ground among the blackened stumps."[36]

In 1932, G. D. Pickford published a classic study of the influence of cheatgrass on these dynamics in the spring-fall grazing lands of the Utah foothills. Livestock in Utah (and in most of the Great Basin) either grazed on desert ranges during the winter or were fed hay on fields; during the summer the livestock went to high mountain ranges. From approximately the middle of April until the latter part of June each year, after the animals ceased to do well on the desert ranges and the ranchers were out of hay, and before the vegetation in the high mountains was ready to graze, the cattle and sheep were turned out on the lower mountain slopes and adjacent foothills. They returned to this same area for a month or 6 weeks in the fall. Pickford described a cycle of (1) cheatgrass invasion, (2) excessive grazing, (3) increased abundance of cheatgrass, (4) increased frequency of wildfires, and (5) continued dominance by cheatgrass in the spring-fall grazing lands.[37] He pointed out that topography and climatic limitations made the spring-fall ranges a finite resource. This resource had been greatly reduced by Pickford's time because farmers were continually expanding the area they put under cultivation. Pickford reported that in 1883, 35 years after the first settlers arrived in Utah, an estimated 100,000 cattle and 450,000 sheep occupied the territory.[38] In 1891 there were about 279,000 cattle and 3,537,000 sheep. In 1931, when Pickford was conducting his investigation of the spring-fall ranges, there were 344,000 cattle and 2,926,000 sheep. He considered that practically all the rangeland in the state had been fully stocked for 40 years.

Very few ranchers have left written records indicating that they recognized the changes that were occurring on the sagebrush rangelands, but they must have realized that something serious was happening to the range. The extreme variation in the climate of the Intermountain Area may have masked these changes on sagebrush ranges. Although the precipitation most years does not favor plant growth, an occasional excellent year comes along. The degraded sagebrush ranges respond dramatically in those wet years, and this growth temporarily masks the effects of degradation.

Utah rancher Glenn Bennion recognized what was happening to the perennial grasses and pointed it out in the *National Wool Grower:* "Sagebrush came when the wasteful, destructive methods of range exploitation, developed as a result of the Government's indefensible free-range policy, destroyed the grass, thus permitting those forms of vegetation that stock can not eat to take the place of grass."[39] Bennion took pioneer stockmen to task for their unknowing destruction of the bunchgrass ranges by prolonged, season-long, excessive grazing and suggested restoring the ranges by reintroducing wildfire coupled with rest from grazing. Bennion blamed the state of the sagebrush/bunchgrass ranges on the federal government's failure to manage the millions of acres outside the national forests that were still open to homesteading but were located in environments too arid to permit rain-fed farming. Pickford found that burning sagebrush/bunchgrass ranges that were protected from grazing resulted in a 32 percent reduction in perennial grasses compared with unburned, nongrazed communities.[40] At the same time, the density of cheatgrass increased by 22 percent on burned, ungrazed communities. One should bear in mind, of course, that Pickford worked for the USDA Forest Service, which was determined to eliminate wildfires on wildlands. Bennion's suggestion that fire was a tool that should be reintroduced to rangelands was bitterly attacked in a subsequent issue of the *National Wool Grower* by C. L. Forsling, the director of the Forest Service's Great Basin Experiment Station. Forsling concluded his rebuttal with: "Generally speaking, fire is an uncertain doctor with a cure more to be avoided than the disease."[41]

Many early stockmen extended promiscuous burning to areas already dominated by cheatgrass, apparently believing that repeated burning did not harm the annual grass and produced better feed the next spring.[42] At first glance such beliefs may seem ridiculous, but the nitrogen released by such fires and the black body effect of the burned area do indeed combine to enhance forage production in the early spring of the year following the fire. It is also easier to burn cheatgrass stands if they have not been excessively grazed than it is to burn degraded big sagebrush/bunchgrass ranges. In fact, the degraded communities without cheatgrass are virtually resistant to burning. What early stockmen did not appreciate was that repeated burning of cheatgrass-dominated rangelands mined nutrients, especially nitrogen, and left the communities open to invasion by other invasive exotic annual plants.

Early in the twentieth century, the ecology of rangelands in the Great Basin was in a tailspin. To be fair, it was very difficult for the range livestock industry to do anything about the destruction of the ranges because the federal lands were both everyone's and no one's. The federal government (Congress and the executive branch, collectively) left the unappropriated lands outside the national forests open to homesteading, but no one managed them. This would continue until the passage of the Taylor Grazing Act in 1934 and the establishment of the USDI Grazing Service.

CHAPTER 4

Scientific Perceptions of Cheatgrass

Cheatgrass had the misfortune to collide head-on with the birth of a new field of applied science that was adopted by the infant conservation movement and became the banner under which scientific range management was to march for a century. Early in the twentieth century, Frederic E. Clements at the University of Nebraska began to inspire students with a zeal for applying plant ecology to problems in natural resource management. Clements was a longtime collaborator with John E. Weaver, also a professor at the University of Nebraska and among the first to offer a course called plant ecology. These men, with very different personalities and skills, formed a powerful team.

Weaver's forte was the exquisite details. He was obsessed with the application of the science of plant physiology to the outdoor world and was noted for his meticulous studies of the root structure and function of the grasses of the Great Plains. Weaver even included a special section "To the Teacher" in his introduction to the textbook *Plant Ecology* giving detailed instructions on how to conduct a laboratory field trip. Weaver explained that even though he had taught ecology for many years, he always followed the same procedures in his preparations. Clements was an idea man who soared with the eagles. It might be cold in the shadow of his wings, but the minds of students were lifted to great heights by his inspiration.

The two pioneers of plant field ecology did not always agree. At a lecture for the botany faculty and graduate students at North Dakota State University, for instance, they had a heated exchange over some obscure detail of the ecology of the northern Great Plains grasslands. Mrs. Clements (also a professional botanist) extracted a collapsible cup from her purse, walked to the laboratory sink in front of the classroom, filled the cup with water, and gave it to her husband. He drank the water, thanked his wife, and smiled at Weaver, and they continued their lecture on a different subject.[1]

In their perennial standard textbook, *Plant Ecology*, Weaver and Clements described plant communities as arising when individual plants come together and interact with each other.[2] They used an abandoned garden to illustrate the process of vegetation accumulating to form a plant community. The first year following the abandonment of cultivation only a few annual weeds whose seed dispersal attributes allowed them to disperse to bare areas grew in the plot. Weaver and Clements explained the subsequent colonization of the plot: "In the struggle for light, water, and nutrients that follows annuals must start anew each year, show their handicap by disappearing, while perennials, which constantly hold the ground and extend their territory, increase in importance. Some are more successful than others and in time occupy the area more or less exclusively."[3]

Expanding this concept to a larger scale, Weaver and Clements cited the research of another pioneer of ecology, H. L. Shantz, who reported that abandoned fields and roads on the Great Plains passed through various stages and after several years could be revegetated with buffalo grass and grama grass.[4] The concept of successive change in the assemblages of plants that occupy a given site is known as "succession." The first North American biologist to recognize the concept of plant community succession was H. C. Cowles, who studied the sand dunes of Lake Michigan at the end of the nineteenth century.[5] In 1916, F. E. Clements published an entire volume devoted to the study and interpretation of plant succession.[6] He included three plates of plant communities in the Carson Desert of Nevada but did not discuss them in the text. Clements did not mention cheatgrass, but he included material provided by H. L. Shantz concerning succession on disturbed big sagebrush and salt desert vegetation in Utah.

The basic concept of plant succession rests on the idea that plants invading bare areas interact both among themselves and with the physical and biological environments of the site. Weaver termed the initial step in this process "aggregation." When sufficient plants have aggregated on the site to exceed the supply of water, minerals, or light, the plants begin to compete for these resources. Some plants die out and are replaced by other plants more adapted to compete on the site. Each successive assemblage of plants is adapted to the conditions created by the previous group of plants, which changed the inherent potential of the site by increasing organic matter, changing the soil pH, providing litter that creates shade, reducing seedbed temperatures, or by altering other environmental factors.

As the succession process continues, increasingly complex assemblages of plants dominated by perennial species replace the original simple, annual-dominated communities. All successional sequences do not end with the same plant community because the potential of the site is determined by the existing climate. Weaver and Clements termed the plant community that is in balance with the climate the "climax community." Perhaps influenced by the vastness and uniformity of the Great Plains, they saw climaxes as existing over enormous areas. This is apparent in the six criteria they used to describe a climax plant community.

1. Climax dominants must all be of the same life form. Trees or forbs cannot be the dominant species of grassland climax. (This may have been the first time the term "forb" was used to describe nongrass herbaceous species.)[7]

2. One or more dominants must range throughout a formation.

3. A large number—and usually the majority—of the dominant genera extend throughout the formation, although they may be represented by different species.

4. When the dominants of two contiguous associations meet in transition areas—called "ecotones"—over vast distances, the two associations should be classified under the same formation but as separate associations. Several of Weaver's students spent a great deal of time in studies of ecotones, especially on rangelands. Perhaps they were trying to understand this principle.

5. Stable "subclimax" communities consisting of associations at the stage just before the true climax stage occur because some repeated event, such as recurring fires, prevents attainment of the climax stage. This principle was to lead to a new theory in plant community ecology, the polyclimax method of community classification.

6. Changing climates exert natural selection pressure simultaneously on plant species and on the communities they constitute. This criterion raised assemblages from a mere list of plants to a living, interacting unit of biology.

If they did not originate plant ecology as a field of science, John E. Weaver and Frederic E. Clements certainly went a long way toward introducing the discipline in North America. Among their most important accomplishments was their mentoring of a series of students who became zealous apostles of the climatic climax theory of plant succession.

Joseph H. Robertson, one of Weaver's students who became a noted range scientist and educator, spent many years at the College of Agriculture at the University of Nevada Reno. He was a noted annotator of book margins (including library volumes), ending each, often acid, comment with a neat JHR. His notes are so perceptive that we have actually cited them in publications. We have a first edition of *Plant Ecology* with a book plate "ex Libris Joseph H. Robertson" that we rescued from a dumpster when the university threw away Dr. Robertson's files after his death. It contains no marginal notations; apparently, Robertson viewed annotating the text of the master as sacrilegious. But the book is in several other ways a treasure. John E. Weaver's chapter tests are carefully attached with glue at the end of each chapter, and old photographs of field trip sites are slipped between the pages. Many of Weaver's and Clements's students approached the application of ecology to natural resource conservation and management in much the same way. Chapter 8 discusses how the results of Joe Robertson's research on cheatgrass cast doubt on Weaver's and Clements's basic concept of community ecology.

Arthur William Sampson was among the most outstanding of the young men influenced by Weaver and Clements. Sampson often credited Frederic Clements with developing his interest in the science of ecology, which he applied to the management of rangelands. Sampson arrived on the scene at the perfect time.[8] A native of Nebraska, he graduated with a bachelor of science degree from the University of Nebraska in 1906, just a year after the infant conservation movement won its first big victory with the creation of the USDA Forest Service.

Created from the USDA Bureau of Forestry, the Forest Service was charged with the protection of timber production and watersheds. Forest reserves formerly overseen by the U.S. Department of the Interior were transferred to the jurisdiction of the Forest Service as national forests. The professionally trained foresters the Forest Service hired to administer these woodlands were graduates of the Yale and Duke schools of forestry or imports from Europe, mainly Germany, where they had been trained primarily as logging engineers. The foresters received a surprise when they examined their new domains: Millions of goats, sheep, cattle, and horses were grazing on them. In the West, livestock grazed more than four-fifths of the national forests' area, and these rangelands constituted 85 percent of the watersheds for the major western rivers.[9]

Chief Forester Gifford Pinchot had been in office only four days when the secretary of agriculture notified him that the jubilant General Lands Office was about to rid itself of more than two decades of grief by bestowing the entire forest grazing problem on the Forest Service.[10] The secretary instructed Pinchot to use "whatever plan, in your judgment, will act for the best permanent use of the range." Pinchot was up to the job. He believed the Forest Service had three essential duties in regard to rangelands on national forests: protection, management, and the effective application of science.

One year after the formation of the Forest Service, the agency hired Sampson, 22 years old and just out of college, to apply protection, management, and science to a critical portion of the new national forest rangelands. Sampson was sent to the high Wallowa Mountains in northeastern Oregon to conduct seeding trials on degraded sheep ranges. At the time, sheep grazed much of the subalpine portions of the Wallowa Mountains; an estimated 252,000 sheep were present in the area in 1906.[11] Forage production had dropped so drastically that many of the sheepmen had proposed closing the range to grazing for a few years. When the sheepmen requested help, Sampson was sent to attempt to seed the degraded subalpine grasslands. He was not successful in his initial assignment, but he gained valuable experience.

In 1912, the Forest Service launched a new system of range research by establishing experiment stations on the national forests. The first was the Great Basin Experiment Station, located in an aspen grove on the western front of the Wasatch Plateau in central Utah at an elevation of 8,500 feet.[12] A. W. Sampson was appointed its first director, and the Great Basin Station was to be the site of his greatest achievements. He worked at a frenetic pace, building and staffing the experiment stations, completing graduate studies for master's (Johns Hopkins) and Ph.D. (George Washington University) degrees, and writing both popular and scientific papers about his research. High in the subalpine bowls of the Wasatch, Sampson formulated an ecological approach to judging the ecological condition of rangelands. In a USDA bulletin published in 1919 Sampson noted that "enterprising stockmen and those concerned with administration of grazing know that the livestock industry has now reached a point where the intensity of use of the forage crop must be governed by a finer discrimination than mere observations of the density of the plant cover and the condition of the stock."[13] He introduced to range science the concept of regular replacement of one assemblage of plants on a given site with another assemblage in response to disturbance or lack of disturbance.[14] Sampson used Clements's term "succession" to describe this process. Clements's theory that the departure from the climatic climax on a given range site could be used to judge the ecological condition of that site was to become the scientific basis for range management in the twentieth century.

Sampson maintained that the fundamental processes controlling succession were simultaneous changes in the substratum; that is, aboveground changes in plant density and species composition were mirrored in changes in the soil throughout the rooting depth. This was straight Weaver and Clements doctrine. Sampson's failure to identify seedbed quality as the most important factor governing successional change in shrub/bunchgrass communities where change was dependent on seed germination and seedling establishment was to have lasting consequences. In grasslands where the dominant species are rhizomatous, sexual reproduction and seedling establishment are not such essential steps.

Sampson's conclusions thundered down from the Wasatch Front with the power of a summer flash flood and flowed across the desert ranges to reach all corners of the western range. Among his edicts was that "grazing may cause either progression or retrogression succession, depending chiefly on the closeness with which the herbage is grazed annually and the time when it is cropped."[15] Sampson was convinced that the western range had been overgrazed during the relatively short period since it had been stocked with domestic livestock. He concluded, "The most rational and reliable way of recognizing the incipient destruction of forage supply is to note the replacement of one type of plant cover by another, a phenomenon which is usually much in evidence on lands used for grazing of livestock."[16]

How does this ecological theory relate to cheatgrass? For several reasons it effectively barred the door against cheatgrass ever gaining acceptance as a forage plant. Cheatgrass is an exotic species, not a native of North America. It is an annual species, not a perennial. It is a lower successional invader species, not a climax species. Weaver's and Clements's basic definition of climatic climax communities classified cheatgrass as an outsider that would disappear if the ranges were properly grazed.

After it was introduced into the Intermountain Region of western North America, probably late in the nineteenth century, cheatgrass remained a very insignificant ruderal, or roadside, species for a quarter century. During the most important period in its evolution as a major weed on the western range, cheatgrass remained a nonentity in science because it existed in a biological and political vacuum—the vacant, unappropriated public lands on the western range, which were largely grazed as community pastures. The only attention cheatgrass attracted was due to the injuries its seeds caused to the mouth and eyes of grazing animals.[17] References to and descriptions of the establishment of cheatgrass are largely anecdotal accounts included in much later publications.[18]

Cheatgrass was not the first *Bromus* species recognized in the Great Basin. Soft chess was apparently much more abundant in former big sagebrush/bunchgrass areas of the Intermountain Region from 1900 to 1915 than it is today, while during the same time period cheatgrass was relatively rare.[19] Did soft chess decline because it was replaced by the better-adapted cheatgrass, or did the originally introduced cheatgrass hybridize with one or more of the other introduced annual brome species and produce progeny adapted to the former sagebrush/bunchgrass rangelands? Interspecific hybridization would not have been necessary for the evolution of an adapted form of cheatgrass, but the possibility should not be overlooked in interpreting how this introduced species came to dominate millions of acres of rangeland.

The 1930s was the decade of great awakening to the potential dominance of cheatgrass. G. D. Pickford wrote the first classic paper of this period, "The Influence of Continued Heavy Grazing and Promiscuous Burning on Spring-Fall Ranges of Utah."[20] Writing nearly two decades after Pickford, Joe Pechanec and George Stewart provided statistics on the spring-fall ranges of southern Idaho that are applicable to those of Utah.[21] On a crescent-shaped belt on the Snake River Plains of southern Idaho, between 700,000 and 900,000 sheep used 18–20 million acres of spring-fall ranges annually. The ranges were used in the spring during and following lambing, a time when the ewes need to eat palatable, highly digestible, nutritious young grasses and forbs to provide milk for the new lambs. In the fall, sheep returning from high mountain summer ranges used the ranges during and following breeding. The availability and quality of forage at this time influenced the size of the spring lamb crop. Pickford pointed out that by 1930 grazing pressure was increasing on the spring-fall ranges in the foothills because former ranges at lower elevations had all been fenced for crop production. Farmers were bringing ditches out of the streams at higher and higher elevations so that more land could be cultivated for irrigated crop production. At the same time, the Forest Service continually set back the date when livestock could move onto the high-elevation summer ranges based on research indicating that native perennial grasses had to restore their carbohydrate reserves in the spring if they were to persist. In the meantime, livestock had to stay on the spring ranges, which expanding farming operations had greatly reduced in area. If soil moisture was exhausted by the time the animals left for the summer ranges, the herbaceous species the animals would depend on when they returned in the fall would be unable to grow. Range livestock were caught in a vicious cycle that, as Pickford recognized, damaged the rangeland resource.[22]

Pickford considered promiscuous burning and excessive, improperly timed grazing to go hand in hand in favoring the dominance of cheatgrass over native perennial grasses, while burning and complete rest from grazing favored the native perennial grasses. Complete rest without burning favored the native shrubs, but an understory of cheatgrass would eventually develop and make complete suppression of wildfires impossible.

In 1939, George Stewart and A. E. Young published an emotionally charged paper entitled "The Hazard of Basing Permanent Grazing on *Bromus tectorum*."[23] At the time, scientific papers concerning range management were generally published either in the *Journal of Forestry* or as USDA Forest Service bulletins. It is somewhat surprising that Stewart, a senior forest ecologist with the Forest Service's Intermountain Forest and Range Experiment Station, would choose the journal of the American Society of Agronomy as an outlet for this material, although Stewart was an agronomist by training.

Perhaps the very strong opinions the authors expressed against cheatgrass and the ensuing controversy the paper ignited had something to do with their choice. Stewart and Young ensured controversy by insisting on using the apparently coined common name "downy chess" rather than "cheatgrass," the name commonly used in the Intermountain Area. Stewart and Young clearly stated two hazards of basing permanent grazing on cheatgrass: (1) the wide variation in forage production from one year to the next, and (2) the uncertainty of any forage for livestock in dry years. Many of their original observations on the history and biology of cheatgrass cut to the heart of the impact cheatgrass was having on rangelands. Their ability to pack so many astute observations into such a short paper was nothing less than brilliant. Further, Stewart and Young were the first to publish the observations that cheatgrass appeared in Utah about 1900 and was an occasional species on roadsides, fence lines, and old stands of alfalfa that gradually spread into degraded sagebrush/bunchgrass rangelands in the foothills. They noted that former bunchgrass ranges that once "produced some harvestable forage during the dry summer months" were bare. By "the end of June only the tough, straw-colored stalks and hard-toothed seeds of downy chess, dead dry, and highly inflammable," remained. They observed that cheatgrass stems are so slender that in wildfires they rapidly burn through, dropping the seed head to the ground; the seeds are not injured by the fire, which by that time has already passed.

Stewart and Young proposed that temperature and moisture influence cheatgrass forage production in two critical periods: (1) the fall, when seed germination ordinarily takes place; and (2) the spring, when flower stalks elongate and most of the herbage is produced. If autumn precipitation is so low that germination does not occur, or if spring weather conditions are dry and cold, the amount of available forage will be very low. If both conditions coincide in the same biological year, there will be virtually no cheatgrass or, very rarely, none at all. A warm, wet autumn coupled with a long, warm, wet spring, on the other hand, will produce more cheatgrass than livestock can consume.

The section of Stewart's and Young's paper dealing with the distribution of cheatgrass at the time they wrote (1937–1938) established benchmarks that held true for 40 years before the annual grass suddenly began a second wave of expansion. "Downy chess has spread throughout Utah in the intermediate foothill zone," they noted, "not being abundant in the high elevations where mountain plants prevail nor in the dry desert valleys where Russian thistle replaces it as the predominant annual on abandoned cultivated land and in spaces of a weakened plant cover that were formerly occupied by native desert perennials."[24] A reprint of the paper from the files of Joseph H. Robertson features a marginal note next to this quotation: "58 words JHR." Stewart and Young noted that day length seemed to control cheatgrass flowering, an astute observation that has largely been lost in the literature on the physiological ecology of cheatgrass.

Cheatgrass had been spreading in the Intermountain West for about 40 years when Stewart and Young published on the hazards of grazing it.[25] They pointed out that cheatgrass was particularly abundant in the foothills of the Snake River valley, especially in a 200-mile stretch centered on Boise, Idaho. It was considered a more recent invader in Oregon, Washington, Nevada, and across the mountains in Montana. Their chronology of the route of invasion of cheatgrass is not correct, but they lacked information in 1939 that is available today. Stewart and Young also attempted to quiet the fears of many stockmen in the Intermountain Area that the dominance of perennial grasses could never be restored once cheatgrass was established. They pointed out that cheatgrass had invaded several areas of good-condition perennial grass ranges where it remained a very minor component of the community. One complex sentence from Stewart's and Young's paper captures the paradox that still haunts the management of cheatgrass-infested rangelands. "Unanswered questions have been raised as to whether perennials can in ordinary circumstances make good headway in stands of downy chess; whether in arid conditions such replacement would not be so slow as to make it inadvisable to sacrifice the spring growth of downy chess in order to give the perennials a chance; and whether, as some think, the downy chess is as valuable for forage as the perennial that can be restored under conditions which are unfavorable to the perennials."[26] A later paragraph continued that line of reasoning: "The many millions of acres of spring and fall range occupied by it [cheatgrass] and the extreme scarcity of perennial grasses make it a highly important feed. In fact, it is so much used during the range lambing season that many stockmen regard it as the principal source of spring feed. This tends to obscure the importance to stockmen of restoring the good perennial grasses which are more palatable, produce higher yields of forage, do not have toothed seeds, and keep soft for 8 to 12 weeks in comparison with 4 to 6 weeks for downy chess."[27] "Keep soft" is an interesting description for forage. Apparently, the authors were referring to the period during which the herbage of cheatgrass and native perennial grasses is green and readily eaten by livestock. The peak green forage period for cheatgrass is the month before seed is set in the spring. If the spring is relatively warm and wet, cheatgrass herbage production is excellent and provides a preferred forage. Stewart and Young did note that the hairs on the leaves of "downy chess" lowered livestock's preference for the forage, but this appears to be a very obscure point. Once the flowering stalks of cheatgrass elongate, the herbage is less attractive to livestock. The same is true for the native perennial grasses; certainly their herbage is less digestible later in the season. Much of the argument comparing the short period of green forage for cheatgrass with that of native perennial grasses is smoke and mirrors. On a true spring-fall range, the grazing animals would be on summer ranges by the time cheatgrass was mature. Only if the foothill ranges were grazed all season long would the maturity of cheatgrass be a factor in livestock preference. When the livestock return in the fall, all the herbage is mature and dry. If early fall rains occur, cheatgrass seeds germinate and the seedlings provide limited green fall forage to mix with the mature herbage, which is softened by the moisture. Stewart and Young did mention a study published by L. C. Hurtt indicating that horses were able to maintain themselves during the summer on a diet of dry cheatgrass.[28]

Stewart and Young were very concerned about the influence of smut (*Ustilago*) infestations on the potential persistence of cheatgrass populations. Smut infection destroys the florets of grasses and inhibits seed production. They noted that smut infestations of cheatgrass were so severe and widespread in the area north of Mountain Home, Idaho, in 1935 and 1936 that virtually no seed was produced.[29] White-faced Herefords were the most popular breed of range cattle at the time, and cows often came off the range with black faces acquired from grazing smut-infested cheatgrass. Stewart and Young thought that these infestations had greatly suppressed cheatgrass and favored the native perennial grasses. Smut infections inhibit seed production of cheatgrass, however, not growth. At the time, no one knew that cheatgrass builds seed banks that persist for 3 or more years. Even the loss of seed production for 2 consecutive years as a result of smut infestation—which would be highly unusual—would not eliminate cheatgrass. Stewart and Young do deserve credit for being the first to mention the occurrence of smut infections in cheatgrass.

The bulk of the actual data that Stewart and Young presented dealt with the large annual variation in the production of cheatgrass herbage. In dry years the cheatgrass produced virtually no herbage while native perennial grasses did produce some, although less than that observed in good years.[30] Stewart and Young presented data for a native perennial grass range in Gem County, Idaho, for the years 1937 and 1938. Precipitation differed from 11–15 inches in the study area in 1937 to 24–28 inches in 1938. Forage production was expressed in very obscure units of "forage acres per 100 acres." The relative heights of the bars in their graph seem to indicate that forage production on the native range exceeded forage production on the cheatgrass range by 1.6 times in a wet year and by 11 times in a dry year. These conclusions should perhaps be considered in conjunction with the fact that Stewart and Young were employees of the USDA Forest Service and were officially writing for that agency.

In 1942, 3 years after Stewart and Young published their paper, another pioneering paper on cheatgrass appeared. *Bronco Grass* (Bromus tectorum) *on Nevada Ranges* was written by C. E. Fleming, M. A. Shippley, and M. R. Miller of the Nevada Agricultural Experiment Station and published as a station bulletin.[31] The authors were scientists funded by the rancher-dominated state legislature. Although we do not question the authors' objectivity, we do note that the environment in which a scientist works, discusses findings with peers, and interprets results influences such interpretations, either consciously or not.

C. E. Fleming was a distinguished range scientist and the developer of the one-night, or blanket, system of range sheep management.[32] A graduate of Utah State and Cornell universities, Fleming joined the Forest Service as a range examiner in 1910 and was appointed the range management staff officer for the Rocky Mountain Region in 1911. He worked as a range scientist for the Forest Service at the Jornada and Santa Rita grazing reserves before becoming the chair of the Department of Range Management at the University of Nevada. M. A. Shipley was Fleming's assistant in range management at the University of Nevada, and M. R. Miller was a chemist. Like

Fig. 4.1. Cattle that grazed on 95 percent cheatgrass range from the time they left the winter feedlot in the spring until August 3, 1941.

Stewart and Young, Fleming and his colleagues called cheatgrass by another name; they used "bronco grass."

The cover of their monograph featured a black-and-white photograph of a group of very healthy looking Hereford cows and large calves congregated near a water trough (fig. 4.1). The caption indicated that the cattle had grazed on a diet consisting of 95 percent cheatgrass from the time they left the feedlot in the spring until the photograph was taken. In their foreword the authors acknowledged the importance of cheatgrass and questioned the wisdom of trying to eliminate it and reestablish native grasses:

> The wide spread and abundant distribution of bronco grass throughout northern and central Nevada seems to have taken place within the period of forty to fifty years. There is no doubt that a combination of favorable climatic conditions with existing methods of range use caused its present abundance. It is found most abundantly on the spring-fall ranges of the foothills and lower mountains where it is necessary to use the forage plants during the early part of the grazing season. It is the grass of the spring ranges. It is also found, but less abundantly, on summer ranges. These summer rangelands lie at higher elevations where moisture conditions are more favorable to perennial plants and where grazing is usually deferred until the plants have an opportunity to make considerable growth. Because of the local abundance and wide distribution of bronco grass on the spring-fall ranges and because of its grazing value at various stages of growth and maturity, it contributes at least as much feed for the grazing of livestock as any other single forage plant found on Nevada ranges. However, many livestock husbandrymen, range administrators, and investigators would like to see bronco grass replaced by an equal quantity of the native perennial

grasses if the change could be made successfully and economically. If a practical method of replacement can be found, this will mean a drastic change in methods of using the spring range which have been established for more than forty years. If the restoration of the old time perennial grasses, the "bunchgrass," is to be brought by any change in methods of range use, the character of the change in method has not yet even been suggested. If the bunchgrass is to be restored through a program of reseeding, then this implies the apparently insurmountable tasks and conditions that are favorable to the growth and reproduction of the perennial grasses while at the same time creating an environment unfavorable to the growth of bronco grass. At the present, we simply do not know of any practical method and we have no information which indicates that bronco grass can be replaced on vast areas of Nevada rangelands which it now occupies. Perhaps a prolonged period of experimental work will show a practical way in which bronco grass may be replaced, but until the way has been found it is necessary to recognize bronco grass as a source of range feed in accordance with the way in which it must be used and the amount and value of the forage which it produces. However, there is another question to be asked. If the old-time growth of perennial grasses, the bunchgrass, were restored would it be as useful as the growth of bronco grass? That is, would it stand heavy spring grazing as well as bronco grass or would it disappear again under the necessary conditions of spring grazing use?[33]

What would Professor Fleming and his associates think if they were to return to Nevada today, 65 years after they prepared their pioneering bulletin, to find that the "prolonged period of experimental work" is continuing and that scientists are still trying to find a practical way to replace cheatgrass? During those same 65 years, the perennial crested wheatgrass was introduced. It meets Fleming's requirements for a species to replace cheatgrass, but modern range ecologists seem to have forgotten the question Fleming and associates raised regarding the fate of the native perennials if they were restored but the absolute requirement for early spring forage was not satisfied by some source other than grazing the ranges.

The major point made by the bronco grass bulletin was that cheatgrass was the most important forage plant on Nevada ranges. Sixty years later, however, federal land managers still do not consider cheatgrass a part of the forage base. Land managers still assume that the exotic cheatgrass will disappear and native perennials will flourish if the range is managed "correctly." The laws of succession "cast in stone" by Weaver and Clements guarantee it. The native perennials will flourish because the laws of succession dictate it. The obvious next step in this narrative is to explore why cheatgrass truncated plant succession on Great Basin ranges and came to dominate millions of acres of rangelands.

CHAPTER 5

Seral Continuum

The First Step

Does the plant community concept of ecology as taught by John Weaver and Frederic Clements apply to introduced weeds such as cheatgrass? The answer is a highly qualified "yes and no." Succession does occur among plant communities dominated by exotic annual species on temperate desert rangelands. In fact, we will show in this chapter that a host of exotic species occupy the same site in succession. Each succeeding assemblage of species modifies the potential of the site, paving the way for the next step in succession. The environmental factors modified by each assemblage of plants are most often seedbed characteristics, because the seedbed is the selective filter that controls the species composition of the next successional stage. The radical difference between the secondary successional process that actually occurs in Great Basin rangelands and the one Clements envisioned is that succession does not proceed through a series of stages, increasingly dominated by perennials, until a stable climax of native species is in equilibrium with the potential of the site. Cheatgrass truncates succession and becomes dominant for prolonged periods. The slightest disturbance is sufficient to perpetuate cheatgrass dominance.

Classic secondary plant succession begins with bare ground, just as it did in Weaver's and Clements's abandoned vegetable garden. The first exotic species introduced into the degraded rangelands of the Intermountain Area turned out to be the primary successional species in the seral continuum that ends in cheatgrass dominance. This species is Russian thistle. Russian thistle is a weed that has been associated with agriculture for so long that its original range is lost in antiquity.[1] It came out of the steppes of Asiatic Russia, where cereal production covered extensive areas. It is one of the few species whose original introduction into North America and subsequent spread are fairly certainly known. The fact that this history was recorded is a tribute to the devastating influence the weed has had on American agriculture, especially on the northern Great Plains. Late in the nineteenth century, the U.S. secretary of agriculture received multiple complaints from the northern Great Plains about a terrible weed that was destroying wheat production. S. W. Narregang, president of the Dakota Irrigation Company, wrote on October 28, 1891:

> I send you here a fair specimen of the Russian thistle. I would say that we first saw it three years ago. Since that time, it has steadily increased, until at present the greater portion of South Dakota east of the Missouri River is infested with the thistle, particularly the strip of counties extending from Eureka, Campbell southeasterly to Sioux Falls, which is covered thickly with this weed. This obnoxious weed has become so formidable in some portions of the state, notably in Scotland, South Dakota, where the Russian[s] formerly settled, that many farmers are driven from their homes on account of it. A man who was there some time ago states that farmers were leaving their land by the dozens simply because of the evil weed.[2]

As complaints mounted, the secretary dispatched Assistant Botanist Lyster Moxie Dewey to investigate the biological nature of the plant and find a means of eradication. Dewey proved to be an astute detective as well as a botanist. He discovered that Russian thistle was first introduced on a farm in Bonhomme County, South Dakota, about 1877. A few seeds of the thistle were mixed with flax seed imported from Europe.[3] One aspect of the introduction had ugly social undertones; many believed that the weed was deliberately introduced by Russian Mennonite emigrants in revenge for perceived social injustices. Dewey went to considerable lengths to dispel that theory.

On the botanical side, Dewey reported that Russian thistle was an annual that completed its life cycle in one year. The most objectionable thing about the plant was its sharp, spinelike leaves, which could lacerate the legs of horses running in pastures. Some farmers bound their horses' legs with leather to protect them.[4] The most remarkable part of the life cycle of Russian thistle occurred in the fall. As the November winds blew across frozen fields, the plants snapped off at the soil surface and tumbled with the wind, racing across the fenceless, treeless plains, scattering seeds with every bounce. The rolling action of the Russian thistle was particularly hazardous during prairie wildfires, when burning plants bounced across fire lines. Russian thistle spread across the plains at about 10 miles per year, making some spectacular jumps as farmers unwittingly helped its spread by sowing contaminated seed grain. The relentless spread spurred citizens to action. E. T. Kearney, a farmer himself, proposed to the North Dakota legislature that the state build a wire fence across the state to hold back the tumbling plants.

What characteristics make Russian thistle such a successful weed? The first key to its success is the plant's tremendous seed production; a very large plant may produce more than 250,000 seeds. How can an annual herbaceous plant with limited photosynthetic potential produce a quarter of a million seeds? Russian thistle solves the problem by putting minimum energy into producing each seed. The seed consists of a miniature plant—radicle (root), hypocotyl (body of the plant), and cotyledons (embryonic leaves)—tightly coiled within a very thin membrane. A second factor that makes Russian thistle such an effective weed is that the small seeds do not dehisce, or drop, from the plant at maturity, but are instead held tightly in the spinescent leaf axil. After the first fall frost, a layer of cells in the stem near the soil surface suddenly disintegrates, releasing the aerial portion of the plant to be driven wherever the wind blows it. The Russian thistle plant's canopy is hemispheric rather than round, so the plants tumble rather than rolling smoothly. Each time the flatter portion of the canopy slams down on the soil surface, the impact forces some seeds on that section of the plant from the axils. This is an excellent seed dispersal system. Although the seeds travel far, Russian thistle seedlings are not very competitive. If the parent plant is to reproduce successfully—that is, produce seedlings that establish, grow, and produce seed themselves—the seeds must reach largely bare areas where competition is at a minimum. Russian thistle cannot occupy stable habitats because of the basic plant ecology principle enumerated by Weaver and Clements. When a Russian thistle becomes established and grows to maturity on a site, the potential of the seedbed to support germination is changed, and there will be succession to another assemblage of plants. A few Russian thistle plants will persist in areas of minimum competition, but within a couple years they will also disappear.

Russian thistle had such great success on the northern Great Plains because homesteaders were still plowing virgin prairie. Typically, they plowed more than they could farm and let the turned-over sod decompose for a couple of years before tilling it or preparing a seedbed for cereal grain production. This vacant disturbed soil was prime habitat for Russian thistle. At the same time, cereal grain production on the plains was not yet perfectly matched to the environment because the hard red spring wheat varieties had not yet been developed. There were plenty of bare areas in cereal grain stands to harbor Russian thistle.

Russian thistle's inability to compete with other exotic annuals or native perennials may suggest that it is a wimpy weed that utilizes a sophisticated dispersal system to passively occupy bare habitat. This is true, but Russian thistle seeds can exploit the harshest seedbeds. In semiarid to arid environments, seed on the surface of seedbeds must take up moisture from the substrate (as a liquid) faster than they lose moisture to the atmosphere (as a vapor). Russian thistle seeds are so successful in part because of their speed in that regard. They have other physiological adaptations as well.

During and immediately after rainstorms in the Great Basin, seedbeds, even those in arid environments, have free water in the substrate and high relative humidity immediately above the soil surface. Russian thistle seeds germinate so rapidly, and the radicle begins taking and transporting moisture so quickly, that germination and seedling establishment can occur during a brief shower on the desert. The membranous seed covering is easily discarded, and the coiled embryonic seedling springs to life as it imbibes moisture. It is a near instantaneous jumpstart. Alvin T. Wallace and his associates, working on the Nevada Test Site of the U.S. Department of Energy, were the first to quantify this mode of germination for Russian thistle.[5] Why were scientists on the site where atmospheric testing of nuclear weapons was conducted interested in Russian thistle? It was the first plant to invade ground zero on nuclear weapon test sites. If the Cold War had ever flamed into a nuclear war, Russian thistle would have inherited the earth.

Another aspect of the inherent physiology of Russian thistle that keeps it near the bottom of the successional scale on temperate desert rangelands is its germination requirements. It is among the last of the exotic invasive species to mature seeds each year, and newly matured seeds will germinate only within a very narrow range of temperatures. The temperature requirement for germination broadens over the winter, and by spring, germination can occur at 32°F or 120°F, or the improbable combination of both temperatures in one diurnal fluctuation.[6] This process of changing germination characteristics after seed maturity independent of environmental conditions of seed storage is known as after-ripening requirements. The seeds may even be immature when they dehisce from the plant. Gradually, the maturity process continues and the temperature-related after-ripening dormancy disappears. Most of the exotic invasive plants that are dominant at higher successional stages can germinate in the fall and complete their life cycle as winter annuals. In an environment such as the Great Basin, where most of the moisture events occur during the cold winter months, this is a decided advantage. By the time Russian thistle seeds are dispersed and free to germinate, the seedbeds where they find themselves are already occupied by the highly competitive seedlings of exotic invasive species that are successful at higher successional stages.

Russian thistle spread from the Great Plains to the Pacific Coast within a couple of decades in the late nineteenth century, in part because of the construction of numerous railroads and irrigation projects. Power for these construction projects was provided largely by horses, which require hay and grain when they are working. Most of the hay and grain they were fed was produced in established agricultural areas and transported to the frontier. Russian thistle thus rode the tide of development across the West.[7]

Russian thistle was well established in the Great Basin by 1900. Although agricultural fields in that region were often separated by vast expanses of desert vegetation, the new weed dispersed rapidly. The road system was still primitive, and the major roads, at least, were graded by horse-drawn implements to improve drainage and smooth the ruts and potholes. Such roadside disturbances created perfect habitat for Russian thistle and avenues for the weed's dispersal. Even before the introduction of domestic livestock and the subsequent reduction in native herbaceous vegetation, naturally occurring vacant space was common in Great Basin plant communities. As the environment became more arid in the lower basins, the amount of vacant space increased. Areas of blowing sand created expanses of habitat suitable for colonization by Russian thistle. No native herbaceous annual had evolved the dispersal system and seed characteristics that allowed Russian thistle to exploit bare seedbed areas.

One of the disadvantages to Russian thistle of producing low-input seeds in abundance is that it does not build seed banks. This is another factor keeping the species at the bottom of the successional ranks. One of the subtle influences of Russian thistle's invasion into Great Basin rangelands has been on the dune characteristics of sand fields.[8] After they have tumbled and dispersed their seed load, dry Russian thistle plants are a serious disposal problem along highways and in irrigation structures throughout the West.

Although Dewey was successful in identifying what was perhaps the first introduction of Russian thistle to the northern Great Plains, it is unlikely that the species was introduced only once. It may have been brought from Asia numerous times. O. A. Stevens, the pioneer North Dakota botanist, told James Young that on his daily nature hike along the Red River he stopped one fall day and tried to visit with a group of Russian settlers camped next to the river. Language proved a barrier to communication, but it was apparent that they were straight from the old country and on the way to homesteads in western North Dakota. The next spring he noted a thick stand of Russian thistle seedlings where the settlers had tied their stock. Seeds may also have arrived from adventive populations in Australia or Chile. If this scenario is correct, it is of more than passing interest because it would indicate that the genetic base for Russian thistle populations is quite broad rather than having passed through a genetic keyhole with a single introduction.

We will jump out of chronological order at this point to discuss a second species of Russian thistle that was not recognized as part of the flora of the Great Basin until the 1970s. Barbwire Russian thistle also originated in Central Asia, where it is the herbaceous annual Russian thistle of desert rangelands. In the Great Basin, Russian thistle had colonized both rangeland and cropland before barbwire Russian thistle was introduced. Now Russian thistle is largely restricted to cropland and barbwire Russian thistle rules the desert ranges.

No one really knows when barbwire Russian thistle arrived in the Great Basin. It took the acid writings of botanist Janice Beatley to convince range managers that a second species of Russian thistle was present.[9] Range managers pride themselves on their ability to identify plants, and they were embarrassed to discover that a new invader had spread across millions of acres undetected. They can at least offer as an excuse the extreme phenotypic plasticity of Russian thistle, which masked the morphological differences of the barbwire thistle. Mature Russian thistle plants come in a bewildering range of sizes, shapes, and colors. The two species overlap in morphological characteristics and probably hybridize. Once the coarser, spinier growth form of barbwire Russian thistle is identified, distinguishing the two species is relatively easy.

We found three major differences when we compared the germination ecology of Russian thistle with that of barbwire Russian thistle.[10] The after-ripening requirements for barbwire Russian thistle seeds are less restrictive than those for Russian thistle, and the former can germinate at a wider range of temperatures sooner after seed maturity. Barbwire Russian thistle also germinates more rapidly at low incubation temperatures than does Russian thistle. In addition, at least a portion of the seeds of barbwire Russian thistle will dehisce beneath the mother plant's canopy without the necessity of tumbling.

Livestock will graze both species despite their spiny leaves. In the winter, Brewer's sparrows and other small seed-eating birds can often be observed feeding on Russian thistle seeds on dry plants along roadsides. Small granivorous (seed-eating) rodents in the Carson Desert of western Nevada eat the seeds of barbwire Russian thistle for extensive periods.[11] When the seeds of the native perennial Indian ricegrass are available, granivorous rodents switch from barbwire Russian thistle to seeds of the native grass. Even in the winter, when Russian thistle plants are mature, stiff, and spiny, cows will utilize them for forage if there is no better alternative. We stopped on the Pyramid Lake Indian Reservation one cold January day and watched a very hungry cow work down a barbwire Russian thistle plant that had lodged against a greasewood shrub. A cow's tongue is a remarkable appendage. This old cow used her tongue to twist and roll the barbwire thistle plant until she got it headed into her mouth root-end first. It required a lot more tongue work and head tossing before she got the entire plant down. Hopefully, gastric juices softened the thistle branches considerably before the old cow lay down to chew her cud.

Another species at the bottom of the seral continuum on Great Basin rangelands is halogeton. Like the two species of Russian thistle, halogeton is a member of the goosefoot family, Chenopodiaceae. Halogeton was first collected in North America near Tobar Siding southeast of Wells, Nevada, in 1934, and rapidly proved to be well adapted to the salt deserts of the Intermountain Area.[12] Halogeton was considered a variant of Russian thistle until 1942, when it was discovered to be highly toxic to sheep. It apparently replaced Russian thistle and extended the realm of exotic annual species into salt desert environments that were too dry and salt-affected for even Russian thistle to grow. When barbwire Russian thistle was introduced, it replaced halogeton on many sites. This replacement illustrates an important aspect of the ecology of exotic, self-invasive annuals first noted by Fritz Went: they can be extremely competitive with seedlings of native perennial species but are highly susceptible to competition from new introductions of their own kind. Went pointed out this concept to James Young many years ago. It is a very simple point, but one not often considered in the management of invasive annuals. The devil is not the one you have, but the one that the current exotic annual allows to invade.

Like Russian thistle, halogeton prefers vacant areas in native vegetation. Roadsides, livestock trails, areas near watering points, and excessively grazed sites all proved susceptible to the poisonous pest (fig. 5.1). Also like Russian thistle, halogeton is not a particularly competitive species and is easily suppressed by perennial species once they become established. In terms of successional stages, halogeton is in a successional stage that parallels Russian thistle but is broader in ecological potential. It occurs in the lower portion of the big sagebrush zone but is not nearly as abundant as Russian thistle in disturbed sites in that zone. Halogeton's seeds are similar to those of Russian thistle, its close relative, but they are produced in even more abundance. Even a tiny halogeton plant with only a couple of leaves and a single flower can produce a seed; large plants can produce hundreds of thousands. Halogeton seeds weigh very little and have an aerodynamic covering, and thus are easily dispersed by wind. In the deserts of the Great Basin, harvester ants are major dispersers of halogeton. It is not uncommon to spot a colony of halogeton growing on a diatomaceous earth bank where no native plant is adapted to grow. Close examination will reveal a harvester ant nest within the halogeton colony.

Fig. 5.1. Like Russian thistle, halogeton is primarily a passive invader of roadsides, live-stock trails, and areas where the native vegetation has been destroyed. Before halogeton lined this road in northern Washoe County, Nevada, Russian thistle would have occupied the same site. Outward from the edge of the road, halogeton shares dominance of the ruderal, or roadside, community with Russian thistle. In many areas of western Nevada, barbwire Russian thistle has largely replaced halogeton in roadside communities.

Unlike Russian thistle, halogeton is not a cropland weed; it is restricted to rangelands and disturbed noncrop areas. Early on during its spread, some people suggested that halogeton might eventually have a range similar to that of Russian thistle.[13] Most scientists scoffed at this speculation. By the end of the twentieth century, however, halogeton was distributed from California to northern New Mexico and north to east of the Missouri River on the northern Great Plains; it had even dispersed across the international boundary to Canada.[14]

Halogeton's ability to build seed banks while Russian thistle does not is a major difference. Dick Eckert used to amuse visitors to the Great Basin rangelands by picking a mature halogeton, threshing the shiny black seed from its papery covering, and wetting the seed with the tip of his tongue. The seed would "germinate," or at least imbibe water, and uncoil before the eyes of the astonished observers. Obviously, such a seed would germinate on its first exposure to adequate moisture, so how does halogeton build seed banks? In a brilliant research effort, M. C. Williams determined that in addition to the quick germinators, halogeton also produces a brown-coated seed that is highly dormant and lasts in seed banks for up to 10 years. These seeds had been previously observed in halogeton seed collections, but because they would not germinate they were thought to be "immature" seeds. Williams showed that day length determines the type of seed halogeton produces, and that the "immature" brown seeds are actually produced before the shiny black seeds.[15]

The largest communities of halogeton occur on sites where the native half-shrub winterfat previously grew. Winterfat is another member of the goosefoot family, although unlike halogeton, Russian thistle, and barbwire Russian thistle it is a perennial. The genus *Krascheninnikovia,* to which winterfat belongs, occurs in both North America and Central Asia. In western North America winterfat is a valuable browse species on salt desert winter ranges. The reasons why halogeton replaces winterfat are complex, but many people believe that excessive grazing usually initiates the process. The loss of a valuable grazing resource and its replacement by a species toxic to grazing animals has created considerable concern.

Halogeton in degraded big sagebrush communities can be biologically suppressed by seeding crested wheatgrass, a perennial. This seeding of a perennial grass to suppress an invasive exotic weed was carried out on a massive scale in the West.[16] Generally, the seedings were highly successful; but on the edge of the salt desert, where moisture is limited, moisture events are often erratic, and salt content of the soils is higher, seeding failures were common. The failures resulted in a bare seedbed with no competing vegetation and did the opposite of what was intended, providing excellent habitat for halogeton and Russian thistle.

Halogeton can truncate succession because it accumulates dissolved salts from the soil profile on the surface of the seedbed. In order to protect its cells from the accumulated salts, halogeton metabolizes the salts into an oxalate, a compound poisonous to livestock. When halogeton plants die and weather, the stored salts leach from the herbage and accumulate on the soil surface, where they interfere with the germination and seedling growth of native perennials.[17]

The most recent addition to exotic invasive species colonizing the rangelands of the Great Basin is annual kochia, yet another member of the goosefoot family. Like *Krascheninnikovia,* the genus *Kochia* occurs in both western North America and Central Asia. The *Kochia* species commonly known as red molly is a half-shrub native to the Great Basin. Although it is more limited in abundance than winterfat, it is also a forage plant on winter ranges. The weedy introduced *Kochia* is annual kochia, a serious pest of cereal crops. Annual kochia should not be confused with the perennial *Kochia prostrata,* which has been deliberately introduced into North America as a valuable forage species that can compete with cheatgrass. On good sites annual kochia may reach a height of 6 feet. It seldom reaches this stature on rangelands, but it is the tallest of the exotic goosefoot species. Like the other chenopods that have found their way onto rangelands, annual kochia is a tremendous producer of seeds. Our research has shown that the seeds exhibit great phenotypic plasticity, and some forms require nitrate enrichment to break dormancy. Annual kochia plants do not tumble like Russian thistle, but they do break off at the soil surface and drift in the wind, often lodging against fences and in depressions. Annual kochia is not widely abundant on Great Basin rangelands. Perhaps its time as a dominant bare ground invader has not yet come, or perhaps it is in a lag period, still evolving ecotypes adapted to the arid Great Basin environments.

The last colonizer of bare ground seedbeds in the Great Basin that we will discuss is filaree, a member of the geranium family (Geraniaceae) that has found a home around the world in environments with a Mediterranean climate. This Eurasian native was actually the first exotic annual to invade Great Basin rangelands. The adobe bricks used to construct missions in Spanish California contained filaree, evidence that some species were established at San Diego before the Spanish arrived overland from Mexico to found the first mission.[18] Long before Spanish expeditions reached California by overland routes, galleons stopped at San Diego for fresh water and apparently introduced filaree at that time. Certainly, it was present when John C. Frémont visited the San Joaquin Valley in 1842. When Frémont first saw the San Joaquin Valley, it was early spring and the entire foothills area was pink with filaree flowers.[19] Filaree is the only exotic plant to have received a name in the Shoshone dialect of the Native Americans of the southern Great Basin. Red stem filaree is the species best adapted to the northern Great Basin.

Unlike the invasive members of the goosefoot family, which rely on superabundant seed production and very rapid germination to become established on harsh seedbeds, red stem filaree uses a strategy of seed self-burial. Filaree seeds fall from the mother plant as fruits encased in a protective body made up of flower parts. The base is composed of an overlapping wrapper of maternal tissue that ends in a very sharp, indurate point. Attached to the top of this covering is a geniculately twisted column of remnant styles. As the seeds mature, the stylar column becomes very tightly coiled as moisture is lost. At maturity, the seeds are dormant because the indurate seed coat does not allow water to penetrate to the embryo. The fruits are dispersed by wind and by animals, to which they attach via the sharp point, boring into wool or clothing. Once the seed reaches an appropriate seedbed and is exposed to diurnal fluctuations in temperature and humidity, the awn begins to coil and uncoil, driving the pointed end of the fruit into the soil. The flexing action of the awn rubs the two enclosing flaps of the fruit across the seed coat until it is sufficiently abraded to allow moisture to penetrate. When the fruit is planted to the depth of the geniculate bend, the column snaps off at the top of the seed covering, leaving the nondormant seed planted at the ideal depth for germination and emergence. Seeds not twisted into the ground remain dormant and enter the seed bank until fortuitous burial occurs.[20] This is obviously a much more complicated seedbed ecology than Russian thistle's strategy of tumbling and casting seeds along the way. Red stem filaree produces far fewer seeds per plant than Russian thistle, but the mother plant invests considerable energy into this highly complex and remarkably successful strategy for establishment on bare seedbeds. Red stem filaree is probably the only exotic invasive annual weed of rangelands that humans deliberately distributed to new sites in the Great Basin. It was considered excellent early spring forage for sheep, and ranchers collected the coiled fruits for distribution on their own rangelands.

This takes us through the exotic annuals that form the first step in succession that eventually leads to cheatgrass dominance. Except for filaree, members of the goosefoot family dominate this stage. The next stage in seral succession introduces the mustard family.

CHAPTER 6

Seral Continuum

The Intermediate Step

The exact nature of the step from Russian thistle to the next level in the succession of invasive exotic plants is not known. John Weaver and Frederic Clements stressed that the stages in succession are largely determined by two factors: (1) how each previous stage or plant assemblage modifies the ecosystem, and (2) by the stand renewal process—the means by which the previous plant community was destroyed.[1] Russian thistle finds a home on bare soil created by severe mechanical disturbance such as road grading and livestock movements along trails or around salt blocks. Road shoulders that are graded every year will never proceed beyond Russian thistle dominance. If the Russian thistle plants mature without additional disturbance, on the other hand, succession will proceed in subsequent seasons. When Russian thistle plants mature, break loose at the soil surface, and tumble off the site, they leave behind a seedbed significantly changed both physically and bio-chemically. The seedbed is still largely bare, but the stubs of the Russian thistle plants provide microtopography that captures subaerially deposited dust and heavier particles moved across the soil surface by wind action. In the Great Basin, the deposition of fine silt-sized particles on the soil surface and the illuviation of these particles into the soil profile after rain are major parts of the soil-building process. An endless supply of these fine particles is constantly being eroded from the playa surfaces.[2]

Stand renewal does not have to be an all-or-nothing event. In a firmly established cheatgrass stand, for example, the area around rodent burrows may continue to support Russian thistle. Wildfire in a cheatgrass stand may burn all the litter, especially the material close to sagebrush canopies, where it burns hot enough to kill most of the cheatgrass seeds in the seed bank. When that occurs, the site will return to Russian thistle dominance. Away from woody fuel, cheatgrass seed banks usually survive and cheatgrass dominance is not interrupted. The stand renewal process that induces regression in succession may vary in spatial distribution and intensity on a given site.

If the bare-ground seedbeds on Great Basin rangelands largely belong to members of the goosefoot family, the next stage in succession is dominated by three members of the mustard family, Brassicaceae. The most widely distributed and dominant mustard on Great Basin rangelands is tumble mustard, a native of Eurasia (fig. 6.1). Very little is known about the timing and mode of introduction of this species, but tumble mustard was apparently introduced relatively recently in California.[3] The oldest California herbarium specimens date from 1910–1918. The second mustard species, tansy mustard, is a native of North America and the rarest of the species in the secondary succession toward cheatgrass dominance. Tansy mustard is highly invasive and may not be native to much of the Great Basin where it is now found. The taxonomy of tansy mustard is very complex, with at least eight recognized subspecies.[4] Tumble mustard is the most frequently found species, but in some sites tansy mustard is dominant. Shield cress, the third mustard species, is also a native of Eurasia. Like tumble mustard, shield cress was first recorded in herbariums in California early in the twentieth century. It is much more salt tolerant than tumble or tansy mustard. In the Great Basin, shield cress is most abundant in the upper shadscale zone and the lower portion of big sagebrush communities, especially on pluvial lake plains, where it is often a roadside dominant species (fig. 6.2).

All three of the Great Basin's mustard species have a mucilaginous seed coat,[5] a character found in many different plant families. Eight percent of the 233 Great Basin species we surveyed had seed mucilage. Interestingly, most of the species with mucilaginous seeds were colonizing weeds. Seed mucilage occurs in so many species in so many plant families that it probably conveys a variety of benefits. A commonly proposed benefit is as an aid in dispersal. Seeds with copious amounts of mucilage are sticky and may

Fig. 6.1. Tumble mustard in full bloom in a cheatgrass community.

disperse to new sites by sticking to animal feet and legs, bird feathers, or vehicle tires.[6] For species of the temperate desert environments of the Great Basin, effects on moisture relations during germination may be a much more biologically significant aspect of seed mucilage.

J. L. Harper and his students were the first to propose the basic concepts of seedbed moisture relations.[7] Harper suggested that every seed has a series of requirements for germination; he termed these "safe site parameters." The number of safe sites in a seedbed determines which seeds and how many seeds of a given species will germinate. Some of the safe site parameters involve temperature interacting with moisture. Most seeds are dormant at maturity by desiccation and can germinate only after they are rehydrated. Three factors control moisture relations of seeds in seedbeds: osmotic potential of the soil water solution, matric potential, and hydraulic conductivity.[8]

Fig. 6.2. Shield cress growing in the transition zone between salt desert and big sagebrush plant communities in the Great Basin.

Osmotic Potential of the Soil Water. As the dissolved salt concentration of the soil water increases, the osmotic potential of the solution becomes more negative and seeds have an increasingly difficult time taking up water from the seedbed. If the osmotic potential becomes even greater, the seeds will desiccate and die without germinating. We previously mentioned that shield cress is adapted for growth on salt-affected soils in the upper portions of the salt desert environments. The exceptional ability of shield cress to grow on salty soils suggests that its seeds can germinate under greater osmotic stress than can the seeds of tumble or tansy mustard. This is not necessarily true, though there is a strong negative correlation between being able to grow in salty soil and being able to germinate at low osmotic potentials. Among cereal grains, barley is the species best adapted to grow on salt-affected soils, yet barley seeds have lower germination under lowered osmotic potentials than the seeds of other cereal grains do. In the Great Basin, the precise relation between germination at low osmotic potential and growth on salty soils applies even within individual species. Basin wild-rye grows in habitats ranging from lake plain playa margins to the upper pinyon-juniper zone high on mountain escarpments. We collected seeds of basin wild-rye over this entire range of variability and found that seeds collected in the pinyon-juniper zone germinated under much lower osmotic potentials than seeds of the same species collected near the playas, where the soils are highly salt-affected.[9]

Matric Potential. Osmotic potential is a chemical relation. Matric potential is a physical relation that refers to soil water molecules' attraction to soil particles. The smaller the soil particles in the seedbed, the greater the force with which they hold water molecules. Sandy-textured seedbeds give up water to seeds during germination much more easily in terms of the energy the seed must expend to imbibe water than do clay-textured seedbeds. Because they hold the water molecules less tightly, sand-textured seedbeds hold much less water against the pull of gravity than fine-textured seedbeds such as clay.

Hydraulic Conductivity. Hydraulic conductivity is the ease with which water can move through soils. The germination of a seed immersed in water is not influenced by hydraulic conductivity, but a seed placed in a mineral seedbed is immediately influenced by hydraulic conductivity. In less than saturated seedbeds, water will flow from the soil to the seed only at points where the seed coat touches the soil particles. The coarser the seedbed substrate texture—and thus the larger the soil particles—the fewer are the seed's points of contact. A seed will have many points of contact in a fine-textured clay substrate. Hydraulic conductivity is also influenced by how densely the seedbed is packed or compressed to eliminate or reduce air-filled voids. The standard farmer's rule for planting seeds is "plant in a firm seedbed." This is why in agronomic—and especially high-value vegetable—crops, soils are repeatedly tilled and compressed to break down clods and prepare a firm seedbed for planting.

At some point lost in antiquity, farmers learned that even in humid climates, seeds placed in the seedbed and covered with a thin layer of soil germinate more rapidly and established more seedlings than seeds broadcast on the surface. Farm implements known as drills were developed to meter the correct number and spacing of seeds, open a small trench or hole in the seedbed, place the seed in the seedbed slit, cover the seed with soil, and firm the surface.

J. L. Harper and his students proposed that seed mucilage interacts with hydraulic conductivity when seeds are buried in seedbeds. The presence of mucilage greatly increases the points of contact between the seed coat and the mineral seedbed. Harper's group carried that concept a significant step further by suggesting that seeds on the surface of seedbeds automatically have reduced points of contact for hydraulic conductivity because only one side of the seed is touching the seedbed. The rest of the seed coat is exposed to the atmosphere, and if the relative humidity is sufficiently low, there is even a moisture gradient away from the seed and into the atmosphere. Harper insisted that the presence of seed mucilage not only increases hydraulic conductivity by increasing contact with the moisture-supplying substrate, but also decreases moisture loss to the atmosphere if the seed is exposed on the surface of a seedbed.

Tumble mustard, tansy mustard, and shield cress all have mucilaginous seed coats.[10] Tumble mustard seeds are quite small, only 0.6–0.8 mm (1.0 mm = 0.039 in) long and scarcely 1.0 mm thick. When they are moistened, a very small amount of mucilage can be detected, but with difficulty. Tansy mustard seeds are slightly larger (0.8–1.0 mm long by 1.0 mm thick), and an easily detectable mucilage forms when the seeds are moistened. Shield cress seeds are much larger (2.0 mm long by 2.0 mm thick), and a large amount of conspicuous mucilage forms on wet seeds. We have observed mucilage bursting out of epidermal cells as the seed coat of shield cress imbibed water.

In an effort to understand the function of seed mucilage, we resorted to some simple but innovative laboratory procedures. We used an uncommon and expensive piece of laboratory glassware called a sintered funnel that looks like a beaker with a funnel spout on the bottom. A disk of sintered (i.e., heated without being melted) glass containing tiny pores is fitted tightly between the beaker portion and the much smaller funnel. Normally a sintered funnel is used to filter solutions; we used it to create a precise environment in which to determine the moisture required for germination. We attached plastic tubing to the bottom of the funnel spout and filled the tubing with water that had been boiled to remove as much air as possible. The holes are so small in the sintered disk that capillary adhesion will support a column of water 200 cm long if you get all the air out of the system, fill the tube, hold your thumb over the end of the tube, and invert the entire assembly to immerse the end of the tube in a beaker of water. Even on a tall distillation rack in the laboratory, this is a tricky process that often requires a mop. The 200-cm column of water exerts a negative force (pull or suction) on the water in the pores of the sinter equal to −0.02 megapascals (MPa, a metric pressure unit equivalent to 1 newton per square meter, replaces the unit bars, 10 bars = 1.0 MPa). To put this level of moisture tension into perspective, free water has a potential of 0. Common herbaceous plants can extract moisture from the soil and have their cells remain turgid to a negative potential of −1.5 MPa. Past this point, known as the permanent wilting point (PW), the plants wilt and will not recover when moisture is added. The PWP was established by using seedlings of the common annual sunflower. Species adapted to extreme desert or salt-affected soil conditions have a much greater tolerance to reduced soil water potentials than other species.

On a bare sintered plate with −0.02 MPa of moisture potential and the top of the funnel open to the atmosphere, only seeds of shield cress could germinate. For seeds of tumble or tansy mustard to germinate, the tube had to be shorter or the top of the funnel closed with a tight-fitting sheet of aluminum foil. Closing the top of the funnel eventually resulted in the atmosphere equilibrating at the water potential of the sinter, or 0.02 MPa. The seeds (a caryopsis) of cheatgrass, in contrast, would not germinate from the bare surface of the sintered plate at 0 MPa if the top of the funnel was left open.[11] Soil coverage, and therefore hydraulic conductivity of the germination substrate, can be approached in the sintered funnel apparatus by adding glass beads with extremely small and precisely controlled diameters (available from the manufacturer of reflective paints for traffic control lines on road highways). Tumble mustard seeds needed a covering of glass beads with a diameter of 0.075 mm or less to obtain significant germination at a moisture tension of −0.02 MPa, an amount within the range of moisture that soils hold against the force of gravity, also known as field capacity. The finer the soil particles (texture), the greater the potential field capacity.

A second way to approach moisture relations during the germination process is to use a soil moisture extraction apparatus designed to determine how much moisture a soil will retain at a given level of moisture stress. The extractor consists of a massive steel chamber constructed of 1-inch-thick steel with a lid secured by a number of large-diameter bolts. The moisture stress is created by pumping compressed air into the chamber. Inside the chamber, ceramic plates with very small holes—similar to the sintered disk in the funnel experiment—are stacked. The back of each plate has a moisture-proof fabric seal from which capillary tubing drains moisture through the wall of the chamber. Saturated soils are placed on the plate and pressure applied. The amount of moisture stress created is proportional to the air pressure.

Before we tested the mustard seeds, we tested germination in relation to reduced osmotic potential by placing seeds in containers with a ground polystyrene substrate moistened with solutions of polyethylene glycol. Polyethylene glycol is a very large molecule that does not pass through the membranes of the seeds being tested, creating an osmotic gradient from the seed to the solution proportional to the increased concentration of the solute. Tumble mustard seeds produced highly variable germination at –1.2 MPa and virtually none at –1.6 MPa. The greatest total germination and the highest germination at reduced matric potentials occurred on a fine clay substrate; germination decreased on silt and sand substrates. We received a surprise when we tried tansy mustard and shield cress in the same experiments. Neither species' seeds would germinate in the extractor, even at 0.01 MPa. We tried charging the chamber with compressed nitrogen gas, and the seeds of both species germinated as expected. Apparently, the seeds of both species are sensitive to partial oxygen tension.

It is hard to find a practical significance to this observation in relation to desert environments, but it may be significant in environments with waterlogged soils. Although seed germination had practically ended at –1.6 MPa, we tried solutions at much lower osmotic potentials to see what the influence would be on the mucilage formation on seeds of shield cress. The conspectus glob of mucilage was present at –3.5 MPa but did not form at –4.0 MPa. In the extraction chamber, the mucilage formed at –0.2 MPa on a clay substrate, but not at any lower matric potentials and not with shield cress seeds on coarser-textured substrates. This suggests that seed mucilage influences water relations only at very high matric potentials. Because the stress becomes greater as the matric or osmotic potential becomes more negative, if there is going to be a real selective differential in field seedbeds among seeds with and without mucilage, it is going to occur while the seedbeds are wet or close to field capacity.

Obviously, the texture of the substrate—and therefore its hydraulic conductivity—has a big influence on germination at reduced matric potentials. In the deserts of the Great Basin, osmotic stress in seedbeds is the product of dissolved salts in the soil water. The specific ion phytotoxicity content of these solutions may be much more detrimental to seed germination than the negative osmotic potentials induced by the salt content.

A principle of germination ecology and seed moisture relations emerged from our observations. At a given level of matric potential, germination is much higher if the seeds are placed in wet soil and immediately drawn down to the matric potential being tested rather than being placed in the substrate already at the desired matric potential. Known among seed scientists as the Young-Evans principle, this fact is difficult to extrapolate to field conditions in the Great Basin. Roughly, it implies that seeding just before a moisture event is better than seeding afterward.[12]

We also investigated seedling elongation of the three mustard species in osmotic solutions and seedling emergence from soil. Under reduced osmotic potentials, the very large seeded shield cress had the greatest seedling elongation at given negative potentials. The very small seeded tumble mustard had greater seedling elongation in reduced osmotic solutions than the larger-seeded tansy mustard. Tansy mustard seeds also emerged in markedly lower numbers after 1.0 cm burial in seedbeds than did the two exotic mustard species. Based on their germination in relation to reduced osmotic and matric potentials, seedling elongation, and emergence, it would seem logical that shield cress with its big seeds and copious seed mucilage would be the dominant plant of intermediate seral stages on Great Basin rangelands. Shield cress remained in the marginal zone between salt desert and sagebrush vegetation types until 2005, when it suddenly spread over thousands of acres of salt desert rangeland as the dominant species. The landscape dominant of degraded big sagebrush communities is tumble mustard, with tansy mustard only an occasional dominant of specific sites.

Moisture relations are a vital part of the germination ecology of temperate desert species, but other factors are important as well. The inherent physiology of the three mustard species differs in that both shield cress and tansy mustard have temperature-related after-ripening requirements that must be met before germination can occur. Newly mature seeds will germinate only at very cool incubation temperatures (5°C) regardless of external influences.[13] Over time, this after-ripening requirement disappears. The seeds of tumble mustard do not require very cool incubation temperatures, giving this species a great competitive advantage for fall germination and root elongation before cold winter temperatures occur.

The number of mustard species present in the Great Basin has increased since 2000. No other invasive seral group has increased in species diversity so rapidly. The next decade may see mustard species replacing exotic species even in other seral stages as well.

CHAPTER 7

Seral Truncation

Cheatgrass truncates secondary succession on temperate desert rangelands before the habitat reaches a natural equilibrium (in the sense of F. E.
Clements). That does not mean that cheatgrass dominance is inherently a
stable environment that will never change. In fact, cheatgrass communities
are susceptible to invasion by other exotic invasive species. R. L. Piemeisel
was the first to recognize that succession was occurring among exotic annuals in the American West.[1] Piemeisel conducted his research during the
1930s on the Snake River Plains of Idaho, where he was employed by the
USDA Bureau of Entomology to conduct research on the curly top virus,
which causes great economic losses to irrigated crops. Plant pathologists had
determined that the beet leafhopper was a vector for transmitting the virus
from plants growing on sagebrush rangelands to irrigated crops. The primary herbaceous species serving as a reservoir for the virus were Russian
thistle and tumble mustard.

Most of the Snake River Plains had been excessively and improperly
grazed by 1900.[2] During the formative years of the irrigation districts, the
homesteaders allowed their horses and livestock to run free on the adjacent rangelands during the winter. The homesteaders in the federally sponsored desert reclamation projects might have been hard-pressed while they
grubbed sagebrush to bring the land under irrigation, but they were in a far

better position than dryland or rain-fed homesteaders. By the 1920s, 2 million acres of homesteads in the sagebrush country of southeastern Oregon had been abandoned.[3] The droughts of the 1930s exacerbated the exodus. Russian thistle colonized the abandoned fields, followed by tumble mustard and then finally by cheatgrass. Writing in 1970, R. F. Daubenmire described old-field succession in an *Agropyron-Poa* habitat in the Columbia Basin: "A small strip of cropland abandoned nearly 50 years ago, then given complete protection, has only a few (very large!) bunches of *Agropyron* scattered about. Since there are no small individuals, these plants were probably established in the first year or two after abandonment, but before the *Bromus* stand reached full density. In fact, *Bromus* may not have reached these *Agropyron-Poa* areas when disturbance was discontinued."[4]

It is very wrong to assume that cheatgrass invaded only areas where the native vegetation had been destroyed by excessive, improperly timed grazing or by cultivation. In the publication summarizing his research in the interior Pacific Northwest, Daubenmire offered the following perspective on the ecological role of cheatgrass: "Although an alien, *Bromus tectorum* must be considered a member of most climax communities in Washington. In undisturbed vegetation the populations are very sparse and the plants dwarfed, sometimes no more than 1 inch tall, and bearing a single spikelet. These plants can be found everywhere including places where there has been no disturbance by man or livestock. The most compelling evidence is its occurrence atop cliff-rimmed buttes that are scalable by humans only with difficulty."[5]

Nor does cheatgrass exist in big sagebrush/bunchgrass communities only because of livestock grazing. The collateral concept would then be "get rid of domestic livestock and cheatgrass will disappear." During the late 1930s and early 1940s, a number of scientists tried to call attention to the hazards associated with the spread and dominance of cheatgrass. The noted naturalist Aldo Leopold wrote a stirring condemnation of cheatgrass under the title "Cheatgrass Takes Over."[6] The crux of the problem as Leopold saw it was whether "cheat" was a curable or incurable affliction. Did it, like other grasses and weeds, serve as a nurse crop for more valuable grasses that would follow? The standard successional theory claimed that the dominant plant in each successional stage modified the environment in a way that permitted or conditioned passage to the next stage of succession.

Many of the early range scientists strongly believed that the cheatgrass problem could be solved through grazing management. In one of the earliest thesis projects on the competitive nature of cheatgrass, S. A. Warg found that bluebunch wheatgrass seedlings were creeping in along the margins of cheatgrass-dominated abandoned fields in western Montana. He estimated that it would take 30 years for the bluebunch wheatgrass to completely invade the field.[7] E. W. Tisdale, who had a distinguished career at the University of Idaho, reported on a study conducted in his native British Columbia whose results indicated that bluebunch wheatgrass seedlings became established in areas of the southern interior of British Columbia that were protected from grazing.[8]

The late 1930s and 1940s were renaissance years in range research in the Intermountain Area. Two of the leaders in this movement were A. C. Hull and Joe Pechanec. Both scientists initially supported the idea that restoring natural succession in areas dominated by cheatgrass was merely a matter of proper grazing management.[9] We do not wish to belabor the point, but it is important to establish the scientific environment that existed during the crucial explosion in cheatgrass dominance. Because leading scientists believed and published papers stating that cheatgrass problems could be solved by grazing management, the idea found its way into university classrooms, and a generation of range managers was trained to believe that this was the technical gospel they were to apply in their jobs. Daubenmire reached his conclusions about the invasive nature of cheatgrass early on in his career when relict sites with perennial grass dominance were easy to find.[10] Later investigators may have missed this important bit of news because of the scarcity of high-condition sagebrush/bunchgrass communities.

One can dismiss the presence of diminutive cheatgrass plants in climax native bunchgrass communities as a mere oddity of no ecological consequence, but there is a grave danger in that conclusion. It is only a matter of time before some form of stand renewal occurs, even something as simple as a single bluebunch wheatgrass plant dying of old age. When wheatgrass seedlings germinate to fill the void in the community, competition occurs between those perennial seedlings and cheatgrass seedlings. Some form of catastrophic stand renewal is always lurking just over the horizon even in the absence of grazing by domestic livestock. When a native perennial grass (or perennial forb or shrub) dies, it leaves a minimound of soil particles subaerially deposited or eroded from interspaces in the plant community and reworked by wind until they were trapped at the base of the bunchgrass.[11] Annual litter fall from the grass adds to this accumulation. A brilliant paper by J. L. Charley and N. E. West demonstrated the micro-patterning of nutrient mineralization in the subcanopy halo area for temperate desert shrubs.[12] Many native perennial bunchgrasses die from the inside out, leaving a picket fence of dry grass culms surrounding the bald dome of the minimound. When the perennial grass dies and the organic matter on and in the soil of the mound decomposes, a very desirable location for seedling establishment is formed. Factors as diverse as wildfires, erosion, explosive populations of voles or jackrabbits, concentrations of free-roaming horses, drought, Mormon cricket bands, and disease outbreaks may induce stand renewal in native perennial grass communities. Once the need for seedling recruitment occurs after stand renewal, the seed bank of cheatgrass plays a potentially dominant role in future succession.

Several researchers have explored competition between cheatgrass and established native perennial grasses. Their results generally indicate that well-established native perennial grasses suppress cheatgrass. This has been interpreted as proof that removal of domestic livestock will ensure the disappearance of cheatgrass. It is at the seedling recruitment stage that cheatgrass competition is such a serious threat, however, not in well-established stands. There is a second aspect of this scenario. The cheatgrass may be present in low concentrations, but every generation it exists in such communities increases the chance that natural selection for site-specific adaptation is occurring. When the inevitable stand renewal process occurs, the cheatgrass will be well adapted to dominate the replacement plant community.

A question arises at this point. If stand renewal is inevitable in plant communities, why has no highly competitive native annual evolved to compete in disturbed sagebrush/bunchgrass communities? Table 7.1 lists the plant species found in such communities. Only two species of native annual grasses are on the list. Six-weeks fescue (*Vulpia octoflora*) is a diminutive annual that occasionally occurs in great abundance but is not a competitive species. James Young proposed many years ago that the lack of a competitive native annual was a result of the lack of large herbivores in much of the Intermountain Area.[13] Bison evolved in the Great Basin, but by the time Europeans arrived the modern American bison had withdrawn

TABLE 7.1 | NATIVE AND EXOTIC ANNUAL SPECIES IN SAGEBRUSH/BUNCH-GRASS PLANT COMMUNITIES OF THE PACIFIC NORTHWEST

NATIVE	EXOTIC
Agoseris heterophylla	Alyssum alyssoides
Amsinckia lycopsoides	Arabidopsis thaliana
Belepharipappus scaber	Arenaria serpyllifolia
Chenopodium leptophyllum	Bromus briziformis
Collimina linearis	Bromus japonicus
Collinsia parviflora	Bromus mollis
Cryptantha flaccida	Bromus racemosus
Descurainia pinnata	Bromus tectorum
Epilobium paniculatum	Camelina pulchella
Festuca pacifica	Draba verna
Gilia minutiflora	Elymus caput-medusae
Helianthus annuus	Erodium cicutarium
Lagophylla ramosissima	Holosteum umbellatum
Lappula redowskii	Lactuca serriola
Linanthus pharmaceoides	Lepidium perfoliatum
Linanthus septentrionalis	Lithospermum arvense
Madia exigua	Myosotis micrantha
Microsteris gracilis	Sisymbrium altissimum
Montia perfoliata	Vicia villosa
Myosurus aristatus	
Oenothera andina	
Pectocarya linearis	
Phacelia linearis	
Plagiobothrys tenellus	
Plantago patagonica	
Plectritis macrocera	
Stellaria nitens	
Vulpia octoflora	

SOURCE: R. Daubenmire, *Steppe Vegetation of Washington*, Technical Bulletin 62 (Pullman: Washington Agricultural Experiment Station, 1976), 80.

NOTE: Plant nomenclature is as originally published. Plants are listed by scientific name because many of the native species have no widely accepted common name.

from all but the extreme northeastern Great Basin. Apparently, the bison moved west over the Rocky Mountains and across the Snake River Plains to the sagebrush steppes of eastern Oregon and the Columbia Basin.[14] Archaeological evidence suggests that during periods in the Holocene era when climatic conditions favored grasses, the American bison population probably expanded throughout much of the northern Great Basin. R. N. Mack and J. N. Thompson addressed the evolution of steppe vegetation in the absence of large-hoofed mammals and concluded that the climate of the modern Intermountain Area—with precipitation occurring as cold rain and snow during the winter and the summers nearly completely dry—was not conducive to the evolution of rhizomatous C_4 grasses.[15] The bison could not survive without these grasses, which are dominant on the American Great Plains.

Some environmentalists eagerly embraced Mack's and Thompson's conclusion as providing evidence that domestic livestock should not be grazing sagebrush/bunchgrass environments. This led to a rebuttal by Wayne Burkhart, titled "Herbivory in the Intermountain West," which claims the Great Basin was a virtual Serengeti Plain bursting with large herbivores before Europeans arrived.[16] He is correct, but it was *long* before European contact, when the Great Basin environment bore little resemblance to the environmental conditions that currently exist.

A much more tenable hypothesis for the lack of native competitive annual species was put forth by Neil West, who pointed to the tremendous seed production capacity of big sagebrush plants as the culprit in the lack of competitive secondary succession species.[17] Big sagebrush is remarkable for certain aspects of its reproductive ecophysiology. Seedlings can flower, and second-year plants can produce a significant number of seeds. Established plants can produce thousands of highly viable seeds year in and year out for a century or longer, and fully stocked stands can produce millions of seeds per acre. The same plants can completely shut down reproduction during extreme drought or when the stand density exceeds the site potential; and they seem to go about the whole process completely backward.

Sufficient moisture for plant growth in the Great Basin occurs during the cold winter months and in the early spring. Native perennial grasses are cool-season species that must grow, flower, and set seed during the spring when moisture is available and temperatures warm up enough for growth. Reproduction is a race against the onset of the summer drought. Members of the rose family such as desert peach and antelope bitterbrush follow the same pattern. Frequently these spring-flowering shrubs lose much of their annual seed production to late-season frost. Big sagebrush plants, in contrast, flower in the fall, often at the height of the annual dry period, and often release their seeds after biologically significant frost has occurred.[18] Big sagebrush plants have two types of leaves. The large, wide, ephemeral leaves, which are very efficient at photosynthesis, emerge in the spring when moisture is available and are dropped as soon as the summer drought begins. The same plants also have small, highly moisture efficient leaves that persist for several years. The inflorescence is both very moisture efficient and very efficient at photosynthesis—

so efficient, in fact, that the plant loses none of its carbohydrate reserves in the flowering process. Big sagebrush is extremely moisture efficient at the time of year when herbaceous plants are dormant (perennials) or dead (annuals). Big sagebrush seeds, which mature in the late fall and early winter, are very small and are loaded with aromatic secondary defensive compounds that apparently protect them from predation by granivores. The pappus is deciduous, so the seeds largely disperse within about 3 feet of the parent shrub. Considering that the average density of an established stand is one big sagebrush plant per square yard, the seed rain is staggeringly dense. Seeds germinate in the top of the seedbed through a unique system of hypocotyl hairs that hold the embryonic seedling in place while the radicle penetrates the moisture-supplying substrate (fig. 7.1). The mass of springtime big sagebrush seedlings have very little chance of establishment in the nearly fully stocked stands in which they were produced. In the early spring, while soil moisture is still available, they are fiercely competitive, effectively eliminating seedlings of native annuals. Dr. West suggested that this apparently excessive production of throwaway seedlings effectively eliminated any selective pressure to evolve native competitive annual species. They simply could not compete.

James Young was formally introduced to cheatgrass in 1965 when he started to work in Reno, Nevada, as a range weed control scientist for the USDA Agriculture Research Service (ARS). Fresh out of Oregon State University with a Ph.D., he was recruited to do a job for which he had not been specifically trained, but he did know what cheatgrass looked like. The cowboys he grew up with called it bronco grass, and he had doctored cows with eyes injured by its seeds and dogs with bronco grass seeds in their ears or between their toes. His father considered cheatgrass a valuable forage. "If we don't have six weeks of north wind this spring," he heard his father say over the years, "the bronco grass feed will really be good."

The first week in Reno was highly confusing because ARS scientists Raymond Evans and Richard Eckert deluged him with publications and progress reports on their research on the control of downy brome. Young's first question was a bewildered, "What is downy brome?" Evans and Eckert explained that the boss in Washington, D.C., insisted on using the accepted common name for weeds, so he had better learn the name downy brome even though every rancher, range conservationist, and scientist in the West called the plant cheatgrass.

Fig. 7.1. Germinating seed of big sagebrush on 1.0 mm grid. The ring of hairs visible on the embryonic seedling emerging from the seed coat are not root hairs but are located on the hypocotyl, where they serve to anchor the seedling to the soil surface and allow the radicle (embryonic root) to penetrate the soil substrate and begin to take up moisture.

Evans and Eckert had established an innovative and extensive research program on cheatgrass control and revegetation.[19] Many herbicides came on the market during the 1950s and early 1960s; all had been screened by the ARS laboratory at Beltsville, Maryland, for their herbicidal activity toward a standard set of weed species (including cheatgrass) and crop species. Evans and Eckert field-tested every one of the herbicides that showed cheatgrass activity.[20] After a series of tests, they settled on two herbicides on which to build seeding programs for perennial grasses. To test these, they had established five experimental sites at five ecologically diverse locations in the Great Basin where seed and seedbed ecology studies of cheatgrass could begin immediately. All of the locations had both undisturbed cheatgrass stands and stands in which the cheatgrass had been controlled and no seed production permitted for 1 and 2 years. Controlling the cheatgrass with herbicides rather than tillage preserved the stratigraphic sequence in the seedbed. To make sure cheatgrass seeds were not blown onto the plots or transported in by rodents, boxes covered with fine-mesh screen were established on each replicated plot at each location.

The experimental sites were at (1) Medell Flat (modern spelling Bedell Flat) north of Reno between the Dog Skins and Peterson mountains; (2) Likely Table in Modoc County in northeastern California, a similar environment but outside the hydrologic Great Basin; (3) Paradise Hill in Humboldt County in northwestern Nevada; (4) Emigrant Pass in Elko County in northeastern Nevada; and (5) Italian Canyon bordering the Reese River valley in central Nevada. All were big sagebrush/bunchgrass potential sites where repeated wildfires had destroyed the woody vegetation and cheatgrass had become the dominant species.

In 1965, when the study was initiated, everyone knew that cheatgrass did not build a seed bank, because freshly harvested seeds commonly had a very high germination rate. For years, though, stockmen and range managers had been puzzled by the occurrence of second crops of cheatgrass following late spring rains. In their comprehensive review, "Cheatgrass (*Bromus tectorum* L.)," J. O. Klemmedson and J. G. Smith offered their opinion that "since germination rate is normally high, it is unlikely that the second crop arises from ungerminated seeds from the previous year."[21] In other words, all cheatgrass seeds germinated with the first effective rain in the fall or early spring and either became established seedlings or died in the attempt. Evans and Eckert suspected otherwise. They had enough experience with herbicidal fallows by this time to believe that at least some of the cheatgrass seeds lasted in the soil for more than 1 year.

The Reno group's approach to studying cheatgrass seedbeds was very simple, although it took some experimentation to settle on a procedure. They began by collecting samples of litter and soil from field sites infested with cheatgrass. Cheatgrass seeds (technically, caryopses) were found in great and varied abundance at all the sites. Some were brightly colored and closely resembled seeds collected at maturity. Others showed various levels of weathering, all the way down to blackened, shriveled specimens that had lost the papery lemma and palea that enclose fresh seeds, leaving the naked embryo, endosperm, seed coat, and pericarp (fig. 7.2). Many of the recovered seeds were darker than cheatgrass seeds collected at maturity. The

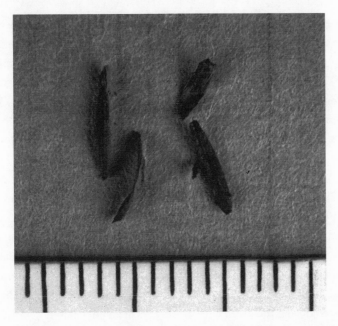

Fig. 7.2. Seeds of cheatgrass recovered from seedbeds in the field. The lemma and palea are no longer covering the embryo, endosperm, seed coat, and pericarp. The black body protruding from the seed at lower right is the synnemata of the seed pathologic fungus Podosporiella verticillata. *Seeds bearing these bodies would not germinate and quickly rotted.*

blackened ones without a lemma and palea had black, peglike bodies protruding from their sides. A plant pathologist helped to identify these bodies as synnemata of the seed-pathologic fungus *Podosporiella verticillata,* which commonly infests seeds of cheatgrass.[22] The caryopses with synnemata were determined to be nonviable. The other seeds were placed in Petri dishes in the laboratory, where they became completely covered with a fungal mass.[23] Viable but dormant seeds are usually surrounded by a clear zone on the germination plate signifying that allelopathic chemicals within the seed are keeping fungal growth at bay. The lack of such cleared areas is usually taken as evidence that the seeds are nonviable.

The tremendous time required to sort through soil or litter samples to find cheatgrass seeds, the number of samples that would have been required to describe the population with any level of precision, plus the difficulty of identifying the seeds once the lemma and palea had been shed and soilborne microorganisms had contaminated germination dishes led the team to use a bioassay system to assess the germinability of cheatgrass seed banks over time. Bioassay sampling is an excellent way to determine the number of viable seeds in the seed bank without having to count them one by one. They dug litter and soil samples from five widely spaced locations in the northern Great Basin. The surface soil litter was handled as a separate layer of each sample. Soil under the litter sample was collected at 1-inch intervals to a depth of 4 inches. The samples were taken to the greenhouse where they were placed in small, well-drained pots and covered with a thin layer of horticultural vermiculite. The samples were watered daily and the emerging seedlings were identified, counted, and removed weekly. The samples were watered for 12 weeks, allowed to dry for 4 weeks, and then rewet and run wet for an additional 4 weeks. After the second wet period, the samples were stored for 1 year and the process repeated. The procedure was repeated for 4 consecutive years. This bioassay technique was patterned on techniques developed during the nineteenth century in England.[24]

It had become apparent that cheatgrass seed banks were variable in distribution at any given location and significantly variable among locations. The large number of samples that would be required to account for this variability reinforced the need to use a bioassay technique rather than trying to recover every seed. Some experimentation determined that a 2×4–inch sampling plot size produced a bioassay sample that fit into a 16-oz cup on the greenhouse bench with a depth that did not interfere with seedling emergence. Each 4×4×2–inch sample dug in the field was placed in a plastic bag and labeled. The wind is virtually always blowing in the deserts of the Great Basin, and chasing little plastic bags became a way of life that lasted for 3 years. The sampling started in May 1965. The current year's population was fully established by May, but no seeds had been produced. Cheatgrass seed banks should have been at their lowest levels. The emerging seedlings were removed at the end of the first week and weekly thereafter for the next 6 weeks. By the end of the 6-week test period the group realized that there was a huge seed bank of cheatgrass seeds left over from the previous year's seed production. Its magnitude was two to three times the size of the existing seedling density.[25] Samples taken over the next 2 years gave similar results for all locations. Obviously, the May seedling emergence was not a fluke. The results were very similar for each of the five locations dispersed over a considerable geographic area. The first point concerning the seedbed ecology of cheatgrass had been resoundingly made: cheatgrass builds biologically significant seedbeds.

Why did it take 40 years of research on the species to discover the presence of the seed bank? Science, especially science applied to natural resources, can be exceptionally blind to the obvious. Those familiar with cheatgrass seeds knew that virtually 100 percent of the seeds collected at maturity in the field would germinate when incubated on moist blotter paper in Petri dishes. Obviously, then, all cheatgrass seeds germinated with the first fall rain. Scientists familiar with the limited literature on cheatgrass physiology also "knew" that exposing imbibed cheatgrass seeds to freezing temperatures killed them; Warg had determined that experimentally.[26] These two "facts" were all that was necessary to prove that cheatgrass seeds did not accumulate in seed banks that lasted for more than 1 year. In retrospect, there was no need for an extensive sampling system and bioassay to prove that cheatgrass builds seed banks. All one had to do was to go out on the range in the late spring before the current year's seed production was ripe, bend over in a cheatgrass patch, and look at a handful of litter. Every handful will contain cheatgrass seeds of a variety of ages and physical conditions. The occurrence of seeds in the litter does not in itself prove that the seeds are germinable, but the sight of literally hundreds of fresh-appearing cheatgrass seeds in a grab sample of litter should raise red flags of scientific interest. The results of the 3-year study were greeted with great skepticism, but subsequent studies confirmed that cheatgrass does build seed banks. Most range managers and scientists accept the fact now, but some still fail to see the obvious.

The study also determined the location of seeds in the seed bank. It soon became apparent that most of the germinable cheatgrass seeds were located on the surface of the soil in the litter. They were not uniformly distributed, however; they accumulated in depressed areas in the microtopography of the seedbed. As we noted earlier, the surface soils of the big sagebrush zone in the northern Great Basin are highly influenced by fine, wind-deposited dust particles from the great pluvial lake plains.[27] Those in the Likely Table site probably also include volcanic tephra from the volcanic cones of the Cascade Mountains. When these silt-textured surface soils become saturated over the winter they are subject to frost action, and they often form polygonal crack patterns when they dry. These cracks are perfect sites for the accumulation

Fig. 7.3. Cheatgrass seeds accumulated in the polygon cracks in a silt-textured surface soil. Such depressed microtopography provides key sites for the accumulation of wind-dispersed cheatgrass seeds.

of wind-dispersed cheatgrass seeds (fig. 7.3). In soils that are not cultivated, there is little opportunity for dispersal of the seeds into the surface horizon of the soil. In the case of cheatgrass on rangelands, the seeds largely remain in the litter or the interface between the litter and the soil surface.

Once it had been determined that cheatgrass builds seed banks, the question of how the initially highly germinable seeds persist in the seed banks arose. That is really the key issue. The answer turned out to be rather complicated. Some of the samples of litter and soil collected in the field over the 3-year study period were saved and allowed to dry, then exposed to moisture. This process was repeated over the next 2 years. Each time the samples received moisture, more cheatgrass seeds emerged. The group's interpretation of this was that cheatgrass seeds in the seed bank had acquired dormancy that gradually broke down. Scientists and range managers alike had a difficult time accepting this idea. The concept of acquired seed dormancy was unknown to most individuals interested in cheatgrass, but it was already well established in science. E. I. Newman was probably the first to recognize that winter annuals are not restricted to simultaneous germination, but actually may benefit from both simultaneous and continuous germination.[28] Virtually all of the current year's production of cheatgrass seeds can germinate if they find safe sites in the seedbed with appropriate temperature and moisture relations. Cheatgrass seeds that are not dispersed to portions of the seedbed that provide safe sites for germination cannot germinate and thus become dormant. This allows them to remain viable but still dormant in the seed bank.

But how can the inherent physiology of a seed be converted from non-dormant to dormant and back to nondormant? Consider the conditions to which the surface of a seedbed in the Great Basin is exposed. High-intensity solar irradiation, wetting and desiccation, freezing and thawing cycles, and microbial action are among the hazards cheatgrass seeds face if they do not find safe sites for germination. Most seeds become dormant at maturity through desiccation. Imbibition of moisture breaks this dormancy. If seeds do not have an impermeable seed coat and can freely imbibe water but are still dormant, something else must be lacking. The most common means of breaking dormancy is nitrate enrichment of the substrate.[29] When the samples were enriched with potassium nitrate solutions, the results were spectacular. Huge flushes of cheatgrass seedlings emerged.[30] The nitrate enrichment explanation also fit nicely with ecological conditions in field seedbeds. In the spring when the seedbeds are warm and moisture is available, nitrification proceeds and nitrate becomes available. Long, wet, relatively warm springs result in maximum density of cheatgrass stands. Obviously, this can be a multifactor interaction. Long, wet springs can create more safe sites for germination through moisture relations, and cheatgrass seeds lose dormancy over time without external stimuli.

Enrichment with gibberellin, a plant growth hormone produced by a fungus, is probably the second most efficient way to break seed dormancy. Samples enriched with minute amounts of gibberellin produced an additional flush in cheatgrass emergence. The response was even greater when nitrate and gibberellin were applied together, although there is still argument about whether the combination is additive or synergistic. No one is sure if gibberellin occurs naturally outside plants as a free chemical in seedbeds, which makes it difficult to assign ecological significance to the dormancy-breaking action observed with gibberellin enrichment.

The research group's field and greenhouse studies in the late 1960s concentrated on defining the physical and biological parameters of germination safe sites for cheatgrass. Fieldwork in fall 1966 at the original experimental sites in and adjacent to the Great Basin provided clues. Close examination of the holes from which samples had been taken revealed that cheatgrass seeds that had blown into the holes were germinating at the bottom of the pits. Cheatgrass seeds lying atop undisturbed soil in between the pits did not show signs of germination. On a few occasions, germination took place in the bottom of the pits before the first fall rain, apparently because diurnal temperature fluctuations caused moisture to condense in the bottom of the pits in amounts sufficient to initiate germination. The tiny depressions had created a microenvironment where it was within the potential of cheatgrass to germinate. These simple observations opened an area of research on the contribution of micro-seedbed topography to safe sites for cheatgrass germination.

Measuring temperature, moisture, and relative humidity on a scale biologically significant to cheatgrass seeds in the field required very sensitive equipment. The emerging field of microelectronics came to the rescue in the form of nickel wire resistance thermometers and thermistors, although the group had to design a specialized switching mechanism and recorder in order to obtain the data on a 24-hour basis. The study sites were far removed from electrical power grids, so the equipment had to be battery powered. Electronic switches were not available at that time, so mechanical ones were used. These mechanical switches proved to be the Achilles' heel of the operation because they would not work at very cold temperatures.

The original five experimental locations provided another clue to the seedbed ecology of cheatgrass when after about a year the herbicide-treated fallow plots lost all their herbaceous litter. The litter consisted mainly of stems from past cheatgrass crops with an occasional fragment of tumble mustard, red stem filaree, and Russian thistle herbage. Cheatgrass did not grow on these herbicidal fallow plots in the field, although seeds were lying on the ground surface and seeds taken from the sites did germinate in the laboratory. Soil samples brought back for bioassay in the greenhouse gave rise to abundant cheatgrass seedlings. What kept the cheatgrass from germinating in the field? Fine-mesh screen cages placed over the herbicidal fallow plots had excluded cheatgrass seeds from blowing onto the plots. An abundant stand of cheatgrass established under these cages even though the litter had decayed on the soil surface. Cheatgrass seeds placed on the surface of sinter plates in the laboratory would not germinate without some form of

TABLE 7.2 | MEAN GERMINATION OF SEEDS OF CHEATGRASS ± ONE-HALF THE CONFIDENCE INTERVAL AT THE 0.01 LEVEL OF PROBABILITY

	PERCENTAGE GERMINATION									
COLD PERIOD TEMPERATURE °C							WARM PERIOD TEMPERATURE °C			
	0	2	5	10	15	20	25	30	35	40
	26±	38±8	48±6	65±6	78±6	87±6	92±6	92±6	87±7	79±10
2		40±8	52±6	69±5	82±5	91±6	95±5*	95±5*	91±6	82±9
5			57±7	74±4	87±4	95±5*	99±5*	99±4*	94±5	85±7
10				79±7	92±5	100±4	100±4	100±4	97±4*	87±7
15					92±6	99±5*	100±5	100±5	95±5*	85±8
20						95±8*	98±5*	96±5*	90±5	79±8
25							89±6	86±6	80±8	69±8
30								73±7	66±8	54±7
35									47±9	36±10
40										13±12

NOTE: Number(s) following the mean is one-half of the confidence interval as determined from regression equations used to develop the response surface.

Asterisk (*) indicates means not lower than the maximum germination (in this case 100%) minus one-half its confidence interval, our definition of optimum germination.

Seeds were incubated for 4 weeks at a wide range of constant and alternating temperatures.

seed coverage. The fact that cheatgrass can germinate over an extremely wide range of constant or alternating temperature regimes (table 7.2) pointed toward moisture relations on the seedbed surface as the limiting factor for cheatgrass germination. These studies were published in two papers that have remained benchmarks in seedbed ecology.[31]

Among the most important sources of germination-enhancing litter on seedbeds are accumulations of cheatgrass seeds themselves. An individual cheatgrass seed with lemma and palea (including awn) intact does not contact the surface of the seedbed efficiently; nor do cheatgrass seeds have mucilage to enhance contact with the moisture-supplying substrate. Individual cheatgrass seeds do not have geniculate awns for self-burial. Despite all these apparent handicaps for germination as individual seeds, collectively, wind-deposited accumulations of cheatgrass seeds effectively provide litter coverage that allows some of the accumulated seeds to germinate. Cheatgrass seeds do not have to have the callus (the sharp, indurate end of the lemma) of the caryopsis—the closest point to the embryo and the normal location of initial imbibition of moisture—in contact with a moisture-supplying substrate in order to germinate. Even suspended in the litter, cheatgrass seeds can imbibe moisture, drop a root to the soil surface, and begin moisture uptake. Even more amazing, if the initial primary root desiccates and dies, additional roots will emerge from adventitious buds and attempt to establish contact with the moisture-supplying substrate.[32]

Succession in the exotic plant communities of the Great Basin rangelands begins with Russian thistle, is followed by the mustard species, and ends with cheatgrass. At each stage of dominance the nature of the seedbed is altered, setting the stage for further succession. The crucial question is why cheatgrass truncates succession for such prolonged periods. The answer to this truncation lies in the nature of the competitiveness of cheatgrass.

CHAPTER 8

The Competitive Nature of Cheatgrass

A "comprehensive" review of cheatgrass published in 1964 describes this annual grass as not highly competitive.[1] In terms of adult herbaceous and woody plants this is true. Most perennial grasses, native or introduced, successfully biologically suppress cheatgrass if they are established, mature plants (fig. 8.1). The important phenological stage for competition is the seedling stage. With hindsight this appear obvious, but it did not become apparent until attempts were made to artificially seed degraded big sagebrush/bunchgrass rangelands.

The classic paper on cheatgrass seedling competition is "Artificial Reseeding and the Closed Community," by Joseph H. Robertson and C. Kenneth Pearce.[2] Joe Robertson, the son of Great Plains homesteaders, was one of John Weaver's students at the University of Nebraska.[3] After graduating, Robertson became one of the brilliant Young Turks employed by the Forest Service's Intermountain Forest and Range Experiment Station during the late 1930s and early 1940s. This extraordinary group, starting with virtually zero knowledge, pioneered the restoration of Intermountain Area rangelands. The Great Basin Forest and Range Experiment Station hired Robertson in 1940 and sent him to two very remote ranger stations in northern Nevada to conduct research on seeding rangelands. At first, range scientists described their research as "reseeding" (note the title of Robertson's and Pearce's paper) until someone suggested they were actually artificially seeding for the first time,

Fig. 8.1. Mature crested wheatgrass surrounded by a space free of cheat-grass. Only in years with very abundant moisture will occasional cheat-grass plants occur in a uniform stand of crested wheatgrass. Observations of similar suppression of cheatgrass by bluebunch wheatgrass led to the conclusion that the annual grass was not competitive, but competition at the seedling level was not differentiated from competition between annual and established plants.

and the "re" was redundant. Considering the seeding attempts' noted lack of success, especially early on, and the number of times that areas had to be reseeded to obtain success, the term "reseeding" is not totally inappropriate. The basic premise of Robertson's and Pearse's publication was pure Weaver and Clements: "The inherent capacity of plants for growth and reproduction leads to competition for every available niche in each habitat and results in a universal tendency for all plant communities to remain closed. This competition has fundamental application in seeding, and indeed in the entire management of rangelands."[4] The authors suggested that the natural tendency of plant communities to remain closed to new members is well established at higher levels of succession. This exposes one of the strikingly ambiguous parts of Clements's succession theory: in order for succession to proceed from at least the lower stages, invasion by a new species is necessary.

Robertson and Pearse pointed out that even great plant formations such as the grasslands of the American Great Plains could be converted from closed climax communities to communities open to invasion by prolonged severe drought. This was the subject of Robertson's Ph.D. dissertation at the University of Nebraska.[5] The authors drew parallels between the effect of a sudden, severe drought on the Great Plains and the prolonged degradation of the sagebrush/bunchgrass habitats of the Intermountain Area. At the time Robertson and Pearse wrote, the bunchgrass component of millions of acres of rangelands in the Intermountain Area had been virtually destroyed by improperly timed, excessive, and continuous grazing. Their comparison began with a most improbable statement: "During gradual and commutative range deterioration, the community remains closed to mass invasion." The authors described the process as follows: "Frequent and continued close cropping of the most relished plants first prevents reproduction and lowers their food reserves. This is followed by a gradual impairment of the root systems. One by one the better plants, now unable to draw moisture from a large volume of soil, or with insufficient reserves of food to maintain respiration through a long period of dormancy, drop from the community" (fig. 8.2).[6] One has to remember the tremendous reproductive capacity of big sagebrush and the longevity of these plants to appreciate this statement. As the perennial grass portion of the sagebrush/bunchgrass communities was destroyed, the density of the big sagebrush component increased. Coupled with continued excessive grazing, this left the sagebrush communities virtually devoid of

Fig. 8.2. *The stages of cheatgrass dominance in a big sagebrush/bluebunch wheatgrass community.* A. *Big sagebrush/bluebunch wheatgrass community with abundant young and vigorous perennial shrubs; wildfires in this community would most likely occur in late August and September.* B. *Big sagebrush community with green rabbitbrush but virtually no perennial grass; this community would burn only under exceptional conditions of drought, high temperatures, and strong consistent winds.* C. *Old-growth stand of big sagebrush with a very abundant understory of cheatgrass; very fast spreading wildfires might occur here from June through September.* D. *Complete conversion of big sagebrush site to cheatgrass with no shrubs; this site is in extreme danger from wildfires throughout the summer.*

the fuel necessary for wildfires to ignite and spread. Wildfires could spread in the sagebrush communities only if the wind was strong enough to bend the flame heads sufficiently to ignite adjacent shrubs more than a yard away. Free from fire as a catastrophic stand renewal process, big sagebrush plants could effectively close a site to perennial herbaceous or woody plant seedling recruitment for a century or more. Continued excessive grazing would result in the death of the native bunchgrasses. Robertson and Pearse described the process eloquently: "Frequent and continued close cropping of the most relished plants first prevents reproduction and lowers their food reserves. This is followed by a gradual impairment of the root systems. One by one the better plants, now unable to draw moisture from a large volume of soil, or with insufficient reserves of food to maintain respiration through a long period of dormancy, drop from the community."[7]

Fitting for what became a basic principle of range restoration, Robertson and Pearse based their conclusion on a vast array of experiments conducted from 1936 through 1940 in Utah, Nevada, and Idaho by the Young Turks working for the Intermountain Forest and Range Experiment Station. Charged with returning productivity to Intermountain Area sagebrush/bunchgrass rangelands, the young scientists conducted seeding experiments in two types of depleted big sagebrush communities: nearly monospecific stands of the shrub, and areas that wildfires had converted to cheatgrass dominance.[8] The species used in these trials were a mixture of native perennial grasses (bluebunch, slender, and western wheatgrasses; the introduced crested wheatgrass; and the exotic annual bulbous bluegrass). The trials were established in the 8–15-inch (20–35 cm) precipitation zone at elevations from 4,700 to 8,000 feet. In some trials the competing vegetation was not eliminated before planting; in others it was partially or completely cleared. Seedings in sites where competition had not been eliminated were doomed to failure. The only perennial grass that had any chance to establish in undisturbed cheatgrass stands was crested wheatgrass, and the stands obtained were not satisfactory. The researchers had many more sagebrush-dominated sites to work with than cheatgrass-dominated sites because of the relative abundances of the two successional stages during the late 1930s. Robertson and Pearse are considered the originators of the closed community concept in regard to cheatgrass, but few remember that at the time they published, degraded sagebrush stands without a perennial grass understory were the predominant closed community for perennial grass establishment in the Intermountain Area.

One of the first studies of the plant physiology and ecology of cheatgrass was conducted in 1938 by Samuel A. Warg as part of his M.S. degree in forestry at Montana State University.[9] A 1934 graduate of Oregon State University's School of Forestry, Warg worked under the direction of the pioneer range scientist Melvin Morris at Montana State. Warg's first experiments concerned germination. He placed cheatgrass seeds in a thin cloth bag that he buried in the surface soil outdoors and left over the winter; he called this his wet freeze treatment. He also placed cheatgrass seeds outdoors over the winter in sealed glass containers; this he referred to as his dry freeze treatment. In the spring, only 3 percent of the "wet freeze" cheatgrass seeds germinated when recovered and tested in the laboratory while 99 percent of the "dry freeze" seeds germinated—the same percentage as seeds stored over the winter in the laboratory. We now know that the "wet freeze" seeds probably acquired dormancy in the seedbed over the winter. Unfortunately, Warg and other scientists who read his thesis interpreted the seeds' failure to germinate as evidence that over-winter exposure in the seedbed rendered cheatgrass seeds nonviable. Warg was adamant in his summary statement that his germination experiments and field observations had shown cheatgrass to be a true winter annual with no spring germination and establishment; therefore, cheatgrass could not complete its life cycle in one calendar year. In fact, his conclusion may have been correct for the plant material that was available in Montana at the time of his experiments.

Warg was among the first to recognize that cheatgrass seedlings were more competitive than seedlings of bluebunch wheatgrass, a native perennial grass. He applied the results of simple seedling development experiments and competition trials in interpreting field measurements of abandoned field succession. Fields abandoned for 1 year were dominated by Russian thistle, tumble mustard (Jim Hill mustard), and some cheatgrass. Fields abandoned for 2 years showed an increase in cheatgrass along with such weeds as prickly lettuce and lamb's-quarters. Fields abandoned for 5 years were dominated by cheatgrass with spot infestations of Canada thistle, sweet clover, and common mullein. Fields abandoned for 10 years or more contained Sandberg bluegrass, bluebunch wheatgrass, and Idaho fescue. The native perennial grasses were most evident along the margins of the fields. Warg believed that native rangelands that had been degraded by excessive grazing but never plowed were much quicker to recover from cheatgrass invasion once grazing was reduced. Warg's research was influenced by Leon C. Hurtt of the Forest Service's Northern Rocky Mountain Forest and Range Experiment Station. Hurtt was perhaps the only early range scientist who publicly stated that cheatgrass was good forage for livestock. Warg's thesis included nearly verbatim the results of Hurtt's experiment with grazing horses on cheatgrass in the Flathead Valley of Montana, citing a newspaper article as the source of the information (Hurtt later published his research as a Forest Service applied research note).[10] Warg noted that the forage value of cheatgrass during the period it was green was as high or higher than that noted for native perennial grasses.

World War II interrupted most of the cheatgrass research. Jerry Klomp later recalled that scientists employed by the Forest Service experiment stations worked evenings and weekends for local farmers to aid in food production efforts to support the war. They were not paid, but they did accumulate leave that they could accrue beyond the statutory annual use-or-lose limit.

After the war, the USDA Forest Service remained the dominant force in range research.[11] Forest Service research produced the 1947 classic "Cheatgrass—A Challenge to Range Research," by A. C. Hull Jr. and Joseph F. Pechanec, which was published in the *Journal of Forestry*.[12] Pechanec was one of the founders of the Society for Range Management and was the society's first president. In the paper's introduction, Hull and Pechanec presented the paradox of cheatgrass:

> The reaction of people to the coming of cheatgrass in Idaho was either one of praise or of condemnation. Cheatgrass was praised for its ability to hold the soil and to produce a large volume of herbage on ranges that had been producing little. On the other hand, cheatgrass was condemned because it fluctuated greatly in forage production and more especially because it encouraged more and larger fires. Consequently, the coming of cheatgrass, which at first tended to protect the soil and provide forage for livestock, later brought recurrent fire which worked hand in hand with overgrazing to destroy perennial grasses and to expose the soil to erosion.

Hull and Pechanec described the occurrence of cheatgrass in southern Idaho as restricted to abandoned farmland, badly overgrazed rangeland, and areas subjected to recurrent wildfires. They concluded that cheatgrass in very small amounts might have invaded properly grazed rangelands, but only if these lands had been previously disturbed; they cited eleven references to justify this conclusion. Although we have read these eleven references many times, we can find no support in them for that conclusion. Further, Hull and Pechanec did not cite Daubenmire's excellent articles published in *Ecology* reporting cheatgrass as an established part of the climax vegetation, even though one of these articles appears in Hull's and Pechanec's bibliography.[13] Like everyone else writing about cheatgrass at that time, Hull and Pechanec considered cheatgrass a problem that could be solved through "proper" management.

Hull and Pechanec considered the desirability of cheatgrass as a range forage plant by assessing six characters: (1) protection furnished to the soil; (2) ability to withstand heavy grazing; (3) season of usability; (4) amount, quality, and reliability of forage production; (5) effect on fire hazard; and (6) the effect of awns and smut on its value. The only way to ascertain how cheatgrass fit these standards was by considering its growth habit, and this the authors did in great depth. No better description of the growth habit of cheatgrass has been published since. Hull and Pechanec described cheatgrass in southern Idaho as primarily a winter annual and noted that if about 2 inches of rain fell in early fall, cheatgrass seeds could germinate and grow a couple of inches before the onset of winter. That occurred about once every 3 years in southeastern Idaho, and once every 8 years in southwestern Idaho. Occasionally, they said, early fall rains prompted cheatgrass germination but lack of further rain resulted in seedling mortality. They cited Piemeisel's *Changes in Weed Plant Cover on Cleared Sagebrush Land and Their Probable Causes* as the source of the latter information, perhaps indicating that such conditions had not occurred within their collective experience.[14]

At Reno, Nevada, fall germination occurs about once every 5 years, but significant leaf growth of cheatgrass in the fall is very rare. If fall germination occurs, five to ten short leaves appear as a rosette flat on the soil surface. By midwinter these leaves will be about the same shade of red that cheatgrass herbage shows at maturity or the onset of moisture stress. Dick Eckert attributed the red color during the winter to nitrogen stress. Others have suggested cold or cold-moisture stress as the cause. Perhaps all three forms of stress are involved. In the spring, the leaves turn green and continue to elongate, then turn red again just before maturity. Cheatgrass plants induced to flower in the greenhouse often turn red just before seed maturity, even though the plants have adequate moisture. In the spring, the seedlings grow rapidly until the soil moisture is almost exhausted, at which point height growth ceases and the leaves curl and become dry. The plants turn purple-red and finally straw yellow. Plants that head out early drop seeds soon after the plants turn red. The cheatgrass dries out completely by early to mid-June.

Hull and Pechanec reported that cheatgrass plants were usually dry by June 5 in southwestern Idaho and by June 15 in southeastern Idaho. Abnormally dry or moist weather could shorten or extend these maturity dates by 4 weeks. Heavy spring rains often caused a second crop of cheatgrass to appear even after the plants had begun to turn red. Second crops also appeared after a late spring frost killed the fall-germinating plants. This appears to be the only original reference in the literature suggesting that cheatgrass is susceptible to spring frost.

Among their other findings, Hull and Pechanec reported that cheatgrass occurred as a single-stemmed plant in thick stands, at densities that could exceed 110 plants per square foot, but could be a well-stooled plant at sites where wildfires or other agents had reduced the density to less than 6 cheatgrass plants per square foot. Over the period 1943–1946 cheatgrass plants averaged 600 per square foot. Piemeisel, working in the same portion of southern Idaho during the very dry years 1933–1935, found an average of 170 cheatgrass plants per square foot.[15]

Hull's and Pechanec's four study sites in southern Idaho yielded 310–690 pounds of cheatgrass seeds per acre, and an average seed rain of 1,650 per square foot. They were the first to estimate the number of cheatgrass seeds per pound at 150,000. Their paper included unpublished data from the files of the Intermountain Forest and Range Experiment Station at Odgen, Utah, indicating that cheatgrass seeds began germination within 1 day and were finished germinating in 2–5 days (depending on the incubation temperature). Our four decades of experience with the germination of cheatgrass seeds at a wide range of incubation temperatures indicate that the 1-day germination is a serious exaggeration.

At the time, cheatgrass was generally believed to be a very shallow rooted plant.[16] Just after the *Ecology* paper came out, however, George Craddock and C. Kenneth Pearse (the Pearse who published with Robertson) published important research results indicating that dense stands of cheatgrass provided more watershed protection on granitic soils than the native bunchgrasses did.[17]

Hull and Pechanec presented the longest record of cheatgrass herbage yields that had yet been published. They also compared the yields of cheatgrass, crested wheatgrass stands, and long-term yields of native perennial grasses from the Snake River Plains. During the period 1940–1946, cheatgrass herbage yields at four sites on the Snake River Plains ranged from 300 to 3,460 pounds per acre. Hull and Pechanec did not report data for all sites for all years, however, and we do not know whether there was no forage production for the missing years or the plots were not clipped. The missing years were not included in the mean herbage production calculations, so we assume the plots were not clipped in those years. The overall average for all locations and all years was 1,260 pounds of herbage per acre. The average production for crested wheatgrass for the same period on seedings located near where the cheatgrass yields were obtained was 1,720 pounds per acre.

Because crested is a perennial grass, one cannot take all the annual herbage production through grazing and expect the stand to persist. We will discuss this in a later chapter on the harvestable amount of cheatgrass annual herbage production. The conservative rule of thumb for grazing perennial bunchgrasses is "take half and leave half" of the yearly herbage production. Annual variation in cheatgrass forage production was a topic of importance in the 1930s and 1940s. It was a major issue in George Stewart and A. E. Young's pioneering paper "The Hazards of Basing Permanent Grazing Capacity on *Bromus tectorum*."[18] Hull and Pechanec found significant variation in annual forage production at their study sites. Cheatgrass production at Arrowrock, Idaho, in 1943 (300 pounds per acre), for example, was one-tenth of the forage production at the same site in 1944 (3,460 pounds per acre) and about 20 percent of the long-term average at the site. Crested wheatgrass production at Arrowrock in 1943 was about 70 percent of the long-term average. In the wet year of 1944, crested wheatgrass production at Arrowrock was 137 percent of the long-term average, and cheatgrass production exceeded 200 percent of the average production. The magnitude in the swing of herbage production was thus considerably greater for cheatgrass than for crested wheatgrass. For native perennial grasses, the variation in annual herbage yield obtained from plots clipped on the U.S. Sheep Experiment Station in Idaho over a 17-year period (including the exceptionally dry year 1934) was 50 percent less than that observed in 7 years of clipping cheatgrass herbage on the Snake River Plains. Hull and Pechanec admitted that Fleming's studies in Nevada had shown that cheatgrass was not as variable as native perennials in forage production but dismissed the findings as a bookkeeping error attributable to the archaic square-foot-density method of sampling.[19]

Burgess L. Kay maintained a nearly 30-year record of cheatgrass production on the Likely Table in northeastern California.[20] His records indicate extreme variation in cheatgrass forage production and more years with below-average than with average forage production. Perhaps the longest record of herbaceous production in the Intermountain Area is one published for crested wheatgrass production at the Point Springs (now Lee

Sharp) Experimental Area in Idaho.[21] Annual production of crested wheat-grass averaged 500 pounds of air-dried forage per acre, but varied from 130 to 1,090 pounds per acre. May–June precipitation was considered to have the largest influence on herbage production.

It seems to us that all of the data indicating that cheatgrass forage production is more variable than that of perennial grasses skirts the vital issue. In dry years, cheatgrass-dominated ranges may produce virtually no harvestable forage while adjacent ranges with native and exotic stands of perennial grass do produce at least some forage. For 75 years this difference has been offered as the shining example of the advantage of perennial grasses over cheatgrass as a range forage base in the Intermountain Area. Perennial grass–dominated ranges may produce more forage than cheatgrass ranges in dry years, but can this resource be safely and sustainably harvested? The season of use is an important factor here. These are spring ranges, which mean that they are used during the drought-abbreviated growing season. All the rhetoric on perennials being less variable than cheatgrass in forage production heard for nearly a century in university classrooms may be meaningless in terms of real life on the ranges of the Intermountain Area.

The annual variability in cheatgrass herbage production obviously has great significance for both livestock carrying capacity on the range and fuel loads for wildfires, but what is the competitive importance of this variation? Its extreme phenotypic variability allows cheatgrass to efficiently exploit an extremely variable climate. Piemeisel considered that neither extreme drought nor extremely abundant population densities threatened the continued dominance of cheatgrass.[22] We concur. We compared the seed production of single-stemmed cheatgrass populations with a density exceeding 1,000 plants per square foot with postwildfire populations of cheatgrass with plant densities of 1 plant per square foot and found seed production to be the same![23] Cheatgrass compensated for the reduction in density with tremendous tiller production.

The winter of 2000–2001 was extremely dry in the western Great Basin, and in the Reno area was one of the driest on record. Very few cheatgrass plants established at lower elevations, and virtually none produced seeds. The winter of 2001–2002 had much more precipitation, but the amount and periodicity of spring precipitation severely limited cheatgrass growth. Most plants headed out early in May at a total height of about 2 inches. The native perennials towered over the miniature cheatgrass plants. Ranchers and researchers commented that it was a "great" year for the perennials and that cheatgrass appeared to be on the way out. The cheatgrass plants may have been tiny, but they did flower and produce seed in abundance to maintain seed banks for succeeding years. At the same time, the perennial grasses had to furnish nearly the entire forage base for grazing animals. W. Ted Hinds recognized the importance of this situation in his excellent monograph on the autecology of cheatgrass when he noted: "This is indeed an adaptive trait for a colonizing annual: maintaining growth potential in succeeding generations without penalty for depauperate conditions in earlier generations."[24]

Hull and Pechanec reported that ranchers considered cheatgrass to be the first grass in the spring to furnish grazeable forage, but their own measurements in 1943–1946 indicated that native perennial grasses started growth before cheatgrass. If this is true, it gives cheatgrass a competitive advantage in areas where excessive very early spring grazing is the norm. Hull and Pechanec did offer the caveat that the perennial grasses started growing before cheatgrass when the latter either did not germinate or made poor growth the previous fall.

The great drought of 1934 was still very much on the minds of the Intermountain Area range scientists in the later 1930s, and Hull and Pechanec looked closely at the response of cheatgrass to drought conditions. They found that cheatgrass-dominated areas returned to full forage production the first year that rains returned to the Snake River Plains while native perennial bunchgrass stands needed 3 or 4 years to return to full production.

Hull and Pechanec destroyed several common misconceptions in the course of their studies. While many range managers considered it impossible to overgraze cheatgrass, for example, Hull and Pechanec reported numerous observations in southern Idaho of cheatgrass dominance being destroyed by excessive grazing, citing work by Piemeisel in Idaho and Daubenmire in Washington to support their observations. Another common assumption was that burning cheatgrass brought on more and earlier feed the next spring. Range sheep operations in particular desired early spring green forage for ewes with newborn lambs. Hull and Pechanec collected data on forage availability on cheatgrass burns for 4 years during the 1940s. They found that early summer burning reduced the density of cheatgrass plants the next spring by 92 percent and reduced early spring height growth by

50 percent.[25] Forage availability at sheep turn-out was often one-tenth that on adjacent unburned range. This work was published in the *National Wool Grower* and was obviously aimed at stopping promiscuous burning. Nevertheless, it seems quite believable, especially if there was sagebrush left in the stand to increase the intensity and duration of seedbed heating. Hull and Pechanec did note that the cheatgrass that had been reduced in density produced numerous tillers, and that herbage production eventually became about equal in cheatgrass stands burned and unburned the previous season.

The next publication of the Intermountain Forest and Range Experiment Station concerning cheatgrass was "Cheatgrass (*Bromus tectorum* L.)—an Ecological Intruder in Southern Idaho," by George Stewart and A. C. Hull.[26] The authors began the paper with a history of cheatgrass on the Snake River Plains of Idaho. Introduced probably about 1900, by the 1940s cheatgrass was the sole dominant on 4 million acres of formerly big sagebrush/ bunchgrass rangelands, the dominant or only understory herbaceous plant on 2 million acres, and had increased from a mere trace to 25 percent of the herbaceous vegetation on another 10–15 million acres of big sagebrush rangeland. When these rangelands eventually burned, that only served to increase the dominance of cheatgrass. A committee formed of Bureau of Land Management natural resource technicians and university professors who met in 2007 to define types of cheatgrass infestations came up with three categories: (1) cheatgrass monoculture (a condition that actually never exists) that requires artificial seeding of perennial grasses; (2) cheatgrass with some perennial native plants remaining, but where artificial seeding of perennial grasses is advisable following wildfires; and (3) big sagebrush communities with a cheatgrass understory where sagebrush and perennial grasses and forbs can be seeded after wildfires. The irony should be evident.

Next in the chronology of cheatgrass comes Lloyd C. Hulbert's exhaustive autecological investigation of the Bromium and Eubromus sections of the genus *Bromus* published in 1955 in *Ecological Monographs*. Hulbert's was the most comprehensive presentation of the physiology of cheatgrass to date.[27] He was a member of the Department of Botany at the University of Minnesota when he published the monograph, but the research was conducted in the Palouse area of eastern Washington and northern Idaho. The grasses Hulbert studied were soft chess, cheatgrass, Japanese brome, ripgut, rattlesnake brome, red brome, chess, and poverty brome. None of these species of brome grasses is native to North America, and all grow as exotics in the Columbia Basin of Oregon, Washington, and Idaho. From the literature, Hulbert developed the following list of chromosome numbers:

Soft chess	28	Cheatgrass	14
Japanese brome	14	Ripgut	56
Rattlesnake brome	14	Red brome	28
Chess	28, 56	Poverty brome	14

Note that the 2n chromosome numbers are multiples of 7, the basic haploid number; most common are the diploid 14 and the tetraploid 28. Hulbert noted that annual brome species are often cleistogamic, or entirely self-pollinated. (We now know that they are largely self-pollinated with environmentally induced outcrossing rather than obligate self-fertilization.) These numbers will be important when we consider sources of variation in the evolution of new ecotypes of cheatgrass to fit changing environments. Hulbert perpetuated a very misleading observation about the flowers of cheatgrass when he stated that he had never observed the anthers of any of the annual bromegrasses to be exerted. In fact, exerted anthers are quite common for cheatgrass plants flowering in the greenhouse, and we have observed them in the field on numerous occasions as well.

Perhaps the strangest misconception concerning the ecology of cheatgrass concerns its root depth. Plant physiologists and ecologists at some point decided—with little or no evidence—that cheatgrass is a very shallow rooted species. Perhaps they found what they wanted to find. Many viewed the native perennial grasses as the rightful owners of the ranges, and cheatgrass the undesirable foreign invader inferior to the native species in every way. The saga of cheatgrass root depth started in 1937 when L. E. Spence studied the root systems of important plants in the Boise River watershed.[28] At the time he published his research, Spence worked for the USDA Soil Conservation Service. When he conducted the research, he was an instructor in the School of Forestry at the University of Idaho and a collaborator with the Intermountain Forest and Range Experiment Station. Spence dug trenches parallel to the cheatgrass plants he wanted to study and excavated the roots from the soil with an ice pick. He found an average of 7 roots per plant, reaching an average depth of 1 foot. In contrast, plants of bluebunch wheatgrass had an average of 176 roots per plant that penetrated the soil to an average depth of 3.5 feet. The evidence fit the perception, so the belief went unquestioned.

A decade later the eminent range ecologist E. W. Tisdale perpetuated the concept when he wrote, "Dwarf, early maturing species, *Poa secunda* and *Bromus tectorum* have shallow root systems, which rarely penetrate below a depth of one foot."[29] A few years later, in 1950, H. C. Hanson added to the misconception: "In spite of its shallow root system, *Bromus tectorum* has been able to invade and compete successfully with shrubs and deep-rooted perennial grasses. In Utah, for example, its roots penetrate to only about six inches while those of the perennial *Agropyron inerme* [bluebunch wheatgrass] and *Helianthella uniflora* [a type of taprooted native perennial sunflower] reach four and five feet, respectively."[30]

Hulbert used a much more precise method to study the nature of cheatgrass roots. He dug trenches parallel to cheatgrass plants, moving a tremendous amount of soil, then washed the delicate roots from the soil rather than attempting to follow them with an ice pick. He also collected volumetric samples of soil from various depths and washed the samples to recover all roots.[31] Employing innovative technology, he injected lithium chloride at various depths in the soil in different trenches and then used a spectroscope to analyze samples of the aerial herbage for the presence of lithium.[32] After an enormous amount of hard labor Hulbert was able to determine that cheatgrass roots extended 4–5 feet into the caliche layer at the depth of the wetting. In the semiarid and arid portions of western North America, annual precipitation is not sufficient to wet soil to the depth of the water table on many sites. The annual depth of wetting level is often marked by an accumulation of calcium carbonate that sometimes forms a cement layer. In the former perennial bunchgrass grassland habitats of the Columbia Basin where Hulbert conducted his root experiments, the depth of wetting is quite deep. In the salt deserts of the Great Basin, where towering mountain ranges cast rain shadows that limit precipitation, the depth of wetting may only be 6–8 inches. Hulbert's work shattered the existing stereotype for cheatgrass roots and paved the way for meaningful research on why cheatgrass is so competitive. Among his contributions was the determination that cheatgrass roots continue to expand during deep winter, even though the aerial portion of the plant remains a small rosette of leaves.

Robertson and Pearse lacked the instrumentation to actually measure the nature of cheatgrass competition with seedlings of native perennial grasses, but the following quotation from their 1946 paper shows that they intuitively understood the nature of the competition.

By virtue of its vigorous spring growth, *B. tectorum* in dense stands draws heavily on the soil moisture provided by winter snows and spring rains by the time it matures seed. Moisture is then not adequate to sustain the growth of the slower growing perennial grass seedlings through the normally dry summer and fall. Any effective summer rains that occur after the *B. tectorum* plants have become dormant are available for perennials, and this may explain the successful reseeding in established *B. tectorum* in occasional years, or in regions with considerable summer rainfall.[33]

Raymond A. Evans entered the field of cheatgrass research in the early 1960s after receiving a Ph.D. in botany from the University of California at Berkeley based on his research on competition among alien annuals. He was perhaps the first participant in the cheatgrass research wars actually trained in annual weed ecology. Evans went to work for the USDA Agricultural Research Service's Range Weed Research Unit at Reno. His first experiment involved seeding crested wheatgrass and cheatgrass at different depths and densities in 3-foot-deep wooden boxes.[34] He filled the boxes with loam soil from a sagebrush area and buried fiberglass soil moisture blocks at depths of 3, 6, 12, and 24 inches in each soil column. In the surface soil of each column of soil he planted 18 crested wheatgrass seeds, approximately three-fourths of the number of seeds per square foot if a seeding rate of 7 pounds per acre is used. Cheatgrass was seeded at rates of 0, 4, 16, 64, and 256 seeds per square foot. The soil columns were saturated before seeding and lightly sprinkled with water until the seedlings emerged, and then no more water was added for the 15 weeks of the experiment, which ran from late November through February, mimicking fall germination. Evans also maintained columns of soil in which nothing was planted. At the end of the experimental period, the boxes without plants had soil moisture levels at the 3- and 6-inch depths near −0.033 MPa; the soil at lower depths remained saturated. Soils in boxes with plants all reached the permanent wilting point (11.5 MPa). The two higher densities of cheatgrass reached the permanent wilting point first, the lower densities second, and the crested wheatgrass by itself last. The 64 and 256 cheatgrass plant densities per square foot greatly decreased crested wheatgrass root and shoot growth and greatly increased mortality of crested wheatgrass seedlings. Even the lowest density of cheatgrass (4 plants per square foot) markedly decreased shoot and root growth and increased crested wheatgrass seedling mortality. Evans found that cheatgrass density as low as 16 plants per square foot caused crested wheatgrass plants to prematurely cease growth and become dormant. If he rewet the plants at the end of the experiment, about two-thirds of these dormant plants would turn green and resume growth. Only 5 percent of the plants in the boxes with 256 cheatgrass plants per square foot turned green after rewetting. In the real world outside the greenhouse, rewetting would seldom occur immediately after the onset of dormancy. The crested wheatgrass seedlings would go dormant in late May or early June and rewetting would not occur for 3 or 4 months. The larger the crested wheatgrass seedling, and thus the more extensive its root system, the greater the chance the seedling would survive the first summer dormant period after germination and establishment.

Evans concluded that soil moisture was the critical factor in competition between cheatgrass and perennial grass seedlings. He spent the next 30 years extending this concept to field situations, using increasingly sophisticated microenvironmental monitoring techniques and weed control methodologies to reduce the competition. Perhaps most important, Evans determined that in all but exceptionally wet years, when spring rains extend into June, almost complete control of cheatgrass—that is, past the 99th percentile—is required for establishment of crested wheatgrass seedlings.

Evans's research efforts were validated in an interesting way. For the next three decades, about 2 years after Evans published the results of an experiment, A. C. Hull would publish on a very similar experiment.[35] We had already met A. C. Hull in his pioneering research in southern Idaho on the nature of the cheatgrass problem. He was transferred from Forest Research to the Agricultural Research Service, but was in the Forage and Arid Pasture Research Unit rather than Range Weed Control. He reproduced the studies very carefully, never greatly disagreed with the conclusions reached by Evans, and usually published in different journals.

During the 1930s, W. S. Chepil studied the longevity of seeds in the soil.[36] Chepil's research was massive in scope and included a multitude of weed species in addition to cheatgrass; because his work was oriented toward agronomic rather than rangeland weeds, range ecologists often overlooked it. Chepil was interested in how long seeds of weedy species could remain viable in the soil and the periodicity of germination of these buried seeds. In rather simple experiments, he determined that cheatgrass seeds remained viable in the soil for at least 2–3 years. Apparently, Chepil had never been exposed to Warg's thesis, and was thus willing to entertain the notion that cheatgrass might build seed banks.

As one would expect of a winter annual that flowers in May, cheatgrass requires both vernalization and long day lengths to initiate flower bud formation. Red brome was the only one of the exotic annual bromes Hulbert tested that would flower from spring plantings without vernalization.[37] Hulbert was alerted to the existence of vernalization and cheatgrass's day length requirements by G. W. Fisher of Washington State University and the pioneer conservation plant material specialist L. R. Schwendiman. Finnerty and Klingman satisfied the necessity for vernalization of cheatgrass plants by germinating seeds at low temperatures.[38]

The next major contribution to the physiological ecology of cheatgrass came from Grant A. Harris of the Department of Forestry and Range Management at Washington State University in 1967. Harris worked in the Department of Forestry and Range Management at Washington State University at Pullman. Obviously, he had a direct connection with Hulbert's earlier work. Harris started by describing the Intermountain Area of the northwestern United States as it had been when it was a vast environment in dynamic equilibrium: "*Agropyron spicatum* (Pursh) Scribn. & Smith (bluebunch wheatgrass) was the major plant species of this region. It flourished on open semiarid sites from the valleys of the Canadian Rockies south to the mountains of the Mexican Sonora, and from the slopes of the Cascade Mountains east to the short-grass plains. In the central part of the Columbia Plateau, and in other less extensive areas, this species was the dominant plant, producing more herbage than all other associated plant species combined."[39] Harris added that in addition to being the dominant plant on millions of acres of grasslands, bluebunch wheatgrass provided understory cover in association with trees and shrubs on many additional millions of acres. He went on to note that "the introduction of European culture into this region with its traditional use of the plow, domestic livestock, exotic plants, and fire control foreshadowed destruction of the pristine ecologic balance forever." Cheatgrass sprouted from the ashes of this ecological disaster. Awareness of the magnitude of the ecological disaster in the Northwest brought the concomitant desire for restoration, Harris said, but restoration attempts by hit-or-miss trials without understanding the basic competitive relations between perennials and annuals were bound to fail.

Harris took a meticulous approach to understanding the competitive relation between cheatgrass and bluebunch wheatgrass. To avoid the genetic variability that Hulbert had demonstrated among cheatgrass accessions, Harris used seed from only one source in all his experiments. He fully subscribed to Robertson's and Pearse's concept that cheatgrass-dominated stands were closed to the establishment of perennial grass seedlings. He wanted to know why. He started with fall germination. At moderately cold temperatures in moderately wet soils, cheatgrass seemed to enjoy a small germination advantage over seeds of bluebunch wheatgrass. Harris suggested that even a slight germination advantage might have great ecological significance in the race to germinate and establish before the onset of winter temperatures. We take a wider view of the genetic potential of bluebunch wheatgrass and cheatgrass seeds to germinate and consider the advantage to be decidedly with cheatgrass.[40] Harris considered the difference in numbers of seeds produced to be an important factor in cheatgrass's competitive advantage. Even a good stand of bluebunch wheatgrass produces far less seed per unit of area than cheatgrass does. Moderate to poor stands of bluebunch wheatgrass under heavy grazing pressure may produce *no* seeds at all. Cheatgrass, even in extremely dry years, produces some seed if the plants have enough moisture to germinate and establish. Furthermore, even excellent stands of bluebunch wheatgrass build very poor seed banks. In fact, of the native perennial grasses in the Intermountain Area, only Indian ricegrass consistently builds large seed banks.[41] In the native perennial grass community that N. A. Hassan and N. E. West studied in Utah, the seed bank contained fewer perennial seeds than cheatgrass seeds in spite of the high cover of native perennial grasses.[42]

Harris also found that cheatgrass had an overwhelming advantage over bluebunch wheatgrass in the rate of root growth.[43] Under standard laboratory conditions, cheatgrass roots grew 50 percent faster than those of bluebunch wheatgrass. In the field in the late fall, when temperatures were near freezing, cheatgrass had an even greater advantage. Early rapid root elongation became a self-propagating advantage for the cheatgrass because the deeper the roots penetrated in the soil profile, the warmer the soil temperatures remained and the faster the roots grew. Cheatgrass roots could continue to grow with soil temperatures in the 2–3°C range, while root elongation for bluebunch wheatgrass stopped at 8–10°C. Harris credited his discovery of this critical factor in the competition between cheatgrass and bluebunch wheatgrass seedlings to R. F. Daubenmire.

When he excavated the roots of native bluebunch wheatgrass plants in Washington, Idaho, and Oregon, Harris found great variability in rooting patterns. Adventitious roots usually ran parallel to the soil surface for 8–12 inches and then dropped down to deeper depths in the soil profile. Some ecotypes had roots that dropped straight down. Harris had no difficulty distinguishing between the roots of cheatgrass and those of bluebunch wheatgrass. Bluebunch wheatgrass has off-white roots while those of cheatgrass are dark brown. Cheatgrass roots are smaller in diameter, weaker in tensile strength, noticeably more diversely branched, and have a greater number of longer root hairs. Harris tried to count the number of cheatgrass and bluebunch wheatgrass roots passing through a 4×4–inch plane located 16 inches deep in the soil profile under a bluebunch wheatgrass stand with sparse cheatgrass. The bluebunch wheatgrass roots averaged 76 per square foot (range 28–180). It was impossible to get an accurate count of cheatgrass roots, but they exceeded 1,800 per square foot. Both cheatgrass and bluebunch wheatgrass have primary and adventitious roots. In exceptionally dense stands, though, the adventitious roots of cheatgrass may never develop.

Harris was the first to compare the internal structure of the leaves and roots of bluebunch wheatgrass and cheatgrass. The two species share the same anatomical structures, although they are quite different in size and shape. Cheatgrass leaves are wider and thinner. Leaves of bluebunch wheatgrass are narrower and obviously involute (i.e., the top surface is rolled inward). Bluebunch wheatgrass cell walls are heavier and the leaves have thicker veins. The rolled leaves and thicker cell walls are probably adaptations to reduce moisture loss during the dry summer. Cheatgrass plants are mature and dead by that time. Harris found similar adaptations in root structures. Bluebunch wheatgrass has tremendously heavier cell structure throughout, and the cell walls in the endodermis are heavily suberized (i.e., corky); these attributes probably help the plant transport water through dry, warm soils during the midsummer drought period. Using radioactive isotopes and autoradiographs Harris was able to determine that cheatgrass roots penetrate the soil below a drying soil front before the slower-growing roots of bluebunch wheatgrass reach the same level. He also determined that cheatgrass stands maintain a heavy demand on soil moisture over a wide range of plant densities. In dense stands, primary and secondary root systems deplete available moisture as they advance. In sparse stands, plants tiller and produce vigorous adventitious roots. If the roots of bluebunch wheatgrass do not find remnant soil moisture in the profile before the summer drought, the seedlings will not survive the summer.

Grant Harris and A. M. Wilson published the capstone research paper on cheatgrass and bluebunch wheatgrass rooting habits.[44] They repeated Harris's original experiments, confirming that cheatgrass roots grow faster than those of bluebunch wheatgrass, and also demonstrated that the exotic annual plant medusahead can grow roots even faster. The crowning point of the research was their finding that the roots of the introduced perennial crested wheatgrass grow as fast as those of cheatgrass. This work has had great importance in the selection of weed control practices and plant material for seeding cheatgrass-infested ranges. Weed control must be nearly perfect if bluebunch wheatgrass seedlings are to become established.

Well-meaning individuals have suggested the need for some treatment that thins the cheatgrass stand and allows the perennial seedlings to establish. Harris's research clearly showed that cheatgrass can compensate for reduced stand density and still close the site. Harris also commented on the suggestion that a bluebunch wheatgrass with seedlings that can compete with cheatgrass should be developed. Such a plant would be extremely ephemeral, both starting and finishing its annual growth very early in the spring. It would also be a perennial with a very short green feed period—the very characteristic that is often used to characterize cheatgrass as an undesirable forage species.

Results of the field microenvironmental monitoring research of R. A. Evans and his associates closely match Harris's findings on cheatgrass–perennial grass seedling competition. Evans's research underscores the importance of temperature and moisture interactions in the field and the periodicity of moisture events in the spring. The emphasis on research concerning the physiological-ecological relations between cheatgrass and perennial grass seedlings has concentrated on the factor that allowed cheatgrass to be so competitive. Taken together, Harris's and Evans's research points toward moisture relations as the form of competition in which cheatgrass wins the battle for survival. The important point here is that in North America, cheatgrass was introduced into a highly disturbed environment. The very nature of the disturbance—plowing, excessive grazing, and more frequent wildfires—changed the inherent potential of the rangeland sites to support plant growth. Accelerated soil erosion and changes in the mineralization of nitrogen as well as the inherent potential of seedbeds to support germination coupled with biological changes within the soil created new and different ecological sites. These anthropogenic changes were superimposed on continual climatic change, which occurs on time scales ranging from annual to geological. The anthropogenic changes did not cease with the introduction of cheatgrass. The interval between wildfires has greatly shortened while the destructive aspects of grazing have often been replaced by grazing management that favors the establishment and maintenance of perennials. Through all of these changes cheatgrass has survived, spread, and prospered. This fact strongly suggests that ability to accommodate environmental change through phenotypic and genotypic variability is more important than any single physiological characteristic. With a constant, roughly uniform environment such as that presented by degraded big sagebrush/bunchgrass plant communities during the late fall to spring months, extreme genetic diversity within a cheatgrass population would be a competitive detriment. A uniform, highly adapted genotype with sufficient phenotypic variability to compensate for local environmental variation would be a great benefit because it would optimize genetic and environmental potentials. An opportunistic species such as cheatgrass needs genetic variability only when the gross environment is significantly changing. The characteristic that makes cheatgrass such a competitive species may thus be the environmentally triggered ability to change genetically.

CHAPTER 9

Genetic Variation and Breeding Systems

"Weeds," Herbert G. Baker was fond of saying, "are excellent subjects for the study of adaptation and micro-evolution, for they are abundantly available, usually grow fast, and reproduce quickly and easily."[1] Baker certainly was not the first to see the value of weeds for genetic studies. In his 1974 review he listed ten more or less comprehensive reviews on the subject. The British botanist E. J. Salisbury was among the first to suggest that some plants are predisposed to be weeds of agriculture, using as an example the number of European agricultural weeds found in the tundras that followed retreating glaciers at the end of the Pleistocene.[2] Baker defined evolutionary success as follows: "The evolutionary success of any organism is to be measured in terms of numbers of individuals in existence, the extent of their reproductive output, the area of the world's surface they occupy, the range of habitats that they can enter, and their potentiality for putting descendants in a position to continue the genetic line through time."[3] R. W. Allard compiled a list of the world's most successful noncultivated colonizing species—or as he put it, plants growing without cultivation (see table 9.1).[4] Some people include plants of uncultivated range and forest lands in this category, but that obviously was not Allard's intention.

Ten of the twelve species in Allard's list were described by Carolus Linnaeus. Another, red stem filaree, was described by Charles L'Héritier de

TABLE 9.1 | WORLD'S MOST SUCCESSFUL NONCULTIVATED COLONIZING SPECIES

COMMON NAME	SCIENTIFIC NAME
Lamb's-quarters	*Chenopodium album* L.
Wild oat	*Avena fatua* L.
Chickweed	*Stellaria media* L.
Yellow oxalis	*Oxalis corniculata* L.
Prickly lettuce	*Lactuca serriola* L.
Red stem filaree	*Erodium cicutarium* L'Hér.
Foxtail	*Hordeum murinum* L.
Purslane	*Portulaca oleracea* L.
Curly dock	*Rumex crispus* L.
Galinsoga	*Galinsoga parviflora* Cav.
Wild radish	*Raphanus sativus* L.
Buckhorn plantain	*Plantago lanceolata* L.

SOURCE: R. W. Allard, "Genetic systems associated with colonizing ability in predominantly self-pollinated species," in *The Genetics of Colonizing Species,* ed. H. G. Baker and G. L. Stebbins, 49–75 (New York: Academic Press, 1965).

Brutelle, another eighteenth-century botanist. All twelve species have been recognized for a long time and were among the first to be formally described under the system of modern plant nomenclature. Yellow oxalis is a notorious weed of greenhouses, and buckhorn plantain is a pest of lawns. Galinsoga is a native of the Western Hemisphere that was accidentally introduced into Europe, where it has become an invasive weed. The other eleven species are native to Europe and Asia and were introduced into North America. Only red stem filaree and prickly lettuce are commonly found growing in association with cheatgrass.

There is little chance of general agreement on any list of the "world's most successful" weeds. Lamb's-quarters and wild oat are excellent candidates, lamb's-quarters in association with concentrations of domestic livestock, and wild oat with cereal production. Russian thistle would seem an obvious candidate for such a list. What did Allard really have in mind? In a subsequent paragraph he made his reasoning very clear: "The above eminently successful colonizing species are distributed in many different families and represent great diversity with respect to morphology and physiological characteristics. There is, however, one feature which the great majority of these notably successful colonizers share: a mating system involving predominant self-fertilization." Allard considered this the central issue to be addressed by the 1964 symposium on the Genetics of Colonizing Species. The concept of invasive species as plants that can actively colonize new environments without the conscious efforts of humans came into common use after the symposium proceedings were published in book form in 1965. Although the authors proposed the term "self-invasive" for such plants, the term has been shortened to "invasive." It is most frequently used to describe exotic species that are weeds. In fact, the species on Allard's list could easily be lumped as invasive weeds.

Allard and his colleagues chose wild oat and soft chess as candidates for genetic studies of invasive species. Soft chess was one of the first exotic species of *Bromus* to be collected in the northern Great Basin, and in chapter 3 we speculated that it was replaced by cheatgrass. West of the Sierra Nevada, soft chess is one of the most abundant exotic grasses in the extensive annual grass–dominated rangelands. It is the co-dominant tall annual grass in many communities with slender wild oat, another exotic invasive species. Cheatgrass is a rare species in the cis-montane grasslands. Allard collected soft chess plants throughout California and grew them in common gardens. Cytologically he found the specimens to be very stable and uniform. In morphological appearance and physiological behavior, however, soft chess was highly variable. When he examined individual progenies in great detail, he found that some were segregating for simple inherited characteristics such as lemma color, pubescence, and hairiness of the rachilla. Selections determined that within-family variability was heritable, hence despite a mating system of predominant self-fertilization the populations were far from homozygous.

What is the significance of self-pollination in colonizing species? G. Ledyard Stebbins, a professor of genetics at the University of California at Davis, was the first to propose and elucidate the importance of this breeding system in populations of colonizing species. In his heyday, Stebbins was a brilliant, often flamboyant lecturer. He introduced his ideas on self-fertilization in a symposium held at Stanford University in 1957 at which he said the following.

> In a population of asexually reproducing organisms, each successive new mutation can produce only one new genotype, and consequently the increase in genetic variability must be very slow, unless the number of individuals is enormously large and reproduction is very rapid in terms of numbers of generations per unit of time. In a population of sexually reproducing organisms, on the other hand, each new mutation

can, after a few generations of outcrossing, give rise to hundreds or even thousands of new genotypes, if the store of genetic variability already existing in the population is great, and if free recombination of genes is possible, sexual reproduction can act efficiently to increase or maintain variability in genetically heterogeneous, cross fertilizing populations. Hence, the most logical explanation for the widespread prevalence of sexual reproduction and obligate cross fertilization in natural populations is that over long periods of time in the geological sense only those evolutionary lines have been successful which have acquired mechanisms enforcing sexuality and cross fertilization.[5]

Although Stebbins seemed to be making a strong case for the evolutionary disadvantages of self-fertilization, he immediately countered with abundant examples of highly successful species that are largely self-pollinated. One of his examples was the scarlet runner bean, which successfully enforces cross-fertilization through self-incompatibility, while the related garden bean and lima bean have been equally successful as self-fertilized homozygous lines. All three species are members of the genus *Phaseolus*. Stebbins mused that such cases of related species opting for self-fertilization might be rare instances of a few homozygous lines remaining healthy despite the ravages of inbreeding depression. Even if they survived, they would represent evolutionary dead ends. He offered an alternative hypothesis: the development of self-pollinated lines is not accidental, and sometimes produces a selective evolutionary advantage. Essentially, Stebbins was suggesting that in certain instances, the self-fertilized plants possess some advantage over cross-pollinated plants. If this is true, then natural selection can direct not only the evolution of visible differences between populations, but also the evolution of population structures. And it might act particularly on those mechanisms that affect the store of variability present in populations as well as its method of release.[6] Stebbins was a noted mountaineer and rock climber, but he did not need to be atop a mountain peak to see a long, long way.

From the plant kingdom Stebbins chose the genus *Bromus* to illustrate variation in breeding systems among closely related species. Smooth brome, a widely planted rhizomatous perennial forage grass, has very large anthers that are normally extruded outside the lemma and palea. Smooth brome plants are self-incompatible. There are perennial species of brome grasses that are self-compatible but are also easily outcrossed. The annual members of the sections Bromium and Eubromus, on the other hand, have small anthers that are rarely extruded and are usually pollinated by direct application of pollen from the anther to the stigma. Among this group of annuals Stebbins specifically mentioned *Bromus tectorum,* cheatgrass. Mark McKone used cheatgrass in studies of the reproductive biology of brome grasses and found it to be largely self-fertilized with anthers that emerge only partially.[7]

Stebbins argued—and backed it up with excellent examples—that self-fertilization is a derived characteristic. Many of the self-fertilized species are annuals, which are considered more specialized than related perennials.[8] The annual species often have highly specialized seed dispersal mechanisms. Additional evidence of self-fertilized species being derived from related outcrossing species is the presence of lodicules on grasses such as cheatgrass. Lodicules are small structures at the base of florets that by taking in water (becoming turgid) force the florets to open and allow the anthers to be extruded. Cheatgrass has these structures, and occasionally they function and allow the anthers to extrude from the floret. Self-fertilizing populations have occurred in normally self-incompatible species within historic times; the cultivated snapdragon is an example. Stebbins considered self-incompatibility breeding systems to require the accumulation of a series of rarely occurring genetic events. Self-compatibility can arise relatively easily from self-incompatible lines or from species that are both self-compatible and outcrossing. The chief barrier to self-fertilization is inbreeding depression caused by deleterious genes that accumulate in homozygous populations.

Stebbins proposed two general categories of selective pressure that favor self-pollinated breeding systems. The first operates on plants colonizing environments where conditions for wind or insect pollination are often unfavorable, as is the case for the multitude of exotic annual grasses that have invaded cis-montane California grasslands. During very favorable periods for plant growth, cross-fertilization can occur. J. R. Harlan reported in 1945 that flowers of California brome with large, exerted anthers were formed during the height of the spring blooming season but were absent if the plants were forced to flower in the fall.[9] Colonizers that disperse long distances and can self-fertilize obviously have an advantage. Stebbins put weeds into this category, using the genus *Bromus* to illustrate his point. The center of origin for the genus is in Eurasia, which is home to many self-incompatible species. Almost all of the exotic self-invasive species of *Bromus* that have invaded western North America are self-compatible. Self-fertilization has been reported for members of eighteen tribes and seventy genera of grasses.[10] Stebbins noted that

> a species that normally behaves as a weed cannot maintain continuous populations of a constant size, since the habitats which it occupies are temporary. Its populations are constantly being destroyed, and must be built up again in newly available locations. Under such con- ditions, a highly selective advantage is given to a genetic type which can quickly build up large populations of well adapted individuals as the progeny of one or two initial colonizers. This ability depends not only upon a high reproductive capacity, but also upon the assurance that the descendants of a single well adapted initial colonizer will be equally well adapted to these same conditions, that is, will have adap- tive properties similar to their parent.[11]

In contrast, a single heterozygous initial colonizer, as would occur with an outbreeding population, is certain to produce segregating offspring with dif- ferent adaptive properties. "Given an initial adaptive advantage, therefore, a homozygous individual belonging to a self-fertilizing line would be most likely to transmit this advantage to its descendents."[12]

From these considerations Stebbins formulated his general hypothesis of the genetics of colonizing species: "Plant populations adapted to certain types of temporary habitats, particularly annual weeds and other coloniz- ers of newly available situations, possess a considerable selective advantage if they remain constant genetically for periods of many generations, but still can produce occasional bursts of genetic variability. The most common genetic systems which bring about such cycles of alternating variability and constancy is that of predominant self-fertilization, with the capacity for occasional outcrossing being nevertheless retained."[13] Cheatgrass truncates succession and remains the dominant plant in Great Basin rangelands for extended periods, which seems to violate Stebbins's concept of transitory colonies of weeds. Cheatgrass is an annual, though, and the climate of the Great Basin is so variable that every year is virtually a new environment in terms of the amount and periodicity of precipitation. Add to this climatic variation, the impacts of grazing regimes, wildfires, and even weed-control attempts, and it is obvious that cheatgrass is a species of transitory habitats that must stretch phenotypic plasticity to cover the minor variations in habitat and periodic hybridization and recombination to meet major changes or to invade new environments.

H. G. Baker spoke to the habitat variation that colonizing species face: "The habitats occupied by plants consist of a mosaic of specific micro-environments associated with obvious features of local geography, such as changes in slope and location of stones, as well as many factors which are not readily apparent to human observers, such as variations in soil chemistry. The simplest and most effective system for efficient occupation of such mosaic environments, at least for sedimentary organisms such as plants, would be to develop an all purpose genotype with infinite plasticity."[14] W. T. Hinds studied cheatgrass plants growing on south-facing and north-facing slopes and found that the former group ran out of water 2 weeks before the latter. Despite this limitation, total energy fixation by cheatgrass did not significantly differ between the two exposures.[15]

C. A. Suneson, who was to barley plant breeders what Stebbins was to geneticists, studied phenotypic plasticity in self-pollinated barley plants. In what he called his evolutionary approach to plant breeding he planted a large number of cultivars of barley in the same field plots.[16] If the environment of the field was held relatively constant, one line of barley always assumed dominance. In a relatively uniform environment, such as degraded big sagebrush/bunchgrass communities repeatedly burned in wildfires, only one genotype is ideally adapted—as long as it has great phenotypic plasticity. The plant that can make endless stable duplicates of this successful genotype wins environmental dominance. Today's society places a premium on diversity. A common assumption is that genetic diversity is always the ideal for which we should strive. On a given constant site, however, there is one ideal genotype—one evolutionary line particularly efficient at seed production, which is the measure of evolutionary success. If this genotype shares the habitat with genotypes less effective at producing seeds, the net reproductive potential is lowered. For cheatgrass, such efficiency is determined by its ability to compete with the seedlings of perennials. If genotypes with reduced competitive ability are also present in the stand, cheatgrass disappears and perennials become dominant. The game cheatgrass plays requires perfection.

That is why numerous competition studies have shown that almost all of the cheatgrass must be controlled if seedlings of perennials are to have a chance for establishment. In a biological sense, cheatgrass cannot afford diversity. It needs, in Baker's words, an all-purpose genotype.[17]

Stephan Novak examined two thousand cheatgrass seedlings from sixty North American populations using twenty-five-loci starch gel electrophoresis and failed to find a single heterozygous individual.[18] He did find a high level of gene diversity among populations. We supplied several of the accessions of cheatgrass used in Novak's experiment and tried to obtain as many specimens as possible from cheatgrass populations that had recently invaded the salt deserts of the Great Basin. Novak used Nei's genetic diversity values to construct a cluster phenogram for cheatgrass.[19] The only cheatgrass accession that made a first-level separation from all other accessions was our collection from "Juniper Flat." Unfortunately, poor recordkeeping prevented us from determining where we obtained this accession. The accession from Emigrant Pass, Nevada, separated at the fourth level. Novak found that this location had the highest expected mean heterozygosity of any population tested. Interestingly, the Emigrant Pass location was always a thorn in the side of R. A. Evans and R. E. Eckert in their control and revegetation studies. They always believed the cheatgrass growing at this location was more vigorous, more difficult to control with herbicides or tillage, and better able to compete with seedlings of perennial grasses. Novak concluded that cheatgrass is nearly entirely self-pollinated and that the variation noted among populations must be the result of multiple introductions from different geographic areas. How did the cheatgrass become genetically variable at the various locations from which it was introduced?

When Lloyd C. Hulbert determined the phylogenetic relations of the annual species of *Bromus* that had been accidentally introduced into western North America, he found that they shared multiples of the same chromosome numbers.[20] Hulbert used common gardens to study variation among twenty-four cheatgrass accessions. All of the accessions from North America were winter hardy in the gardens in Washington and Idaho where he conducted his research. An accession from Israel was not winter hardy and suffered 95 percent winterkill. There was about 3 weeks' difference in time of maturity among the accessions, and about a 25 percent difference in height between the tallest and shortest accessions. Regarding specific morphological differences, Hulbert found that pubescence on the lemmas varied continuously from scaberulent to thickly pubescent.

More recent studies have identified cheatgrass ecotypes that are resistant to sulfonylurea herbicides.[21] Some individuals in a stand of cheatgrass growing in a Kentucky bluegrass seed field in eastern Oregon were not killed by applications of a herbicide widely used for control of annual grasses in winter wheat fields. The progeny of these plants were resistant to herbicides of the same herbicidal family. This stretches the "multiple introduction" explanation for variation in morphological characteristics for cheatgrass beyond the breaking point. Molecular genetics techniques should be able to determine the existence and potential consequences of environmentally induced outcrossing in cheatgrass populations. Perhaps neither numerous introductions nor hybridization is the explanation for the observed variability in cheatgrass populations, and future scientists will expand the known universe of invasive weed genetics with a completely new explanation.

We proposed many years ago to take Allard's and Stebbins's basic concepts concerning the genetics of largely self-pollinated species a step further, stimulated by our observations of areas in the Great Basin where cheatgrass-infested ranges burned in wildfires. As we have noted, cheatgrass plants on sagebrush/bunchgrass rangelands can reach densities of several hundred or even a thousand plants per square foot. At such extreme densities, inflorescences may consist of a single fertile floret containing a single seed. When such a cheatgrass-infested area burns, almost all of the cheatgrass seeds in the seed bank beneath shrub canopies are killed by the intensity and duration of the heat generated by the burning woody material and subcanopy litter. Many of the cheatgrass seeds on the surface between the shrubs are killed as well, but apparently a more important factor is that burning consumes herbaceous litter accumulations in the shrub interspaces, reducing the safe sites available for cheatgrass seed germination.[22] Postwildfire densities of cheatgrass may drop to one plant per square foot in the former interspaces between shrubs. These relatively widely spaced cheatgrass plants have a comparatively huge amount of resources available for growth. Much more soil moisture is available per plant, and the burning process has released nutrients.[23] Such plants respond with exceptional growth in characteristics including culm height, panicle size, and number of tillers. Prolonged tiller production extends the flowering period for weeks compared with the day or two that plants in dense stands flower. These well-supplied plants produce tremendous amounts of seed. Populations with one plant per square foot can equal or exceed seed production of preburn communities that had several hundred plants per square foot.[24] The second growing season following the wildfire, cheatgrass seems to explode across the burned rangelands with large, very vigorous plants. We propose that these plants may be expressing hybrid vigor.[25] Hybrid vigor is the opposite of inbreeding depression, which is expressed in inbreeding populations. Largely self-pollinated plants such as cheatgrass are highly inbred. Further, selection for fitness has occurred within inbred lines for specific sites. The phenotypic plasticity expressed by cheatgrass plants growing in greatly reduced population densities following wildfires opens the door for potential hybridization and the crossing of inbred lines. This would account for the apparent burst of vigor in cheatgrass populations the second growing season following wildfires; genetic segregation would occur in that generation. Any genotype that finds an ecological niche can lapse into self-fertilization and produce stable duplicates of itself. Molecular genetics techniques will soon determine whether this hypothesis is fact or fiction, but molecular genetics is a scientific tool to be used in the general understanding of complex biological processes, not an end unto itself.

The spread of cheatgrass into the salt deserts of the Great Basin is one of the most sudden and most extensive range increases of an exotic invasive weed ever to occur in North America. Before the 1980s, cheatgrass was almost completely confined to the sagebrush zone of the Great Basin. This zone is characterized by loam to sandy loam–textured surface soils with well-developed argillic horizons. Precipitation ranges from 8 to 14 inches annually. The predominant woody species is big sagebrush, and the understory when Europeans arrived was characterized by various species of perennial bunchgrasses. The zone below the sagebrush zone, which is characterized by woody species of chenopods rather than sagebrush, has the potential to develop into salt desert.[26] W. D. Billings pointed out in the mid-twentieth century that despite its name, much of the salt desert zone does not have salt-affected soils. Sagebrush does not grow at lower elevations in the basins of the Great Basin because of the aridity (4–6 inches annual precipitation). At one time we assumed that it was just too dry for cheatgrass to grow in these arid areas. In the early 1980s, however, we began to observe cheatgrass growing in shadscale communities in the arid zone. By 1985 there was sufficient cheatgrass to fuel huge wildfires in the salt deserts north of Pyramid Lake and in the Slumbering Hills–Silver State Valley area northwest of Winnemucca, Nevada. In 1999, about 850,000 acres of rangeland, approximately half of it salt desert, burned in the Winnemucca area during one week of firestorms. Prior to the invasion by cheatgrass, the salt desert communities were considered immune to wildfires because they lacked herbaceous fuel.

Did cheatgrass suddenly develop the inherent potential to exploit salt desert environments, or did the climate of the salt deserts suddenly change to encompass the potential of cheatgrass? If the genetic potential was supplied by means of a newly introduced genotype, how did that genotype become so widely dispersed in such a short period? If the new inherent potential was the result of hybridization and recombination or mutation of genes controlling some key feature of moisture relations, these must have recurred many times in many places simultaneously. Despite the magnitude, speed, and consequences of cheatgrass invasion of salt desert environments, we know of only one scientific paper dealing with the phenomenon.[27]

In visualizing the dynamics of the breeding system of cheatgrass and genetic change, it is worthwhile to consider the population sizes involved. Even a very minimal population of one hundred plants per square foot produces populations exceeding four million plants per acre, and millions of acres of rangeland are infested with cheatgrass in the Great Basin. The flip side of these extremely dense populations is that during times of drought the populations decline to zero over millions of acres for multiple years. Cheatgrass populations renew from seed banks after droughts, but do all available genotypes persist equally in seed banks? Drawing a large population down to virtually nothing and then allowing for near unlimited expression of individuals can be a powerful evolutionary mechanism. It is possible that a single homozygous genotype may be the only one represented among the millions of cheatgrass plants now growing on vast expanses of degraded sagebrush-bunchgrass rangelands. If this genotype is generally adapted to the particular environment in terms of soil characteristics, amount, and periodicity of precipitation, it is the genotype best suited to utilize the environmental potential.

We have repeatedly stressed that growing conditions on Great Basin ranges are highly variable from one year to the next. In addition, although the soils on many Great Basin landforms, such as lake plains and alluvial fans, are relatively uniform, there are always microdifferences in seedbed characteristics. How do the fittest cheatgrass genotypes that reproduce by self-fertilization adapt to environmental variability? The answer is through phenotypic plasticity. Anurag Agrawal's review and synthesis paper on phenotypic plasticity suggests that "single genotypes can change their chemistry, physiology, development, morphology, or behavior in response to environmental clues."[28] Evolutionary biologists often dismiss phenotypic plasticity because it has no genetic basis, but Agrawal embraced the adaptive plasticity hypothesis that phenotypic plasticity evolves to maximize fitness in variable environments. Essentially, phenotypic plasticity allows cheatgrass to be "stable" in an inherently unstable environment. When the extremes in variability in a given environmental setting exceed the phenotypic plasticity of cheatgrass, it is time for hybridization.

Stand renewal in cheatgrass stands is through burning in catastrophic wildfires. Such wildfires reduce population densities, allowing more environmental potential per individual cheatgrass plant. This enhanced environmental potential induces an expression of phenotypic plasticity that enhances the chances of fertilization with wind-transferred pollen from cheatgrass plants with different genotypes. Fertilization produces seeds with the most effective form of evolutionary fitness advantage, heterosis or hybrid vigor. In a paper on the selective advantage of immigrant genes in hybrid populations, Elbert and colleagues offered this conclusion, "An important effect of hybrid vigor is the genetic rescue of populations from extinction because it may influence extinction and colonization dynamics of the whole metapopulation."[29]

CHAPTER 10

Control of Cheatgrass and Seeding Prior to Herbicides

In chapter 3 we mentioned that the editor of a Carson City newspaper wrote an editorial in 1886 asking the state to take action to restore overgrazed rangelands. If the problem was apparent in 1886, why did it take until the mid-twentieth century for successful restoration seedings in the Great Basin? Four major problems hindered the development of seeding technology: (1) the leading conservation agency—the USDA Forest Service—believed that ranges could be restored through management without seeding; (2) the equipment necessary to control competing woody vegetation and to seed degraded sagebrush rangelands was not available; (3) early seeding efforts on rangelands, which largely had been conducted by the Forest Service, had almost entirely been perceived as failures; and (4) it took a long time for the majority of ranchers, and even longer for politicians, to be convinced that something was wrong with the western range.[1]

The Forest Service was created in 1905 to administer the nation's existing forest reserves, formerly managed by the U.S. Department of the Interior.[2] These lands were generally higher-quality rangelands in terms of environmental potential for plant growth; much of the land was higher-elevation summer rangeland. Prior to the passage of the Taylor Grazing Act in 1934 and the presidential decree closing the remaining vacant federal lands in the West to homesteading, no U.S. government agency managed the desert and foothills ranges.[3] Afterward, the Grazing Service was charged with managing the semiarid to arid winter and spring-fall ranges. These areas had suffered grazing abuse for three decades after the national forests were established and closed to unmanaged grazing and had less environmental potential for plant growth than the mountain ranges managed by the Forest Service did.

Certainly, some individual ranchers exercised high-quality stewardship over the grazing lands located on vacant federal lands that they considered their "home range." They were handicapped in their efforts to pursue natural resource management, however, because the vacant rangelands were legally open to anyone to graze, and several ranches shared most ranges. In the Great Basin, virtually the only fenced rangelands were areas where hay was produced. The individuals who practiced conservative grazing management were likely to see the forage eaten by their neighbors' livestock.[4] Combined with the growth of the tramp sheep industry on the western range during the early 1900s, this situation was the prelude to cheatgrass dominance of millions of acres of rangelands.

In hindsight, Americans' lack of consensus and the government's failure to come up with a policy for scientific management of the vacant public lands is puzzling. The issue was widely discussed in the press. Glenn Bennion, writing in 1924 in the *National Wool Grower,* blamed the government's "indefensible free-range policy" for the badly degraded rangelands. "Sagebrush," he wrote, "came when the wasteful, destructive methods of range exploitation, developed as a result of the Government's indefensible free-range policy destroying the native grass and allowing sagebrush to become dominant, thus permitting those forms of vegetation that stock can not eat to take the place of grass."[5] Bennion had a simple plan to restore degraded bunchgrass: Burn the ranges during the hot summer months, rest the burned areas from all grazing until the grasses had a chance to recover, and then use moderate stocking rates with seasonal, managed grazing. He offered evidence that many sagebrush-bunchgrass ranges still had sufficient remnant stands of native perennial grasses to restock the ranges if the competing shrubs were removed and grazing was managed. Burning was not a new idea. Throughout the West, the earliest ranchers used promiscuous burning to reduce woody vegetation on rangelands. Early conservationists called such burning wanton destruction of natural resources, but perhaps it was merely prescribed burning for which the "doctor" had not yet written the prescription.

The basic problem that stifled restoration of big sagebrush rangelands in the Great Basin was the overabundance of big sagebrush.[6] Would-be range improvers faced the daunting task of removing tons of woody material from each acre to make room for restoration seedings. Essentially the range rehabilitators faced the same problem that had plagued thousands of homesteaders, especially those trying to bring land under irrigation. Homesteaders sometimes resorted to flooding to kill sagebrush, which cannot stand wet feet, but most often that was not an option.[7] Usually, the would-be homesteaders had to resort to hand grubbing, which was backbreaking labor. Often they saved the sagebrush and used it for household fuel. A southern Idaho homesteader who could switch from using sagebrush for the cookstove to lodgepole pine was considered a success. The range managers did not have the option of flooding, either; they had millions of acres to treat rather than portions of a 160-acre homestead.

What was wrong with Bennion's suggestion of burning during the hot summer months to eliminate competition from big sagebrush? The first problem was a general lack of herbaceous fuel to spread the fire from sagebrush plant to sagebrush plant. Herbaceous fuel simply did not accumulate on grazed ranges. Cheatgrass had not yet become a widespread dominant, and the extremely heavy grazing pressure biologically suppressed the little that was present. Second, the entire forest and range conservation movement in the western United States was based on the total exclusion of wildfires from wildlands.

The first publication written about controlling big sagebrush, titled "Eradication of Big Sagebrush (*Artemisia tridentata*)" and published by the Forest Service, did include prescribed burning as an alternative treatment for removing shrubs in preparation for seeding.[8] The authors stated that prescribed fire could control 100 percent of old and young sagebrush plants. They either did not know or failed to say that the rockiness of the soil was also a factor. Among the disadvantages of fire, they included loss of 30–40 percent of the remnant native-perennial grasses, no control of sprouting shrubs, and a slope limit of 30 percent (perhaps slopes steeper than 30 percent were considered an erosion hazard). The estimated cost of planned burning was 25–35 cents per acre, probably for firebreaks and standby crews during the actual burning. The authors ended with a cautionary statement, "Great care needed in use. Not advisable on soils that will blow badly."

Prescribed burning for sagebrush control was not practiced on publicly owned rangelands until late in the twentieth century, and then only on a very limited basis.

The third obstacle to restoration of sagebrush/bunchgrass rangelands in the 1920s and early 1930s was the reported lack of success experienced by Forest Service scientists in early trials. The attempts of Arthur W. Sampson, one of the founding fathers of range science, to revegetate sheep-denuded subalpine ranges in the Wallow Mountains of northern Oregon generally failed.[9] Sampson's work in the Billy Meadows in the Wallow Mountains did establish two basic rules of range seeding: (1) control competition from weeds, and (2) protect seedlings from excessive grazing. In a review of early range research published in 1937, W. R. Chapline described the Billy Meadows research plots as the "crucible where future range scientists learned their trade."[10] By the second decade of its existence, the Forest Service was the federal government's lead agency in rangeland research.

J. S. Cotton of the USDA Bureau of Plant Industry conducted trials similar to Sampson's in northern California and Washington as early as 1902.[11] C. L. Forsling and W. A. Dayton characterized those attempts and numerous later ones at a variety of locations as failures.[12] The forage species seeded to replace the degraded native meadows were largely exotic perennials that had been successful in the humid eastern United States. The failure was attributed to lack of adapted species, lack of suitable tillage equipment to reduce competition from weeds in the meadows, and failure of the introduced species to stand up to continued heavy grazing. If scientists of the agency charged with revegetating the ranges said seeding of the mesic portions of the western range was impossible, it was considered to be impossible. This put a definite damper on any attempts to revegetate the arid to semiarid degraded sagebrush/bunchgrass rangelands. The early seeding experiments were conducted in mountain or subalpine meadows, which bear no environmental resemblance to big sagebrush/bunchgrass communities. Why did these meadow-seeding failures completely stop research on seeding degraded sagebrush/bunchgrass sites? Apparently, the researchers thought that if they could not seed a wet meadow, they certainly could not seed a semiarid desert.

The major obstacle in restoring degraded big sagebrush/bunchgrass ranges was the lack of an adapted perennial grass to replace native perennials. Various scientists collected seeds of the landscape's dominant bluebunch wheatgrass, an obviously valuable forage species found from northern Michigan to Alaska and south to western South Dakota, New Mexico, and California.[13] Attempts to establish bluebunch wheatgrass on degraded sagebrush/bunchgrass rangelands usually ended in failure, though. Seed germination of many collections was low, and seedling vigor was weak. There was no native plant seed industry at that time, and seeds of the native perennial grasses were impossible to buy. The remnant stands of bluebunch wheatgrass were often located on steep, rocky slopes where mechanical harvesting was impossible.

The final impediment to restoring western rangelands was apathy, both public and political. The western range was endless, bountiful, eternal, and—most important—free; there was no valid reason to spend taxpayers' money on rangelands. P. Beveridge Kennedy, writing in 1901, warned about the danger of government playing too large a role in attempts to restore degraded sagebrush rangelands.

Perhaps one group of men holding one portion of the public range by force of arms may decide also not to overstock their range and to improve it by sowing seed of valuable grasses and forage plants. This is not likely to occur, because the reseeding of a large tract is a costly undertaking, and one still so largely an experiment whose results cannot clearly be foreseen that no stockman will likely undertake the reseeding of lands not his own. How, then, shall the open ranges of the public lands be made fully productive again? A socialist has suggested that this should be undertaken by the general government as a public work to be paid for by taxation of the whole people; that seed should be collected, enormous grass farms planted, and that the seed raised on these farms should be sown far and wide on the ranges, and that the cost of all this enormous undertaking should be borne by the general government.[14]

The droughts and the economic and social upheaval of the 1930s furnished the catalyst for change. The northern Great Plains were almost destroyed by the droughts. Millions of acres of virgin prairie had been plowed for wheat production to meet demands during World War I. Much of this land was marginal at best for farming and should never have been brought into agriculture.[15] When the droughts came, many homesteads on these marginal lands were abandoned, and extensive areas became subject to wind erosion. The subsequent dust storms had a tremendous impact on public opinion concerning conservation of natural resources. The result was the formation of the USDA Soil Conservation Service. The conservation practices this agency introduced included plant material centers that produced plants and seeds for use in conservation plantings. President Franklin D. Roosevelt's New Deal established agencies that provided manpower for conservation projects. The new Civilian Conservation Corps provided young men to fight wildfires, build check dams and access roads, and attempt restoration seedings.

The native prairie grasses on the northern Great Plains were largely very difficult to establish by direct seeding. Desperate to find a forage-producing grass that would stabilize abandoned cropland, the Soil Conservation Service turned to crested wheatgrass, an import from Russia. Crested wheatgrass was to have a profound impact on the sagebrush ranges of the Great Basin. The plant was first imported into the United States in 1898 by N. H. Hansen of the South Dakota Agricultural Experiment Station while he was employed by the USDA as a plant explorer in Russia.[16] Apparently, none of the plantings from the original collections survived. Hanson sent additional seeds in 1906, but these suffered the same fate as the first batch. Crested wheatgrass plantings were very successful at the Belle Fourche Experiment Station at Newell, South Dakota, and at the Northern Great Plains Experiment Station at Mandan, North Dakota, and these locations supplied the seeds for plantings across the northern Great Plains and well into Canada.

The first known range seeding of crested wheatgrass in the Intermountain Area occurred in 1932 on Herman Winter's farm near American Falls, Idaho, and on the USDA Sheep Experiment Station near Dubois, Idaho.[17] The stand on the Sheep Experiment Station was moderately grazed and after 30 years produced more than a ton of air-dry forage per acre. The very early crested wheatgrass seedings in southern Idaho were highly successful because they were seeded on well-prepared seedbeds in abandoned cereal grain fields using grain drills.[18] The Rural Resettlement Administration bought submarginal homesteads in the Intermountain Area during the 1930s and started seeding 57,000 acres of former cropland in Curlew and Black Pine valleys in Oneida County, Idaho. Similar large-scale seedings of former croplands were conducted in the Crooked River area of eastern Oregon. The Rural

Resettlement Administration used the farm equipment of the former home-steader to establish these successful crested wheatgrass seedings.

The use of exotic plant material to restore degraded Great Basin range-lands has been highly controversial from the very beginning. The first restoration efforts used native species only. F. H. Hillman, the first botanist employed by the University of Nevada, understood the challenge of "domesticating" the fine examples of needlegrasses he saw on his nature walks along the Truckee River.[19] Several of the species he thought could be important for use in seedings—such as desert needlegrass, Thurber's needlegrass, and needle-and-thread grass—were either unavailable in the seed trade or were rarely used in plantings.

In a 1901 paper discussing range improvement P. B. Kennedy railed against those who thought an exotic plant would instantaneously restore the overgrazed ranges. "When the average man begins to think about restoring depleted ranges," Kennedy wrote, "he is apt to imagine that somewhere on earth, in Australia, or South Africa, or even in Siberia, there must be some wonderful grass or a salt bush, or something else which can be made to grow on his range up on the dry mountains and down in the wooded, shady valleys, furnishing abundant food for all his stock. This is just as possible as it is to find a patent medicine at one dollar per bottle which really will cure all diseases."[20] Kennedy's mention of Siberia in this regard is a remarkable coincidence. Hansen collected most of his crested wheatgrass specimens from a Russian experiment station in European Russia, but at least some of the specimens came from Siberia. Kennedy was perhaps the first scientist on the western range to write about the importance of nongrass herbaceous species. Early scientific papers on rangelands classed herbaceous species that were not perennial grasses as "weeds." Kennedy was also among the first to recognize the importance of browse species on rangelands.[21] At least half a century passed between Kennedy's initial comments concerning the value of forb and browse species on rangelands and the contemplation of artificial seeding of these species. Grass was the king of revegetation attempts for a long time. In the century since Kennedy forcefully expressed his feelings on the subject, the controversy concerning exotic versus native species has intensified. Current proposals are to plant only native species that are genetically identical with those that originally occupied the site. This fear of "genetic pollution" must be classified as "voodoo genetics" because it does not consider the processes of exotic weed invasion and the dominance that changes the inherent potential of the site.

We have extreme examples of such change on sites infested with medusahead in northeastern California.[22] The sites consist of deep vertisol (self-churning) expanding-lattice clay soils that originally had discontinuous mounded surface soil horizons of wind-deposited silt. The original vegetation was Lahontan sagebrush growing on the mounds, with squirreltail and Sandberg bluegrass bunchgrasses in the interspaces. Early spring grazing by cattle and sheep depleted the perennial grasses, and cheatgrass invaded the site. Dry cheatgrass accumulated and provided fuel for recurrent wildfires, which killed the Lahontan sagebrush. Wind eroded away the silt-textured mounded surface soil, leaving the clay beds exposed. The shrinking-swelling clay made an extremely harsh seedbed that is outside the potential for establishment of the previously existing native plants and was subsequently open to the invasion of another invasive weed associated with clay-textured soils, medusahead. If an inch of fine sand had been applied to the surface of the seedbed, at least some of the native species would have grown.[23] This is an extreme example, but all sites that have suffered degradation of the natural plant community after invasion by exotic species share some aspects of this change in potential. The plants native to the site are not necessarily adapted to occupy the site after the self-invasive exotic weeds are controlled.

One of the few studies comparing the survival of seedlings from locally collected and commercially obtained seeds of bluebunch wheatgrass and Indian ricegrass conducted in the Great Basin showed no statistically significant difference between the seed sources.[24] It would have been very interesting if a cultivar of these native species that had undergone selection for seedling establishment had been included in the experiment. The controversy is not going to disappear, because the individuals who favor completely passive management of rangelands will continue to oppose the use of any seed not native to the site in restoration attempts. Unfortunately, this attitude dooms many restoration efforts to failure and directly contributes to further degradation of rangelands. It assumes that all the genetic variation that has arisen through intra- and interspecific hybridization and mutations has had the opportunity to disperse to all sites. Considering the vast range across western North America of species such as bluebunch wheatgrass, such an assumption is patently unrealistic. The genetics of rangeland plants is dynamic, not static.

Those who insist on using only plants from that particular site are also assuming that climates are perfectly stable. As the current climate changes should have made clear to everyone, this is an unwarranted assumption. If plant communities are to persist, they must continually evolve to adapt to changing conditions. This is especially true for communities that have been invaded by highly competitive exotic species such as cheatgrass. In an ever-changing environment, a touch of hybrid vigor is the spice that leads to evolution and adaptation to new environments.

Crested wheatgrass was not the first exotic species deliberately introduced into the Great Basin. Filaree was brought in much earlier, deliberately collected and planted on degraded spring ranges to provide early spring forage during lambing time. Even A. W. Sampson wrote an article about seeding "alfilaria." A folk story tells that cheatgrass was deliberately spread by an itinerant peddler who sold seeds of "hundred-day grass."[25] We once heard someone comment that the peddler's market was short-lived because the cheatgrass probably moved faster than he did.

At the time crested wheatgrass was introduced into the Great Basin, the primary requirement was the potential to establish and persist through droughts and grazing. Improved seedbed preparation and planting technology promoted the establishment of crested wheatgrass, which persisted under everything but the most brutal grazing practices. Most important, crested wheatgrass provided forage in the early spring when it was desperately needed and native perennial grasses were most easily harmed by grazing. It presents some management problems, especially when large plants with seed heads are allowed to persist as wolf plants. Cattle will avoid these plants unless grazing pressure is heavy.

Some environmentalists oppose its use because it is an exotic plant, but perhaps if crested wheatgrass were viewed as an environmental "band-aid" that closes ecological voids, stops accelerated erosion, and biologically suppresses a host of exotic self-invasive weeds, the species would be acceptable to a wider spectrum of individuals concerned about Great Basin rangelands. Maybe environmentalists would be more likely to accept planting crested wheatgrass to biologically suppress cheatgrass if the perennial grass was not such a good forage species.

Site selection and preparation were crucial factors in obtaining a good stand of crested wheatgrass. L. A. Stoddart noted that "rangeland, like farmland, produces better when properly planted and when the soil is good and water is plentiful. No grass is known which possesses miracle qualities. Poor land management, non-productive soil, or arid climate are factors which increase chances of failure in range seeding. No grass will grow when thrown out upon the land without preparation or attention, nor will any known grass produce green forage on sterile and dry soil."[26] Stoddart suggested that deep, well-drained, salt-free soils on valley bottoms where the presence of tall big sagebrush indicated soil fertility were the best sites to seed crested wheatgrass. He considered 10 inches of annual precipitation the absolute minimum for successful seeding.

For better or worse, crested wheatgrass was the plant chosen to revegetate the degraded Great Basin rangelands. The next step involved developing the equipment for controlling sagebrush, preparing seedbeds, and seeding crested wheatgrass on often rocky, previously uncultivated rangelands. The scale of this undertaking was enormous; eventually, millions of acres in the Intermountain Area were seeded. By the time the program began, the situation was desperate. In 1939, George Stewart reported that 112 million of the 145 million acres of rangelands in the Intermountain Area required immediate remedial action if the region's five million range sheep and one million range cows were to survive.[27] Track-laying tractors made vegetation conversion on such a large scale possible.[28] These machines were rugged enough to work under rangeland conditions and had the horsepower to pull implements strong enough to defeat dense stands of big sagebrush.

The Intermountain Forest and Range Experiment Station started hiring young scientists to work on range seeding during the 1930s. Men such as A. W. Sampson, J. T. Jardine, and W. R. Chapline essentially developed range management techniques as they went along.[29] Their work was nothing short of brilliant. The second generation hired during the 1930s—Joseph Pechanec, A. C. Hull, Jerry Klomp, A. Perry Plummer, and Joe Robertson—were an exceptional group as well. Most of them had very strong personalities, and they were seldom in total agreement, but they were innovators who thrived on accomplishment. Pechanec, Hull, Plummer, and Robertson wrote the original paper on "eradication" of big sagebrush.[30] Although they were pioneers in range improvement research, all were classified as forest ecologists. As its name implies, the Forest Service was an organization for forest management. Range management was very much a secondary activity.

The young scientists established nursery trials with native and introduced plant materials throughout the Intermountain Area and started the search for equipment that could be used to convert degraded big sagebrush sites to perennial grasses. They jointly instigated the Interagency Range Seeding Equipment Committee to develop range weed control and seeding equipment.[31] Initially, they tried to plow under the big sagebrush using wheatland disk plows pulled by track-laying tractors. In the earliest trials, a man sat on the back of the tractor and hand broadcast crested wheatgrass seed before the disk turned the big sagebrush under. The disk plow worked fairly well on rock-free sites, but every time it hit a rock or large sagebrush trunk the disk, or even worse, the shaft on which the disk turned, broke. Robertson and Plummer reported on their firsthand experiences with the plows. "Frames of wheatland plows occasionally break, and wheel axles bend, as for example, when a fast moving wheel drops into a badger hole. . . . [C]astings break so frequently, especially with a new plow on rocky or gullied land, that some foremen have considered it necessary to keep an acetylene welding outfit and a supply of scrap steel on the project."[32] Electric arc welders were rare in the remote sections of the Great Basin in the 1930s. Most ranchers were not hooked up to an electric power grid, and very few of those who were had enough amperage for welding.

Joe Robertson was trying to disk-plow big sagebrush in Ruby Valley when the disk frame on his plow broke. He spent a couple of days disassembling the frame into pieces small enough to be loaded into the back of his Forest Service–issued Ford pickup. Dr. Robertson knew that a service station in Wells, Nevada, had a new arc welder, so he took the broken frame there. When he arrived in Wells, the station owner said he thought he could weld the frame, but Joe would have to pump gas and wash car windshields while the owner welded. One of the first cars that stopped for gas was driven by the manager of the Utah Construction Company Ranches, which constituted about 3 percent of the land area of Nevada. The manager, who had met Dr. Robertson at a meeting at the Forest Service Regional Headquarters in Odgen, Utah, recognized him and asked, "Things a little slow today in Forest Service range research?" More than a little embarrassed, Dr. Robertson explained that he was pumping gas while the station owner was welding the frame on his disk plow. Sensing an opportunity to sell the crested wheatgrass program, he explained what he was trying to do in Ruby Valley. Rather than support, though, he got a tongue-lashing from the ranch manager, who bluntly told him that he had carefully read all of A. W. Sampson's publications, and Sampson had proclaimed it impossible to seed meadows in national forest lands, much less the sagebrush hillsides. Robertson's project, the rancher continued, must be another of those New Deal waste-the-public's-money ideas from Washington, D.C. Being a pioneer range improvement scientist in the sagebrush was not a bed of roses. The death of Robertson's tractor driver from exposure at the Arthur trial site drove home the point.[33]

Another technique used for clearing big sagebrush for seeding was railing, pulling one or more railroad rails behind a track-laying tractor.[34] Railing was not very efficient because it tended to leave sagebrush in piles that made seeding difficult. The plots the Forest Service scientists initially established for nursery trials were called "rod plots" because they were 1 rod (16 feet) square. These may seem tiny, but the scientists had to grub the big sagebrush by hand and construct their own fences to exclude cattle and hungry jackrabbits. Their budgets were only a few hundred dollars per year, and that did not buy very much fencing material. The scientists were frequent visitors to the recycling yards (junk piles) of large ranches, where they gleaned fencing material. Dr. Robertson never passed up the opportunity to retrieve a discarded tire along the Nevada highways. He might be able to trade it for 25 cents' worth of gas at a service station.

World War II dramatically changed the fortunes of the sagebrush range scientists. The War Production Board in Washington, D.C., demanded red meat and wool from the western range to feed and clothe the millions of Americans entering the armed services.[35] The Forest Service had answered the same call from the Hoover Commission during World War I with disastrous results. Livestock numbers greatly increased on the national forests, but beef production apparently declined. Chapline noted that "many of the more favorable National Forest grazing areas again became seriously depleted and often practically devoid of palatable grasses. It took more than 30 years to overcome much of the damage."[36] It was these damaged national forest grazing lands that the Forest Service researchers were still trying to rehabilitate when Walt Dutton, chief of Range Management for the Forest Service, explained to the board that millions of acres of the western ranges were producing forage well below its potential because of past abusive grazing practices. Forest Service researchers had promising results indicating that these lands could be restored to productivity by seeding, but they needed funding for their research. The War Production Board responded with $75,000 for range research on seeding and $100,000 for large-scale trials. Congress initially did not want to fund the research portion of the request because the government was unwilling to wait for research to be complete. The Forest Service pointed out that research on range seeding had been under way for 10 years with very promising results. Chapline remembered that the House Agricultural Appropriations Committee was impressed with the work Hurtt had done in Montana with President Hoover's emergency employment funds in 1930, and released the money.[37] That is how Joe Robertson got the funds to do the seeding at Arthur, Nevada.

Assigned to find areas in the national forest on the Ruby Mountains of Nevada where seeding crested wheatgrass could be demonstrated, Robertson reported that the steep mountain slopes made poor locations for seeding and generally still supported remnant stands of native perennial grasses.[38] What these native plant communities needed to recover was rest from grazing in the early spring. The mountains were surrounded by thousand of acres of degraded big sagebrush/bunchgrass communities highly suitable for seeding crested wheatgrass, however; most of this land was administered by the USDI Grazing Service. Robertson's supervisors bought the idea. The site chosen was actually a mixture of private land and Grazing Service land, so the project became a three-way cooperative undertaking.

Frustrated because his wheatland plow continued to break almost every day, Robertson recalled seeing a reference to a "stump-jump plow" in the proceedings of the 1939 World Wheat Congress. It was similar to a wheatland plow, but each pair of disks was independently suspended on a spring-loaded arm and could rise over obstructions. Robertson called the plow to the attention of the Forest Service administration, and eventually one was imported from Australia. Interestingly, when we interviewed the surviving members of the 1930s research group during the 1970s, each was certain the stump-jump plow had been his idea. They remained highly competitive until the end. The imported plow was tested and redesigned to meet the specific needs of the western ranges, and was manufactured by the

Fig. 10.1. Rangeland plow developed to plow degraded stands of big sagebrush for the seeding of crested wheatgrass working on large-scale experimental plots at Cain Springs, Nevada.

Inter-agency Rangeland Seeding Equipment Committee.[39] The "rangeland plow" (sometimes called "brush-land plow") was the committee's first big success (fig. 10.1). The plow weighed 7,000 pounds and required a 40 horsepower tractor to pull it (quite a lot of power for those days).[40]

The area seeded at Arthur was rested for two years and then grazed by four hundred cows and calves for three weeks every spring thereafter. The delay gave the native perennial grasses on the national forest land where the cattle would otherwise have grazed a chance to grow and renew their carbohydrate reserves. The Arthur seeding and the other pilot seeding trials conducted by the Forest Service during World War II helped to dispel the prevailing attitude that sagebrush ranges could not be seeded to restore perennial grasses. The Forest Service claimed a 90 percent seeding establishment rate with the pilot program. Ranchers were so impressed with the results of the seeding demonstrations that they asked Congress to increase the appropriation for range seeding to $500,000 per year and Congress agreed.[41]

Equipment failure remained a problem. The rangeland plow was very good at knocking down and uprooting big sagebrush plants, but it left a very rough seedbed spotted with sagebrush trunks and rocks. The drill used for seeding crested wheatgrass was the standard grain drill found on cereal grain farms throughout America. It used two flat disks of lightweight steel that were slightly offset to open a slit in which the seed was deposited. When such drills were used to seed big sagebrush sites that had been plowed with a rangeland plow, someone had to follow behind to pick up the parts as they broke off. In 1951, for example, Floyd Iverson, the regional wildlife and range officer for the Pacific Northwest Region of the Forest Service, was on a routine review of the Fremont National Forest in southeastern Oregon. During a review of rangeland seeding practices, John Kucera, a Fremont forest range and wildlife officer, told Iverson that his crew was breaking three or four drill arm assemblies in each 8-hour shift. Iverson suggested that a new drill for rangelands would be helpful, and Kucera agreed to try to develop one if funds could be found for it. From this encounter grew the best-known product of the Range Seeding Equipment Committee, the rangeland drill.[42]

Thus, apparently by chance, a number of events that greatly influenced seeding of degraded sagebrush rangelands occurred in the area north and east of the Ruby Mountains in Nevada. It was not chance that northeastern Nevada was the center of the range cattle and sheep industry in Nevada, though. The vast areas of big sagebrush/bunchgrass rangelands were ideal for grazing, and the 10,000-plus-foot mountains provided runoff for irrigated hay production.

The new developments came just in time. In 1934 a Forest Service range scientist and his summer assistant were collecting and pressing plants for the Forest Service herbarium. While working south of Wells, Nevada, about 50 miles north of Arthur, they collected a fleshy annual herb near a railroad siding called Tobar. The plant, which was new to both of them, looked like a strange Russian thistle.[43] Identification took several years, but the plant was finally determined to be halogeton, apparently another of the multitude of exotic annuals that had been accidentally introduced into the Great Basin from Central Asia. In 1942, sheepman Nick Goicoa lost 160 ewes out of a band grazing on winter range near the Tobar siding. When Dr. C. H. Kennedy, the district veterinarian for the Nevada Department of Agriculture, investigated, he found the rumens of the sheep full of a plant he could not identify. He sent samples to the University of Nevada, where comparisons with herbarium material proved it to be halogeton. A chemist at the University of Nevada analyzed halogeton samples and found that the herbage contained oxalates equivalent to 19 percent anhydrous oxalic acid.[44] Feeding trials confirmed that halogeton could be highly poisonous to sheep.

The range sheep industry was at its peak in the Great Basin, and was considered the second most important livestock enterprise in the entire Intermountain Area. After reports came out that halogeton was poisonous and range managers and ranchers learned to identify it, the plant was associated with numerous large-scale deaths of sheep that had previously gone unexplained. Halogeton spread at an amazing rate across the Great Basin, and eventually across all of the Intermountain Area. The huge losses of sheep gained attention in the national press and even a feature article in *Life* magazine.[45] Weed-control districts, counties, states, and finally the federal government appropriated considerable amounts of money to try and control halogeton. A new herbicide known as 2,4-D seemed to show promise for halogeton control. It had been developed during World War II and was revolutionizing agronomic and rangeland weed control.[46] The herbicide killed halogeton, but it also killed the native salt desert shrub vegetation, creating more habitat for halogeton.

Joe Robertson's research opened the door to a solution. Dr. Robertson had adapted a three-row vegetable seed planter to tow behind his 1930s pickup for seeding range revegetation plots. After a wildfire south of Wells, Nevada, near the site where halogeton was discovered, Robertson towed his drill around the burned area, leaving behind three rows of crested wheatgrass. He got good establishment of the perennial grass the next year, and in subsequent years noticed that where crested wheatgrass became established, halogeton was excluded. This result combined with the observations of many other range managers and scientists led to the seeding of crested wheatgrass to biologically suppress halogeton. When the Grazing Service was merged into the Bureau of Land Management and Marion Clausen became the director, he enthusiastically supported the seeding program because he realized that the real disease affecting Great Basin rangelands was more livestock on the ranges than the forage base could support. Halogeton got all the publicity, but it was merely the symptom of overgrazing, not the disease. Seeding crested wheatgrass had a double advantage: it biologically suppressed halogeton while at the same time treating the basic problem by increasing the available forage base. Transferring research technology to local ranchers was an important part of the success of crested wheatgrass seeding on depleted sagebrush/bunchgrass ranges. The Forest Service published a pamphlet titled *How to Reseed Range Lands* for each state in the Intermountain Area.[47] Many of these bulletins featured cartoonlike panels on the front cover with captions such as "Choose Favorable Sites," "Plan Early," "Buy Suitable Seed," "Make Correct Mixture," "Use Effective Tools," and "Manage Wisely."

By 1950, range managers in the Great Basin had crested wheatgrass seed, the equipment to convert degraded big sagebrush communities to crested wheatgrass, and the funds to plant it. Probably more important than the actual funding, which came from the halogeton suppression program, was establishment of the concept that federal funds could be used for range improvement on lands administered by federal agencies. The size of individual seedings ranged from as few as 2 acres to as many as 15,000.[48] In Nevada, about 1 million of the estimated 19 million acres of formerly big sagebrush/bunchgrass range was converted to crested wheatgrass. In the 1980s, this million acres of crested wheatgrass supplied 25 percent of the forage base for the Nevada range livestock industry.[49] Currently, most of that area has been reinvaded by big sagebrush. The presence of the sagebrush has dramatically reduced the production of herbaceous vegetation but has apparently created habitat with a mixture of shrubs, perennials, and annual forbs and grasses approaching the environments that existed at the time Europeans arrived.

How much of that 1 million acres of crested wheatgrass could be successfully seeded with a perennial grass today? Probably none. The early crested wheatgrass seedings were done when big sagebrush closed communities to herbaceous seedling recruitment; the development of the rangeland plow solved this competition problem. And the grazing pressure was so extreme on Great Basin rangelands before crested wheatgrass was added to the forage base that cheatgrass was biologically suppressed. In the 1960s, virtually every grazing allotment on publicly owned rangelands in the Great Basin came under some form of grazing management. Usually it involved rest-rotation grazing, with at least three pastures involved. One pasture was heavily grazed early in the season, one was grazed only after seeds were ripe, and one was completely rested. The following year, the pasture grazed after seed ripe was rested, the one grazed early was grazed after seed ripe, and the rested pasture was grazed early in the rotation. At lower elevations where few native perennial grasses remained and site potential was limited, rest-rotation grazing resulted in complete dominance by cheatgrass.

The general reduction of the number of grazing animals—particularly sheep—permitted on publicly owned rangelands coupled with the end of community grazing allotments and the instigation of single permittee allotments had a dramatic influence on the ecological condition of rangelands. According to the ecological theory accepted in the 1950s, the grazing reduction and management should have resulted in a dramatic improvement in range condition. The reduction in the numbers of livestock on Great Basin ranges was probably even greater than changes in permits indicated because the management agencies did not have good estimates and control of the number of animals on Great Basin ranges until well after World War II.

The dramatic increase in cheatgrass after 1950 was at least partially due to relaxation of the biological suppression from grazing. The most significant expression of this relation was the accumulation of cheatgrass herbage in degraded big sagebrush stands in sufficient quantities to readily ignite and burn in wildfires. In addition, cheatgrass may have been evolving and adapting to a wider variety of habitats. It invaded and dominated crested wheatgrass seedings that failed to establish, especially in areas at the lower edge of the environmental potential for big sagebrush where shadscale intermingled with Wyoming big sagebrush. Lee M. Burge, a pioneer in halogeton control on Nevada rangelands, claimed that the only sign of Bureau of Land Management crested wheatgrass seedings on the margins of the salt deserts was the stands of halogeton arranged in rows left by the drill openers. Failed seedings also provided habitat for cheatgrass. Some sites were seeded again and again in futile attempts to establish crested wheatgrass stands. Once the big sagebrush was plowed and the first crested wheatgrass seeding failed to establish, cheatgrass ruled the site unless some form of effective weed control was practiced.

Another source of cheatgrass dominance stemmed from herbicide applications to release perennial grasses from competition from dense stands of big sagebrush. If 2,4-D was applied properly at the correct time and on the correct site, the results could be spectacularly successful. Selection of the site for herbicide application was critical. There had to be a sufficient density of perennial grasses to preempt the environmental potential released by the killing of the sagebrush. It became a standard rule that you had to be able to step from one remnant perennial bunchgrass to another for the release to be successful. If too few perennial bunchgrasses were present, killing the shrubs opened the site to cheatgrass.

The rules of seeding evolved as crested wheatgrass seed and the mechanical tools for site preparation and planting became available. One of the first rules range managers learned was to protect the seedling stand from grazing, usually until after seed ripe in the second growing season. Crested wheatgrass seedlings might establish under grazing, but their subsequent vigor was very low. This meant that fences had to be built to protect the plants. If the enclosed area did not contain stock water, a water source had to be developed. Failure to obey the rules resulted in additional areas dominated by cheatgrass and subject to recurrent wildfires.

Two persistent myths reporting easy, low-cost methods of controlling cheatgrass have interfered with control efforts. The first myth is an apparently never substantiated story that burning cheatgrass in the early summer before the current year's seed crop is ripe will effectively control the annual grass, making it possible to seed the site successfully the following year. That is wishful thinking. We now know that cheatgrass builds huge seed banks that contain two to three times as many seeds as there are established plants in any one year. Burning has no effect on them. The second myth is that if cheatgrass is heavily grazed in the spring of the year that crested wheatgrass is seeded and the livestock are then removed, crested wheatgrass seedlings will establish. In fact, fall-germinating cheatgrass plants develop a root system that exploits the soil profile throughout the entire depth of wetting. Even cheatgrass seedlings that do not establish until the spring have very rapid root growth. With soil moisture the limiting factor, cheatgrass will win out over the perennial seedling every time. The seedling morphology of the two plants favors cheatgrass as well. Overwintering cheatgrass seedlings from fall germination are rosettes of short leaves that lie flat on the soil surface; wheatgrass seedlings grow straight up in the air and are always grazed first. Experience has taught us never to say never, but if prescribed fire or early spring grazing is going to help perennial seedlings to establish, it has to be applied in some innovative manner that has not yet been tried under experimental control.

A. C. Hull suggested tillage as the best way to control cheatgrass and establish seedlings of crested wheatgrass.[50] Virtually all of the cheatgrass seed bank is located on the surface of previously untilled rangeland seedbeds, and plowing with a moldboard plow completely turns over the seedbed and buries the seeds too deep for emergence. Unfortunately, rocks and the power and fuel requirements for moldboard plowing make this form of tillage impractical for many rangeland sites. Soil conditions in northern Nevada also affect the way in which severe tillage operations such as moldboard plowing control cheatgrass. Fred Peterson, formerly a professor of soil science at the University of Nevada and state soil morphologist, was the first to see the connection between the big sagebrush belt that extends across northern Nevada diagonally from southwest to northeast, and the zone of deposition of silt-textured soil particles that eroded from the pluvial lake basins of northwestern Nevada after the Pleistocene. These subaerially deposited surface soils make excellent substrates for big sagebrush/bunchgrass communities—and for cheatgrass once the native species are displaced. We compared different methods of tillage for cheatgrass control on such soils on the western flank of the Santa Rosa Mountains. Tillage using either a moldboard or a large disk plow destroyed the structure of the silts and led to the formation of a versicular crust on the soil surface in which grass seedlings were unlikely to establish.[51]

The golden age of crested wheatgrass seedings in the Great Basin extended from the close of World War II through the 1950s, largely funded by halogeton suppression programs. Agencies were so busy implementing projects and generally so successful with seedings that new weed control and seeding technology was not in great demand. The fact that new seedings had to be protected from grazing for a year or two played into the hands of the management agencies; each new fenced pasture gave them more leverage in controlling livestock on the previously open ranges.

But cheatgrass infestations required new approaches and new research. New approaches often meant new scientists. Richard Eckert Jr. joined the USDA Agricultural Research Service in 1958 to work on the seeding of degraded rangelands. Raymond A. Evans worked on range weed control in conjunction with seeding. Eckert and Evans formed an uneasy partnership that was to last for some 30 years. The two men were opposites in many ways, but both were extremely productive and innovative. They probably never realized that their research contributions would become the script for modern herbaceous range weed control. Throughout their careers, Evans and Eckert reached out to other colleagues to help attain their goal of cheatgrass control and seeding. Evans cooperated with the dynamic Burgess L. Kay, a range seeding specialist with the University of California at Davis. Eckert cooperated with Harry Summerfield, a brilliant and eccentric soil scientist who was a direct descendant of Chief Winnemucca of the Northern Paiute. These new weed warriors were to have access to a weapon that had been largely unavailable to the Young Turks of the 1930 and 1940s: herbicides.

CHAPTER 11

Control and Seeding with Herbicides

Weed control has been a concern of farmers since agriculture began, but herbicides came into wide use only in the first half of the twentieth century. Salt (NaCl) was widely used for weed suppression in the nineteenth century. The first half of the twentieth century produced new herbicides and new methods of administering them.[1] Scientists tested a variety of compounds. Several formulations of arsenical herbicides, often referred to as "blackberry killers," were used in brush control. Dr. Alden Crafts, the noted weed scientist and teacher at the University of California at Davis, told his introductory weed control class in 1955 that the only discomfort he ever suffered while applying herbicides occurred when he was experimenting with sulfuric acid application for weed control in onion fields. Very light viscosity petroleum-based oils were widely used for weed control during the 1920s and 1930s. Among the early supporters of research on the control of halogeton was the Richfield Oil Company, which manufactured weed oil.[2] Sodium chlorate was used as a soil sterilant, especially for noxious perennial weeds. In 1939, 2.8 million pounds of it was used in Kansas alone for weed control.[3] The borate-chlorate mixtures tried in the 1930s gave even better total vegetation control. Ammonium sulfonate was widely tested as a herbicide before World War II and was actually used in an attempt to control mesquite on rangelands.[4]

The really big breakthrough in herbicide development occurred in the 1930s when Fritz Went discovered the phenoxy compounds that regulate plant growth.[5] The secret discovery and development of one of these herbicides—2,4-D (2,4-dichlorophenoxy acetic acid)—during World War II initiated a revolution in chemical weed control. Before the war, farmers spent an estimated $1.5–2.5 million annually on herbicides. Within two decades after the war, herbicides constituted a $250 million annual industry in the United States.[6]

P. W. Zimmerman and F. Wilcoxen of the Boyce Thompson Institute produced and tested 2,4-D as a plant growth regulator in 1942 following the actual synthesis of the compound by R. Pokerav in 1941.[7] E. J. Kraus, head of the Botany Department of the University of Chicago, conceived the idea of using massive doses of growth regulators as chemical weed killers that would essentially cause plants to "grow themselves to death." He communicated this idea to John W. Mitchell and Charles L. Hammer, two of his former doctoral students, who were working as plant physiologists at the USDA Bureau of Plant Industry Station at Beltsville, Maryland, conducting secret research on herbicides as defoliants to use on jungle islands held by the Japanese in the Pacific. Mitchell and Hammer made the first public suggestion of using 2,4-D as a herbicide in June 1944.[8] In August 1944 the public became excited about the largely unproven herbicide when Hammer and H. B. Tukey reported that field bindweed sprayed with a dilute solution of 2,4-D died within 10 days.[9] The characteristic that made 2,4-D unique as a herbicide was its physiological selectivity: it acted only on broadleaf plants. It could be applied to cereal grain crops (grasses) without harming them, but it would selectively kill competing broadleaf weeds. Similarly, it could be applied to excessively dense stands of big sagebrush to release remnant stands of native perennial grasses.[10] More than a half century after its introduction, 2,4-D remains one of the most widely used herbicides in the world. The problem with controlling cheatgrass to establish perennial grasses is that it requires killing a grass to grow another grass. The basic problem Eckert and Evans faced in the early 1960s was the lack of a truly selective herbicide—one that would kill cheatgrass and not injure the seedlings of perennial grasses. The problem remains unsolved to this day.

American chemical companies were the first to take advantage of the explosion in herbicide chemistry during the postwar years. Later, as Europe recovered from the war, international companies joined in the agrochemical business. There was big money to be made in the discovery, manufacture, and sale of herbicides. The USDA Agricultural Research Service (ARS) saw the future of American agriculture in agrochemistry. W. B. Ensis, the head of the Crop Protection Branch of ARS, was one of the memorable characters employed by ARS in those days. Dr. Ensis was Napoleonic in stature, a dynamic leader, and possessed a volatile temper. James Young's first day of work for ARS in Reno happened to coincide with a visit by Dr. Ensis. The next day, Ray Evans and Dick Eckert joked that after surviving that first day, the next 30 years would be all downhill. Weed control research on a national level in ARS was under the direction of W. C. Shaw, another powerful personality. Critically wounded on Utah Beach during the Normandy landings, Shaw was evacuated to England where doctors told him he would be completely paralyzed for the rest of his life. Shaw defied their diagnosis, perhaps from the sheer intensity of his personality, and recovered. He was working for North Carolina State University when 2,4-D first became available, and he raced from farm to farm across the state establishing experimental trials with the new tool for weed control.

Range weed control was headed nationally by Dayton Klingman, a midwesterner whose brother also became a leader in university research and teaching on weed control. Dr. Klingman had taught at the University of Wyoming before joining ARS. Dick Eckert's boss, Dayton Klingman's counterpart, was Wesley Keller, an agronomist from Utah. Klingman and Keller ran the research operations of a far-flung empire from their offices in Beltsville, Maryland. Once a year they would descend on their individual field locations to examine the projects and results. The field scientists both dreaded and anticipated their visits, which, though traumatic, offered the opportunity to plead for more funds. Eckert and Evans made their request in memorable fashion. When Klingman first visited their field sites, the two researchers shared a World War II surplus pickup that leaked oil fumes into the cab. They sat Klingman between them in the truck cab and took him on the roughest, dustiest roads they could find in the Nevada desert on a hot summer day. Clad in the obligatory white shirt, tie, and suit, Klingman must have felt like a smoked bureaucratic turkey at the end of the day. That winter, Evans and Eckert both received new vehicles. That was the good news. The bad news was that the new vehicles were what the General Services Administration (GSA) termed "sedan deliveries," station wagons with the rear seat replaced by a sheet of plywood. A vehicle less adapted to field research in Nevada could not be imagined.

The ARS ran a huge program in Beltsville screening new compounds for herbicidal characteristics on an array of weed species and an equally diverse array of crop and forage species. At Dayton Klingman's insistence, cheatgrass and the wheatgrasses were included in the screening program. Chemical companies submitted newly synthesized compounds either as identified materials or as numbered compounds to protect proprietary rights. The results were public information available to anyone on request, and were required reading for ARS weed control scientists. New compounds flowed out of commercial chemical research laboratories in increasing numbers between 1950 and 1955. Not only were variants of existing compounds discovered, but also entirely new families of chemical compounds with new herbicidal activities, including chloracetic and propionic acids (the range herbicide dalapon),[11] phenyl and chloro-phenylurea (the range herbicides monuron, diuron, and fenuron), and chlorobenzoic acids (the range herbicide dicamba). Most important for cheatgrass research, the s-triazine compounds (range herbicides atrazine and simazine) were discovered in 1955–1958.

Herbicides enter plants either through the aerial portions (primarily leaves) or through the roots. Thus, herbicides are classified as either foliar or soil active. A few herbicides can gain access to the vascular systems of plants by both means. The first growth regulator herbicide, 2,4-D, enters plants primarily through the leaves. Most of the triazine herbicides are soil active. If a foliar herbicide is to be used on an annual species such as cheatgrass, it must be applied to the weed postemergence; that is, after the weed emerges from the soil. Soil active herbicides can attack the weed during the germination period, or preemergence.

Researchers are still looking for an herbicide so physiologically selective that it will kill cheatgrass and not harm perennial wheatgrass seedlings in the same stand or will kill ungerminated cheatgrass seeds in the seed bank without harming those of wheatgrass. The closest we have come to such selectivity was with a herbicide called siduron. Developed as a turf herbicide, siduron killed cheatgrass seedlings and left wheatgrass seedlings apparently unharmed in tests performed at Beltsville.[12] Dayton Klingman rushed the new chemical to all locations where research was being conducted on cheatgrass control and reseeding for field trials. When it was tested in the field, siduron was very effective on cheatgrass but killed all the wheatgrass seedlings as well. Those results were not popular back in Maryland. The field researchers suggested that the greenhouse experimenters at Beltsville had been sniffing too many herbicides, and the Beltsville staff were sure the cowboys in the field were incompetent. The field tests continued, on orders from Dr. Klingman, until a very wet spring occurred and siduron appeared to work in the field. On closer examination, however, the wheatgrass plants proved to have virtually no roots, and those that were present were very short and greatly thickened. Greenhouse trials finally determined that siduron inhibits root elongation. The inhibition was sufficient to kill cheatgrass seedlings, but seedlings of perennial wheatgrass survived if they were kept nearly continuously wet.[13] It was close but no cigar. Siduron has no use in an environment where moisture is the environmental parameter limiting the establishment of seedlings.

Evans and Eckert entered the cheatgrass–crested wheatgrass program like two whirlwinds. The cooperative studies on halogeton that Eckert published with Floyd Kinsinger had already shown his exceptional research abilities.[14] Evans jumped right in with an elaborate greenhouse study that proved that near total control of cheatgrass was necessary to give perennial wheatgrass seedlings a chance to establish.[15] Whether it was through Klingman's and Keller's leadership or the need to cooperate to use their tiny budgets most effectively is not clear, but Evans and Eckert formed a remarkable and productive research partnership. With the cooperation of Burgess Kay, Evans had a ground sprayer built at the University of California at Davis. It had a boom with only three nozzles and rode on two bicycle wheels. An air-cooled engine drove the centrifugal pump. A calculated dilution of herbicide was placed directly into the pump chamber, and as the sprayer was pushed forward at a constant calculated rate, the herbicide concentration in the pump chamber was diluted by water sucked in from a tank mounted on the sprayer frame. Using the gallons per minute of spray issuing from the nozzles, which was calculated by the size of the nozzle orifice, and the pressure produced by the pump coupled with the spacing of the nozzles and the forward speed, it was possible to determine a rate of herbicide application. For this sprayer the rate was never constant, but declined as the original solution was diluted. Essentially, the sprayer allowed a continuous dilution of herbicide material over a broad range of rates. This mode of application saved hours compared with establishing individual fixed-rate plots. Eckert adapted four Plantet Jr. Vegetable Seeders to a tool bar that fit the three-point hitch of a World War II surplus Fordson tractor. They needed some form of tillage to serve as a comparison to herbicidal weed control treatment. A light, two gang offset disk harrow about 5 feet wide stretched the power of the ancient tractor to the limit, but it worked. A half-ton two-wheel-drive pickup and a single-axle tilt-bed trailer with no brakes hauled the contraption from site to site.

Murphy's Law is the most basic rule concerning broadcast spraying herbicides at remote locations in the Great Basin: if it can go wrong, it will. Fatigued metal breaks; nozzles clog. One early morning in mid-March at Paradise Hill in Humboldt County, northern Nevada, we could not get the spraying started because the nozzles were plugged. All our efforts to unplug them failed. We were trying to spray at the crack of dawn because as soon as the desert warmed, even in the early spring, thermal winds would rise and interfere with the spraying. There was bright early spring sunshine and no wind yet. We were working comfortably in short-sleeve shirts. Finally, some genus walked around to the shady side of the truck and came back with the information that the air temperature was still below freezing. Apparently, the venturi effect as the spray solution passed through the nozzles cooled the solution below the freezing point so ice crystals clogged the orifice. We had no way of knowing that because by the time we removed the nozzle, the ice would have melted. It was one of those mornings when everyone glared at everyone else because we were collectively too dumb to identify the problem.

Lacking a selective herbicide that would control cheatgrass and allow perennial wheatgrass seedlings to establish, Evans and Eckert had to come up with a strategy that would fit the materials available. Soil active herbicides that would kill cheatgrass left behind an active residue in the soil that would kill wheatgrass seedlings. Eckert and Evans considered fallowing—allowing the herbicide-treated area to remain bare for a season after the herbicide application—so the toxic residues would leach away or be broken down by sunlight and/or microbial degradation. Fallowing is an ancient practice developed in semiarid environments at the limit of rain-fed crop production. The key to fallowing is keeping the land free of all vegetation so that the soil stores moisture and accumulates nitrates. One of the myster-ies of cheatgrass control research is why the Young Turks of the 1930s did not suggest mechanical fallows. A. C. Hull Jr. came very close when he and George Stewart (an agronomist by training) reviewed tillage methods.[16] One of the early leaflets on seeding abandoned croplands infested with cheatgrass suggested fallowing abandoned wheat-fallow land to reduce cheatgrass the year before seeding wheatgrass.[17]

From 1958 through 1966, Eckert and Evans conducted research trials with eighteen soil active herbicides at seven locations in Nevada and northeast-ern California. They evaluated the herbicides for (1) control of cheatgrass, (2) control of associated weeds, and (3) influence of herbicide residues on wheatgrass seedlings.[18] The first and third points are obvious, but even today the spectrum of weed control is often overlooked. Our earlier discussion of succession among annual exotic species made the point that if cheatgrass is removed, the lower successional stages have a chance to dominate the site. Evans and associates evaluated the impact of soil active herbicides on the expression during the fallow year of Russian thistle, red stem filaree, tumble mustard, and tansy mustard. Timing is the critical issue in this evaluation. If there is not enough herbicide residue or the herbicide does not have the spectrum of activity to control these species, they will grow luxuriantly dur-ing the fallow year and no moisture or nitrate will be stored in the soil; fur-ther, they may persist into the wheatgrass seeding year. Some method had to be found to control these weeds during the fallowing.

Evans and his associates quickly identified the triazine herbicides as pos-sibilities. They tested four different triazines but settled on atrazine as the most promising. They summarized their results with atrazine as follows: "Activity of atrazine was long enough for consistent season long control of downy brome [cheatgrass] under conditions that existed on all years of the study. Atrazine also controlled a wide spectrum of weeds. In fact, with only one exception, it controlled all weeds found in these experiments."[19] The one exception was Russian thistle growing on a plot after a summer thun-derstorm dropped significant precipitation.

Atrazine was applied in the fall so that winter moisture would move it into the surface of the soil where cheatgrass seed banks were located. At the time these experiments were conducted atrazine was formulated as a wetable powder; that is, it did not dissolve in water, but remained in suspension during the application process. A relatively large nozzle orifice was required to apply the suspended solids, and that meant a concomitantly high volume of carrier (water). This was a problem in desert environments and greatly increased the cost of the application. Evans developed the methodology necessary by using whirl chamber nozzles, which allowed low-volume application of atrazine.[20]

In terms of chronology, Eckert and Evans had already published the herbicidal fallow technique for cheatgrass control and seeding before Evans published the overall evaluation of soil active herbicides. Dick Eckert was the lead author of an article introducing the term "chemical fallow," although Harold Alley in Wyoming and Rod Bovey in Texas had already published on the concept.[21]

Eckert and Evans conducted their herbicide experiments at Medell Flat, Likely, Paradise Hill, Italian Canyon, and Emigrant Pass. Emigrant Pass was the only site where the atrazine fallow technique did not consistently give excellent results. It was also a site where S. J. Novak and colleagues found a markedly different genetic variant of cheatgrass.[22] The problem at Emigrant Pass was competition from cheatgrass in the seedling year.

The basic sequence of events for the atrazine fallow was to apply 1 pound per acre of active ingredient of atrazine in the mid to late fall. The next year was the fallow year, with nothing growing on the treated area. The fall of the fallow year, 1 year after the herbicide was applied, the plots were seeded to perennial wheatgrasses. The next year was the seedling year for the wheatgrasses. The area was not grazed until after seed ripe the second year after seeding. Worried that residues of the herbicide would affect the seedling wheatgrass, Eckert developed analytical procedures to determine the levels of atrazine residues in seedbeds. Results of 5 years of work, with Jerry Klomp collecting the numerous field samples, produced an apparent paradox: there were potentially toxic levels of atrazine in the seedbeds, but wheatgrass stand establishment was generally excellent.[23] Eckert suggested that reducing the rate of atrazine application from 1 pound to 0.5 pound per acre active ingredient might reduce the residue problem. Weed control was less consistent with the lower rate, but applying the herbicide under optimum conditions—suggested to be wet soil, in the fall, just before another moisture event—might improve the efficacy to the point that the lower rate was acceptable. Considering the erratic nature of Great Basin weather conditions (an area where Eckert was an acknowledged expert), it was a very difficult recipe to follow consistently.

Atrazine's relative insolubility contributed to the presence of residues of herbicide activity during the seedling year. Back in the 1930s, A. C. Hull had reported that crested wheatgrass seedling establishment was better if a semi-deep furrow drill was used for seeding.[24] William McGinnies, another of the eccentric characters who worked on the revegetation of rangelands, greatly amplified Hull's research. Bill McGinnies's father had been one of the leaders in the conservation movement during the dust bowl days. Late in Bill's career, James Young served on an evaluation panel and attempted to justify a national award for McGinnies's research work, especially his demonstrated expertise in restoring rangelands disturbed by strip mining for coal during the first energy crisis. Bill's publication list was impressive, but one of the panel members pointed out an oddity: Dr. McGinnies was the sole author of every work. He was truly a lone wolf in research.

McGinnies determined in replicated experiments that grass seedlings established much better from furrows than from flat seedbeds; he credited the idea to a bulletin published by the Young Turks and independent research by A. L. Nelson.[25] Measurements of soil moisture indicated that the micro-environment at the bottom of the furrow held more water than the surface.[26] The superiority of furrow seeding was even more pronounced on atrazine fallows. There was less evidence of herbicide injury (chloric and necrotic leaf tips) on seedlings growing in furrows. It gradually became apparent that the furrowing action was moving the surface soils contaminated with atrazine away from the seeded row and the seed was being planted below the area where the residues occurred.[27] Laboratory experiments later demonstrated that the loss of litter on atrazine fallows eliminated the safe site the cheatgrass seeds need to germinate.[28] Raymond Evans was later to measure the environmental attributes of furrowed seedbeds with microenvironmental monitoring.[29]

At the time Eckert and Evans started their research on herbicidal fallows, most researchers believed cheatgrass could not build persistent seed banks because the freshly harvested seeds showed no dormancy. Because they could create clean fallows with herbicides where cheatgrass seedlings subsequently grew, Eckert and Evans strongly suspected that seed banks existed. This is where James A. Young entered the cheatgrass research picture and demonstrated that acquired dormancy allows cheatgrass to develop seed banks.[30]

The gypsum resistance blocks they used indicated to Eckert and Evans that the fallow fields were accumulating moisture during the treated year.[31] They thought nitrates were accumulating as well but had no actual data. Eckert followed through with a detailed study of nitrate accumulation on check (cheatgrass-infested) and herbicide-fallowed soils. The results clearly showed that both mechanical and herbicidal fallows resulted in nitrate accumulation.[32] The evidence was clear: a moderate-sized furrow greatly enhanced wheatgrass seedling establishment. The problem was that the experimental furrows were made with fixed-shank shovels. In research plots carefully selected to avoid moderate or large rocks this was fine, but in general use on rangelands such furrow shovels would break every time they struck a rock. The original rangeland drill was developed with each opener individually mounted on a hinged assembly so it rode up and over obstructions. The opener was a cupped single disk that scratched out a very shallow furrow with only one near-vertical side.[33] Bill McGinnies's research, however, indicated that the steep-sided furrow should be 2–4 inches deep (fig. 11.1).[34]

At the height of the golden age of crested wheatgrass seedings a lot of technicians were out in the field practicing range seeding. Some of them became legends for their success rate with crested wheatgrass seedings as

Fig. 11.1. Track-laying tractor pulling three rangeland drills in a single hitch. The three drills probably cover a 25-year span in age. The drill arms on the older drills have been bent (intentionally or accidentally) so that the openers create furrows. The older drills probably leaked at least half as much seed as they put in the furrows.

they mastered both the science and practical art of seeding. Richard Holland, who finished his career with the Carson City District of the Bureau of Land Management (BLM) in Nevada, was one of these consistently successful rangeland seeders. Dick chain-smoked Raleigh cigarettes (the brand with the highest tar and nicotine content) and enjoyed a quart of good bourbon as a daily tonic. A rumor circulated that he had made a pact with the devil, promising to go quietly when his time came if he was allowed to enjoy himself growing crested wheatgrass while he was alive. Events on a winter day in 1973 seemed to disprove that story. His crew was unloading a rangeland drill north of Reno to begin seeding the huge Hallelujah Junction wildfire. The Washoe Zephyr was howling down the eastern slope of the Sierra Nevada, and when the crew removed the chains binding the load on the truck trailers, the wind blew the drills off the high truck beds and sent them crashing to the ground. The rangeland drills were extremely strong. If any part was damaged in loading and unloading accidents, it was usually the opener arm assemblies. Holland's drills had some disk openers that ran perfectly straight and made no opening and others that were severely bent and made only a small furrow. This type of damage was common; many of the older rangeland drills had the same kind of variation in openers. Someone noticed that shallow furrows made by some openers often had no seedling establishment while deeper furrows made on the same drill pass had good seedling stands. Out came the cutting torches to modify the original disk openers. The new openers would even work on firm seedbeds if a few hunks of scrap iron (old car blocks) were hung on them.

Eckert was very interested in developing a furrowing drill for seeding wheatgrasses both for the favorable microenvironment aspects it would create in the seedbed and to keep atrazine residues away from the crested wheatgrass seedlings. He teamed with Jerry Asher of the BLM's Elko District to test drill arm assemblies fabricated by the Forest Service Equipment Development Center under the direction of engineer Dan McKenzie. The deep furrow arms had 24-inch-diameter disks instead of the standard 20-inch disks. The disk angle could be adjusted vertically and horizontally, and the arms could be loaded with 400 pounds of weights. The drill was tested on a site with clay-loam to loam surface soils with rocks 5–15 inches in diameter. The heavy-duty drill arm assemblies consistently made furrows 3 inches deep and 4 inches wide.[35] Asher and Eckert found that crested wheatgrass seedlings in the deeper furrows had greater establishment rates and were consistently larger. Testing in several adjacent states produced excellent results as well, but the deep furrow drills were never popular. Although the 12-inch spacing between rows was optimum for forage production, the drill rows had to be farther apart to accommodate the larger disk. The wider spacing did not fully close the site to cheatgrass or medusahead. The deep furrow drills were also unpopular because they were more expensive, the arms were extremely heavy to lift, and the resulting seedbed looked like a war zone.[36]

We do not want the story of Richard Holland and the Hallelujah Junction wildfire seeding to go unfinished. As is typical with bureaucracies, funding did not materialize for the wildfire rehabilitation until it was too late to seed in the fall of 1973. This appeared to be bad luck for Holland because wildfire rehabilitation seedings in the fall in northwestern Nevada always had a better chance of seedling establishment than spring seedings.[37] James Young established several research seedings on the burn that fall. Severe wind erosion cut the emerging seedlings off at the soil surface, and all the plantings were complete failures. The erosion effectively blew away any seed reserves of cheatgrass that remained after the fire. Holland began seeding in the early spring, and for five consecutive Fridays, biologically significant rainfall shut down the seeding process at noon. By Monday morning, the sand-textured, windblown surface soils were dry enough for seeding to resume. The result was a seeding spectacularly successful in density and species diversity. It never hurts to be lucky as well as skilled in seeding rangelands.

Eckert and Evans went from testing the herbicidal fallow technique in small plots in the late 1950s to 1,000-acre test plots in the 1970s. In the larger experiments, the herbicide was applied aerially and the seeding was done with rangeland drills.[38] Along the way, in cooperation with the herbicide manufacturer, Eckert and Evans evaluated the fate of herbicide residues. All requirements for registration of atrazine for herbicidal fallows on rangelands were met or exceeded. Within the natural constraints of the Great Basin environment for seedling establishment, grasshopper infestations, absolute droughts, and trespass cattle grazing, the atrazine herbicidal fallow was an unqualified success for seeding former big sagebrush/bunchgrass rangelands that had converted to cheatgrass.

The basic atrazine fallow technique was embellished for special applications. Combining small amounts of the broadleaf herbicide picloram, which had both foliar and root uptake by some plants, provided a means of controlling Scotch thistle in seedbeds at the same time annual grasses were being suppressed.[39] The basic herbicidal fallow could be used to aid in the establishment of transplants of the valuable native shrub antelope bitterbrush.[40] This progressed to weed control–revegetation systems for big sagebrush rangelands that resulted in revegetation with perennial grass–native shrub mixtures.[41]

Eckert tried to release submarginal stands of crested wheatgrass from competition with cheatgrass by applying atrazine over the established plants. The crested wheatgrass plants had no physiological resistance to the herbicide, but their deeper roots and the relative insolubility of triazine herbicides made it possible to remove cheatgrass without damaging the crested wheatgrass plants.[42] Eckert hypothesized that during what would normally be called the fallow year, the established but suppressed crested wheatgrass plants would benefit from reduced competition and produce abundant seed. The seeds would disperse and germinate during what would be considered the seeding year in the atrazine fallow system. Indeed, Eckert showed that seedbed concentrations of atrazine during the seeding year were not lethal to crested wheatgrass seedlings. In a companion study, Young and Evans investigated the influence of apparent carbohydrate reserves on resistance of established perennial grasses to atrazine applications.[43] Reducing carbohydrate reserves by clipping the herbage of perennial grasses in the early spring greatly increased the plants' susceptibility to atrazine. This suggested that differential rooting was not entirely responsible for crested wheatgrass plants' resistance to atrazine; the physiological status of the plants played a role as well. By definition, these restoration treatments were designed to release stressed crested wheatgrass plants from competition with cheatgrass. The question then became, were these stressed plants automatically more susceptible to atrazine?

At the same time Eckert and Evans were developing and testing the atrazine fallow methodology, they were developing a parallel technology with a contact herbicide named paraquat. Evans actually published the first paper on paraquat before Eckert published the original paper on herbicidal fallows to control cheatgrass.[44] Burgess Kay played a big role in the paraquat study because he had a highly successful program under way controlling weedy annual grasses and seeding annual legumes such as sub and rose clover on annual-dominated rangelands in cis-montane California. As 2,4-D had been, paraquat was a unique herbicide at the time of its development. Developed by the Imperial Chemical Company in the United Kingdom, paraquat was a contact herbicide that killed any green plant tissue it touched when applied in the correct concentration at the proper stage of plant phenology. Paraquat was unique in that it deactivated as soon as it touched soil particles and left behind no soil residue. This meant that it could be applied during the seeding process.

The first time James Young worked with paraquat was on the Likely Table in northeastern California on a bitter cold March day. The experimental plots belonged to Burgess Kay, and it was Young's first opportunity to meet him. They sprayed plots with paraquat and then spent the rest of the day seeding the plots and building fences to protect them. The north wind was blowing that afternoon, and they were all chilled to the bone by the time they left the Table. The group stayed that night in Alturas, California, in a utilitarian motel. The electric wall heaters were not turned on until someone actually occupied the room, and it took a while to get the frost off the walls. The group gathered in Raymond Evans's room because he had a quart of Jim Beam bourbon. Burgess Kay was relaxing in the room's only chair with a dark glass of whiskey when James Young entered. Young asked why the outside of the glass was purple. Kay looked at the glass and then answered, "I guess I should have washed the paraquat off my hands." We did not know at the time that paraquat is one of the very few herbicides extremely toxic to mammals. Paraquat breaks down when exposed to sunlight, so it is shipped in heavy, dark brown plastic gallon bottles with tight screw caps. Farm workers thought the discarded bottles made excellent water bottles for use in the field. The label instructions now advise paraquat users to rinse empty containers three times and puncture them before discarding them in the proper receptacle. Such lessons were learned the hard way.

In contrast with the fall seeding that was done with atrazine fallows, paraquat control of cheatgrass required springtime herbicide application and seeding. Once a significant amount of cheatgrass seeds in the seed bank had germinated in the fall or early spring, the paraquat was applied, followed by immediate seeding. The cheatgrass seeds were in the seed bank because they had acquired dormancy. The dormancy would break down over time, but the breakdown could be accelerated by nitrate enrichment of the germination substrate. There was usually a flush of germination in the early spring, especially if no fall germination had occurred. This flush was conditioned by cheatgrass seeds from the previous year's seed crop that had not acquired dormancy, seeds from the seed bank that had lost their acquired dormancy over time, and seeds brought out of dormancy by the nitrification stimulated by spring temperatures. Killing this flush of germinated cheatgrass thus created a gap in cheatgrass germination that could be exploited to successfully establish a perennial grass seedling before the onset of the summer drought. Evans considered this spring seeding system to be feasible only in years when 3 or more inches of precipitation fell from March through May. Because fallowed soils store moisture, the fallow fields had a greater environmental potential to compensate for dry springs.

Paraquat's environmental interactions proved interesting. Burgess Kay had been very successful in controlling medusahead in California with paraquat as a preparation for seeding annual clovers. He and James Young attended a range weed conference in Vale, Oregon, and on a field trip visited some plots established by Bob Turner as part of his graduate studies at Oregon State University. While showing the group his research Turner casually mentioned that paraquat did not kill medusahead. This insult to his favorite herbicide brought an irate response from Kay. He and Young later tried an experiment to determine why Turner's results differed from Kay's. Accessions of medusahead seed gathered from a wide expanse of its range were planted in common gardens at Reno and the University Farm at Davis, California. Turner was right. The same rate of paraquat that killed every medusahead plant in every replication of every accession at Davis failed to kill a single plant at Reno.[45] In fact, Kay and Young later found paraquat to be an excellent herbicide to keep cheatgrass out of their medusahead common gardens at Reno.

Charlie Robocker and Dillard Gates had earlier tried using a foliar herbicide to control cheatgrass before seeding while working for the ARS in eastern Washington.[46] They had used dalapon rather than paraquat, though, and had not been successful. Unlike paraquat, dalapon does not deactivate once it touches the soil, and it killed the seedlings of perennial wheatgrasses. They later had success with dalapon in medusahead control and seeding by using it as a soil active herbicide to create a fallow.[47]

A persistent problem with the paraquat method of cheatgrass control in Nevada was the failure of aerial applications to control weeds. Simultaneous treatments with a ground spray using the same amount of carrier at the same location consistently gave excellent control. In cis-montane California, on the other hand, paraquat was and is applied by aircraft for weed control and as a preharvest desiccant. An added difficulty with the Nevada research was the scarcity of aerial applicators. Almost every farm in the Central Valley of California had one or more aerial pesticide applicators in close proximity. Nevada agriculture was rather behind California in technology at the time, and aerial applicators were very difficult to find. When 2,4-D was being applied to thousands of acres of sagebrush in Nevada, aerial spray applicators had to come in from out of state. They ferried in their aircraft, refueling facilities, water trucks, and mixing facilities. The pilots were unique individuals. Evans was stuck with one who claimed to have survived aerial combat against the terrible Hun in the big war. Naively, Evans assumed that he meant World War II. The pilot kept his plane at an old World War II emergency landing field in the Reese River valley outside Battle Mountain, Nevada, where he kept a collection of old military planes from which he salvaged parts to more or less keep a surplus Navy torpedo bomber flying to drop borate solutions on wildfires. All summer long, he drank beer in the shade of the wing and watched hydraulic fluid drip from abundant leaks as he waited for the BLM to call him for big-money wildfire runs. Winter and spring were slow times in the wildfire suppression business, so he was available for aerial coyote hunting and crop dusting. His crop-dusting aircraft was a fabric-covered Stearman biplane with a huge radial engine and an open cockpit. We never saw the pilot fly his borate bomber, but in the Stearman he wore a leather helmet, antique goggles, and a flowing white silk scarf. The aircraft was old, too, but it had been designed for aerial application of pesticides and had an excellent reputation.

The research plots that Evans and Eckert had on loan from the BLM were not located next to landing sites, so the pilot had to mix the paraquat solution in Battle Mountain, load the plane, and fly to the research site. Applications were always timed for early morning when the winds would be light. The research team would be in place with safety equipment and flags to show the pilot where to apply the paraquat solution. Unfortunately, he generally failed to show up. His absentee record had earned him the nickname Phantom of Battle Mountain. It was enormously frustrating for the researchers. After a bad night's sleep in a seedy motel in Austin, they would get up at 5:00 AM, have a terrible breakfast at the dirtiest café in Nevada, and then stand for several hours staring at the northern horizon waiting for the Phantom to show. It was difficult to maintain a positive attitude. His excuses were wonderfully original. He had been on his way but looked down from the open cockpit and saw his wife having an automobile accident. He had looked down from the open cockpit as he circled the airfield and saw a cowboy in a big hat get out of a pickup in front of his trailer home at the airport. On one memorable occasion, though, he did show up at the experimental site in the Reese River Valley.

The ARS project did not rent vehicles from the GSA motor pool at Reno. The GSA would not rent them to the ARS because the researchers hauled dangerous pesticides. On this occasion, though, Evans needed another vehicle to haul the entire crew to central Nevada. The motor pool manager had one available, but he was adamant that the car have no trace of dangerous chemicals when it was returned. Evans threatened his team with dire consequences if they carried herbicides in the rental vehicle. There was no way to control, however, what might come down from above. The Phantom had received detailed instructions regarding the drop. After he applied the paraquat solution to the areas marked by people waving flags, he was to circle to the upslope portion of the large exclosure, where fixed flags would show him where to spray all the solution that remained. This was standard operating procedure. That morning, the researchers were overjoyed when the Phantom showed up; he was even early. He buzzed the plots once to get oriented and then laid down the herbicide from such a low altitude that the plane's wheels knocked down the 36-inch lath used to mark the plots. The flagging crew hit their bellies. It was dead calm, and they could see the paraquat droplets settle out of the air onto the cheatgrass leaves. If aerial application of paraquat was ever going to work in the Great Basin, this had to be the time. The Phantom stood the biplane on its side and came roaring back to accept accolades for his flying skills. He apparently remembered the remaining herbicide solution about then because he made one more turn, lined up the plane, opened the tank drain, and made a perfect drop on the GSA rental vehicle. That was when the researchers discovered that the distinctive bright pink paraquat solution oxidizes instantly to insoluble purple splotches when it strikes a hot metal surface painted standard GSA gray. The purple pox was a permanent feature of the vehicle. Evans had survived bitter fighting in the

Battle of the Bulge during World War II, and it was the Phantom's good fortune that Evans did not have access to antiaircraft weapons that morning in the Nevada desert.

The lack of aerial applicators continued to be a problem in research and implementation of research in the central Great Basin. When Evans's team undertook large-scale testing of the atrazine fallow techniques at the Gund Ranch experiment station, they developed the prototype sagebrush sprayer so ranchers would have a means of applying herbicides at remote locations in the desert.[48]

New herbicides continued to appear during the 1960s. Karbutilate was first used to control small pinyon and juniper trees that escaped mechanical clearing efforts in woodlands, and then for herbicidal fallows for control of cheatgrass.[49] Karbutilate was much more soluble than atrazine, and it leached down to the rooting zone of rabbitbrush and sagebrush where atrazine had no effect. A single application was sufficient to convert a degraded sagebrush stand with abundant rabbitbrush and a cheatgrass understory back to a perennial grass stand. Burgess Kay had developed a similar idea much earlier when he sprayed degraded sagebrush-rabbitbrush stands with 2,4-D and seeded with a rangeland drill through the dead brush,[50] although he did not have to deal with cheatgrass.

After a great deal of testing and collection of residue and environmental data, the company holding the patent on karbutilate seemed very enthusiastic about a potential rangeland market. James Young presented the karbutilate research results at a national meeting of the Weed Science Society of America. The hierarchy of the agricultural division of the chemical company sat in the front row and applauded loudly. In fact, they had already decided that building a new plant to manufacture karbutilate was too expensive and they had sold the patent to the company that manufactured atrazine, which quietly filed it away, eliminating any potential competition to atrazine in rangeland markets.

Young and Evans replaced karbutilate with atrazine and 2,4-D in experiments at the Gund Ranch, but the results of that study were never published. Instead of applying the atrazine in October, it was applied in late May or early June when the sagebrush and associated species were most susceptible to 2,4-D. This meant the atrazine had to lie on bare soil or litter until fall rains leached it into the soil to kill the germinating cheatgrass plants. The researchers expected a big loss in atrazine efficacy from photodegradation, but it was not apparent. They did discover that dragging a rangeland drill through standing dead brush requires a lot more horsepower than seeding a burned or plowed site.

How many acres of cheatgrass-dominated former big sagebrush/bunchgrass rangelands were converted using atrazine fallow or spring paraquat applications? Outside of large-scale experimental sites, virtually none. Several interacting factors contributed to this failure to apply the research. Rachel Carson's *Silent Spring* aroused a general public outcry about the potential dangers of widespread use of pesticides in agriculture.[51] It made no difference that most of the environmental disasters she described involved insecticides, not herbicides. Second, the passage of national environmental laws brought democracy to natural resource management, especially on public lands. Prior to the passage of these laws, management strategies were commodity driven, with the ranching, timber, or mining interests providing most of the input. Afterward, creating more forage for livestock on Great Basin rangelands was no longer a primary goal. Public rangeland management agencies did not drop the use of herbicides because they were afraid of the environmental consequences of using pesticides; they dropped them because they were afraid of comments from a highly vocal but not necessarily knowledgeable portion of the general public. Congress stopped appropriating money for the improvement of publicly owned rangelands to avoid criticism from environmental groups.

The private ranching sector should have welcomed the new technology as a means of converting their own rangelands from cheatgrass to perennial grasses, but in fact they seemed apathetic. Why should they spend their own money on range improvement when federal agencies had been doing it free? If ranchers want to make a case for continued use of public rangelands for grazing, they should make their privately owned rangelands models of environmental sustainability. Only a few have done that. Peter Emerson Marble converted tens of thousands of acres of family-owned lands at the base of the Ruby Mountains to crested wheatgrass, but even this extremely progressive agro-businessman (Harvard graduate school) used mechanical rather than herbicidal tools for the conversion.

Public land management agencies responded to the new national environmental review regulations not with compliance but by trying to avoid them. Coupled with reduced funding for range improvement, this policy replaced range improvement with grazing management. The obligatory photo of the last big seeding found in every BLM office, flanked by sheaths of crested seed heads, was tossed on the trash heap. Employees like Dick Holland who had spent their careers restoring degraded rangelands were discarded along with these shrines to the false god. BLM employees sat up nights trying to memorize the pasture sequences in the Gus Hormay three-pasture rest-rotation grazing system.[52] Rumors abounded that some had to have it tattooed on the inside of their wrist.

The wave of agrochemistry that was to lift agriculture from the drudgery of the hand-held hoe to the heights of technology crashed on the rocky reef of pseudo-environmentalism born in the wake of Rachel Carson's brilliant exposition of the impact of chemical residues in the environment. Naysayers pointed out numerous errors in Rachel Carson's science, but she made the irrefutable point that pesticide residues affect the environment on a prolonged basis. Business card printers did a brisk business as employees of the major chemical companies who once handed out impressive cards with the slogan "The Company That Invented DDT and Saved Millions of Lives After World War II" sought a new theme. In fact, DDT did save millions of lives.

Public land managers who were quick on their feet caught the new wave of grazing management and rode it for much of their careers. The politics of bureaucratic survival called for saying and doing as little as possible. They learned never to propose a seeding to increase forage supplies because government agencies, environmentalists, and archaeologists would descend en masse demanding a full environmental impact statement. Even wildfire burns were not seeded. When public pressure dictated seeding some very important habitat, the seed mixture was composed of species that had no chance of establishment and was seeded by aerial broadcasting on unprepared seedbeds. Young managers who tried seeding and failed were careful never to try again. This malaise lasted from the 1960s through the early 1990s, when the results of decades of passive management came crashing down on public rangeland managers, wildlife habitats, the ranching industry, and the general public. Wildfires fueled by cheatgrass denuded millions of acres of rangelands, destroying homes, costing millions of dollars to fight, and placing an army of firefighters in harm's way, sometimes with fatal results.

A new rallying cry rose across the West: "We need a method for controlling cheatgrass on rangelands so that we can reseed perennial grasses to suppress wildfires!" Some among the management agencies may have remembered the work of Evans, Eckert, and Kay, but by then it was too late. The patent had run out on atrazine, and it was manufactured by numerous smaller companies on a worldwide basis. Renewing the registration for its use on rangelands would probably have cost at least $20 million. There was no demonstrated market for this use of the herbicide, so no one had bothered to renew the registration. The triazine herbicides had become linked with groundwater contamination in the Midwest and with some forms of cancer, and they were blamed for the demise of frogs around the world. Paraquat was still available as a restricted use pesticide, but the weight of obligatory spring seeding and no viable aerial application procedure kept it on the ground.

Glyphosate is the new "miracle" herbicide. Like paraquat, it bonds to soil particles in an inactive form and is applied in the spring after cheatgrass has germinated; but unlike paraquat, it does not kill cheatgrass instantly. In the prolonged cold springs of the Great Basin, it takes a very long time for cheatgrass plants treated with glyphosate to die. The race to establish seedlings of perennial grasses before the summer drought does not halt while glyphosate-treated cheatgrass dies. Several accelerators said to increase the speed of action of glyphosate have been marketed, but apparently they had never been tested in cheatgrass control programs. Perhaps glyphosate applied in mid-spring to establish a summer fallow followed by seeding in the fall is a valid option.

A new family of herbicides, the sulfonylurea compounds, seemed to be showing promise. The sales department of the manufacturing company and technicians working for a major federal public land management agency proposed one compound as the ideal cheatgrass herbicide. In field tests, however, sulfometuron methyl did not have the wide weed spectrum of efficacy or the fallow year summer herbicide active residue levels to control broadleaf weeds on fallows. Its effect was often to convert cheatgrass stands to tumble mustard, Russian thistle, and annual kochia patches that negated the moisture and nitrate storage of the fallow. Herbicide residues during the seedling year were a problem as well, as was the failure to determine the correct rate of application. Several Intermountain Area states permitted special-use labeling for sulfometuron methyl for the control of cheatgrass on range-lands but did not issue permits for aerial application. Applying soil active herbicides on irregular landscapes is a tricky process. If the boom passes overlap, some strips have double the recommended amount of herbicide and may have excessive herbicide residues on the seedling year for peren-nial grasses. If any strips are missed, cheatgrass grows, mining moisture and nutrients from the herbicide-treated areas and producing a superabundant seed crop that is dispersed across the supposedly fallowed area. We faced this problem years ago with atrazine and developed a foam marking system that made it much easier to get even coverage.

Perhaps the most significant difference between atrazine and sulfome-turon methyl fallows in cheatgrass-infested sites involves the litter that con-ditions germination of cheatgrass seeds. This litter does not consistently dis-appear from sites treated with sulfometuron methyl during the fallow year. If the litter is present, cheatgrass seed in the seed bank can germinate and establish during the seedling year for perennial grasses. When we made the original observations about the disappearance of cheatgrass litter on atrazine fallows and the subsequent influence of safe sites for germination, we had no idea that this phenomenon had any relation to the herbicide itself. It may not have a consistent relationship, but it certainly deserves more research. Would the litter disappear from a fallow created by glyphosate during a summer fallow?

The use of sulfometuron methyl to create fallows on cheatgrass-infested sites ended after the herbicide was apparently improperly applied to some 26,000 acres of largely burned cheatgrass range in southern Idaho. The label for sulfometuron methyl features a bold warning: "Do not apply to powdery dry soils subject to wind erosion." The burned soils in Idaho were severely wind-eroded during the next winter, and herbicidal symptoms appeared on crops grown under irrigation in fields downwind from where the sul-fometuron methyl was applied. The level of damage has been estimated to be more than 100 million dollars. The manufacturing company voluntarily withdrew its registration for use of the herbicide for control of cheatgrass on rangelands. Unfortunately, the incident may have ended all herbicidal weed control on rangelands.

While the dust was still settling from this disaster, a herbicide from another family was proposed for cheatgrass control and seeding. Imazethapyr uptake is both foliar and soil active, and wheatgrass seedlings are said to be relatively more tolerant of this herbicide than cheatgrass is. If so, imazethapyr could be applied in the fall or spring and the treated area immediately seeded to perennial grasses. The key issue is the relative tolerance of cheatgrass to imazethapyr. The application rate has to be carefully calibrated to kill cheatgrass but not harm the seedlings of perennial species. The inherent variability associated with rangelands infested with cheatgrass, in terms of both the amount and periodicity of moisture events and soils, may make such precise rate determinations difficult. Development of new herbicides is so expensive that a clear market must be demonstrated before any chemical company is willing to undertake such a project. At the same time, ensuring environmental and human health safety, which is responsible for much of the development cost, is absolutely essential. The sulfometuron methyl mess did nothing to solve and much to hurt efforts to control cheatgrass. Perhaps the future holds the possibility of nontraditional biological control with plant pathogens or genetically engineered wheatgrass plants that are resistant to glyphosate or some future herbicide. One thing is certain: the cheatgrass problem is not going to go away by itself. As each new promising herbicide has appeared, the ever-increasing frustration with cheatgrass and associated resource damages has led range managers to jump on the bandwagon without scientifically approaching the proposed control method. This is apparent in the sulfometuron methyl case. This type of approach can certainly lead to the loss of tools in cheatgrass control efforts that we cannot afford to lose.

CHAPTER 12

Revegetation Plant Material

To this point we have referred to crested wheatgrass in general terms as if it were a homogeneous species. In fact, the taxonomy of crested wheatgrass is quite complex. Dr. Douglas Dewey of the Agricultural Research Service (ARS) spent much of his career developing a genomic system of classification for the perennial species of the grass tribe Triticeae (for classification purposes, large plant families are subdivided into tribes).[1] There are about 250 perennial species in the Tribe Triticeae, including many of the world's important forage and crop species. Annual grasses within this tribe include the crops wheat, barley, and rye. As grasses within the Triticeae evolved through inter-specific hybridization, they may have acquired different genomes (the term "genome" refers to one haploid set of chromosomes) and then gone through further hybridization. Understanding the genomic structure of the grasses of Triticeae has made it possible for plant breeders to select new plant material from among the members of the tribe for cultivar development. Throughout his career, Dr. Dewey was besieged with demands for new plant material for revegetation of Great Basin rangelands, but he refused to be distracted from his taxonomic research. As he mastered the international literature of grass genetics and classification (much of which is in Russian), he revised the scientific names of many of the major grasses of the western range. For example, the extremely widespread dominant bluebunch wheatgrass, originally described by Pursh from material brought back by Lewis and Clark, went from *Agropyron spicata* (Pursh) Scribner & Smith to *Pseudoroegneria spicata* (Pursh) Löve. The good old cowboys working as technicians out on the range howled, "Sacrilege!" when that happened. They had recorded the shorthand identification "AgSpic" (abbreviating the genus and species names) a million times. Dr. Dewey became quite defensive during the later years of his career, especially when addressing audiences containing substantial numbers of range managers, but students are now writing *Pseudoroegneria spicata* on the inside of their wrists so they can pass laboratory examinations in range plant identification courses.

Dewey summarized the problems he encountered in classifying crested wheatgrass as follows: "The taxonomic history of crested wheatgrass reflects widespread confusion and uncertainty. Any expectation of a simple and neat solution to crested wheatgrass taxonomy is unrealistic because the problem is inherently complex, and subjective judgements must be made. Multiple chromosome races and the ability of crested wheatgrass taxa to hybridize are the root of the taxonomic problems. All crested wheatgrass taxa can be hybridized with each other, and many taxa hybridize with ease and produce fertile offspring."[2]

Crested wheatgrasses occur at three "ploidy" levels: diploid (2N = 14), tetraploid (2N = 28), and hexaploid (2N = 42). That is, individuals may have two, four, or six sets of chromosomes. The tetraploids are the most common and span the entire natural distribution range of crested wheatgrass in Europe and Asia. Diploids occur from Europe to Mongolia as sporadic islands in the general distribution of the tetraploids. Hexaploids are very rare, occurring only in northeastern Turkey and northwestern Iran.[3] Tetraploid strains are the most commonly seeded types in North America, where they are represented by the group called Standard crested wheatgrass. Standard is not a plant classification or formal taxonomic name; it was made up by the Montana Seed Growers Association to describe the type of crested wheatgrass they were harvesting and selling.[4] Under the broadly used category "Standard" are the cultivars Nordan, Summit, and Ephraim. When plant breeders formally release plant material for use in agriculture, they describe it and assign a name; these releases are known as cultivars. In scientific writing, cultivar names are enclosed by single quote mark, but technically, the use of single quotation marks along with the word "cultivar" is redundant. The diploid forms of crested wheatgrass are grossly classified under the trade name Fairway, which is equivalent to Standard. There are two cultivars, 'Parkway' and 'Ruff.'[5]

Dewey proposed the following scientific names for the introduced crested wheatgrasses found in North America: (1) diploid forms of *Agropyron cristatum* are fairway crested wheatgrass (Dewey suggested not capitalizing "fairway" to help avoid confusion with the cultivar name, and we follow that usage from this point on), (2) tetraploid forms of *Agropyron desertorum* are standard crested wheatgrass, and (3) tetraploid forms of *Agropyron fragile* are Siberian crested wheatgrass. *Agropyron fragile* is the accepted species name and is a synonym for *Agropyron sibiricum*.[6] There is obvious scientific justification for Dr. Dewey's suggestion that "Siberian crested wheatgrass" is the proper common name for this group, but common practice is to use simply "Siberian wheatgrass."

Most of the early seedings in the Great Basin were made with standard crested wheatgrass harvested from successful seedings on the northern Great Plains and sold in the commercial seed trade. J. R. Carlson and J. L. Schwendiman estimated that 17 million pounds of common crested wheatgrass (largely standard crested wheatgrass) harvested on the Great Plains of Canada and the United States was shipped to seed abandoned or retired cropland.[7] W. J. McGinnies suggested that farmers who had converted marginal cereal grain land to crested wheatgrass should keep one-third of their acreage in reserve each year. If a drought occurred, the reserve was available to graze; if rainfall was adequate or above average, they could make good money by harvesting seed from the reserve fields.[8] Crested wheatgrass seed production on the northern Great Plains began when the Rural Resettlement Administration started seeding abandoned croplands as part of the Agricultural Adjustment Act of 1933.[9] The first seed planted under this program was harvested from native prairie stands, but virtually no successful seedings were obtained with native species. At that time, crested wheatgrass was the only species that could be artificially seeded. The drought was so severe in 1934 that no seed was produced on the northern Great Plains, and crested wheatgrass seed was obtained from the far northern prairies of Canada where the drought was less harsh. Canadians in the western Prairie Provinces found crested wheatgrass a savior during the great droughts of the 1930s. From Saskatoon,

Saskatchewan, to Manyberries, Alberta, it was the only perennial grass that had a chance for establishment on abandoned croplands.[10]

Favorable weather returned in 1935, and many of the 1934 seedings established. Very favorable growing seasons in 1937 and 1938 produced sufficient crested wheatgrass seeds for extensive seedings. This grass in turn produced the seeds for the extensive seeding done in the Great Basin after World War II.

The first cultivar of crested wheatgrass developed in North America was 'Fairway,' released in 1927 by the Saskatchewan Research Station.[11] 'Fairway' was the result of a mass selection from plant material introduced in 1906 from Samara Province in western Siberia. The mass selection criterion was for fine, leafy plants. 'Fairway,' a diploid, was more winter hardy in Canada than the common standard crested wheatgrass. 'Fairway' was widely planted in at least a few trials in the Great Basin but did not establish as well as standard crested wheatgrass. The seeds of 'Fairway' are smaller than those of standard crested wheatgrass, and the seedlings are not as vigorous. Some people made the mistake of seeding mixtures of standard crested wheatgrass and 'Fairway.' The 'Fairway' grew well on sites with higher precipitation but quickly disappeared under grazing because livestock preferred the finer-textured herbage of the cultivar over the coarser standard crested wheatgrass.

'Parkway,' an *Agropyron cristatum* cultivar developed in Saskatchewan and released in 1969,[12] is a sixteen-clone synthetic derived from 'Fairway' after several cycles of selection for forage and seed production. 'Parkway' is more vigorous and 2–3 inches taller than 'Fairway' and produces at least 10 percent more seed and forage production. It is less leafy than 'Fairway' but more resistant to lodging. It is used for hay production on the plains but to the best of our knowledge has never been widely planted in the Great Basin.

'Ruff' was released by the ARS and the Nebraska Agricultural Experiment Station in 1975. Under Nebraska conditions it produces forage equal to 'Nordan' crested wheatgrass and is more of a sod former than 'Fairway.'[13] We tried this cultivar at several locations in Nevada, but seedling establishment was poor and plants had poor survival during droughts.

'Ephraim' is unique for several reasons. All of the other cultivars of *Agropyron cristatum* originated from plants brought back from Russia in 1906 by N. H. Hansen. Thus, they all originated from a very narrow gene base. 'Ephraim' originated from Plant Introduction (PI) 109012 collected near

Ankara, Turkey, by the Westover-Enlow Expedition in 1934.[14] This was the first and apparently only venture into the selection and release of cultivars of introduced perennial grasses for Great Basin rangelands by the Forest Service's Shrub Sciences Laboratory in Provo, Utah. Subsequently, some scientists from this laboratory became extremely vocal opponents of the seeding of introduced perennial grass cultivars. If any selection was performed in developing 'Ephraim,' it was based on rhizome development. 'Ephraim' was planted at many locations in the Great Basin, especially as part of mining reclamation. It was included in the seeding mix planted at Flanigan, north of Pyramid Lake on the California-Nevada state line, after an extensive area of big sagebrush with an understory of cheatgrass burned in a 1985 wildfire. The big sagebrush was growing on stabilized sand dune fields from pluvial Lake Lahontan. Annual precipitation at this very arid site averages 4–5 inches, with infrequent much wetter years and not infrequent years with near zero precipitation.[15] The big sagebrush apparently survived there because of the unique moisture relations of the deep sands. The perennial grass native to the site was Indian ricegrass. The Bureau of Land Management BLM elected not to seed Indian ricegrass because of the cost of seed and the poor record of success with this species. As it turned out, much of the sand was revegetated by Indian ricegrass from seed banks present on the site. Cheatgrass was not a problem, apparently because the extreme wind erosion depleted much of its seed banks. A few 'Ephraim' plants established on the sands, and once winter grazing was permitted on the site again, cattle preferred the 'Ephraim' forage because its green leaf growth occurred during late winter when Indian ricegrass plants were completely dormant. In the late 1980s the site experienced 3 consecutive years of near complete drought, but the 'Ephraim' plants had elliptical halos of seedlings stretching to the northeast from each mother plant. This would be a most interesting location for plant breeders to select plant material for use in improvement programs. It also provides a good lesson in the fallacies associated with generalization about plant material adaptation. Generally, cultivars of *Agropyron cristatum* are not as well adapted to the drier portions of the Great Basin as are cultivars of *Agropyron desertorum*. Obviously, the "success" of 'Ephraim' at Flanigan leaves such generalities open to question.

The first selected cultivar from *Agropyron desertorum* was 'Nordan,' released in 1953 by the ARS in North Dakota.[16] 'Nordan' was selected from plant material established at an evaluation nursery at Dickinson, North Dakota, in 1937 and probably originated from plant material introduced from Samara Province in western Siberia in 1906. 'Nordan,' developed by G. A. Rogler, originated from seven superior plants and went through two cycles of mass selection for short, compact, dense, awnless, seed heads; uniformity; and improved forage production.[17] It took considerable time for this improved cultivar to reach the Great Basin. Eckert and Evans, who initially started experiments in cheatgrass control and seeding with standard crested wheatgrass, had switched to 'Nordan' by the mid-1960s. The seeds of 'Nordan' are significantly larger than those of standard crested wheatgrass, and the seedlings are more vigorous. The introduction of 'Nordan' had a marked impact on crested wheatgrass seedings in the Great Basin. Eckert and Evans virtually ensured this by conducting tours of their research plots for land managers.

Kay Assay, who was assigned by the ARS to develop new wheatgrass cultivars, wanted to hybridize the three ploidy levels of the crested wheatgrass complex. Dr. Assay thought that "it would be helpful for breeders to investigate the potential benefits of genetic introgression among related taxa. Combining the genetic resources of the three ploidy levels of the crested wheatgrass complex would substantially expand the genetic diversity available at any one level."[18] He faced a major problem because the diploid, tetraploid, and hexaploid forms presented barriers to hybridization and the production of fertile offspring. Diploid forms of crested wheatgrass, as represented by 'Fairway,' produced forage superior to cultivars of the tetraploid standard crested wheatgrass but lacked the seedling vigor for consistent establishment on harsh rangeland sites. The limited amount of genetic variability available to plant breeders in the diploid form precluded making significant advances in seedling vigor within the diploid forms. Assay and Dewey worked together to bridge the ploidy barrier between tetraploid and diploid forms of crested wheatgrass through induced polyploidy.[19] The process involved treating 'Fairway' diploids with colchicine.[20]

Dr. Assay used the diploid-tetraploid hybrids they obtained to establish a massive breeding program. He selected eighteen clones from an original population of 8,000 hybrid plants on the basis of seedling and mature plant vigor, seed and forage yield, and resistance to environmental stress and plant pests. The selection was carried out both in the nursery and at a variety of field locations, including sites infested with cheatgrass and halogeton.[21] The new cultivar was released as 'Hycrest.'[22] 'Hycrest' had an immediate impact on crested wheatgrass seeding in the Great Basin. The number of acres seeded declined greatly during the 1980s and 1990s because public land management agencies had lost interest, but much of what was seeded was 'Hycrest' (fig. 12.1).

'Summit,' released by the Saskatchewan Research Station in 1953, was selected from clones that originated from the Omsk Experiment Station in western Siberia and has not been widely used in the Great Basin.[23]

When Dick Eckert and Ray Evans were conducting their cheatgrass control and revegetation experiments with perennial grass, they first used standard crested wheatgrass but switched to 'Nordan' when it became available. They used their four-opener experimental drill to seed two rows with crested wheatgrass and two rows with 'Amur' intermediate wheatgrass. (Intermediate wheatgrass apparently received its name because it was taller than crested wheatgrass but not as tall and coarse as tall wheatgrass. At the time, intermediate

Fig. 12.1. 'Hycrest' crested wheatgrass seeded north of Winnemucca, Nevada, near the original Paradise Hill study site of R. A. Evans and R. E. Eckert. Crested wheatgrass provides plant material critical for the effort to biologically suppress cheatgrass.

wheatgrass was known as *Agropyron intermedium*; after the revision of the Tribe Triticeae the name became *Thinopyrum intermedium*.)[24] Planting this grass in the 8–12-inch precipitation zone was heresy; everyone knew it could not persist in areas with less than 12-plus inches of precipitation. 'Amur' had the additional problem of being much more preferred as forage than crested wheatgrass. Eckert and Evans had very consistent results: the intermediate wheatgrass always established much better than crested wheatgrass. They attributed this to the much larger seed and much more vigorous seedlings of intermediate wheatgrass. Once established, intermediate wheatgrass is rhizomatous. In the wetter areas of northeastern California, it is the preferred grass for rangeland seedings, but ranchers complain that it becomes sodbound and drops in apparent productivity.

Eckert and Evans had a large-scale experiment located near Cane Springs on Highway 95 between Winnemucca and Orvada, Nevada. The site had been plowed and seeded but was considered a failure; occasional large wolf plants of standard crested wheatgrass stood out in a sea of cheatgrass. Mechanical summer fallow, atrazine herbicidal fallow, spring-applied paraquat, and control plots were compared. Half of each treatment was seeded to 'Nordan' crested wheatgrass and half to 'Amur' intermediate wheatgrass. The treatment plots were each 40 acres in size with three replications. The mechanical tillage and seeding was done with commercial equipment. The long direction of the plots ran up and down a gentle slope, and the contractor seeded on the contour. Excellent stands were obtained in both herbicide treatments. The mechanical fallow was a failure because a very loose seedbed resulted in the seed being planted too deep to emerge. Stoddart, Robertson, Hull, Hyder, and McGinnies all wrote papers on the danger of seeding wheatgrasses in loose, fluffy seedbeds;[25] it is considered the major reason why rangeland seedings fail. Neither herbicide treatment killed the wolf crested wheatgrass plants. During the seedling year, when the plots were free of cheatgrass, these plants had tremendous seed production. Initially, the stands of seeded intermediate wheatgrass were noticeably better than the crested seedling stands. Within 2 years, however, the intermediate had disappeared and was naturally replaced by an excellent stand of crested wheatgrass. The general interpretation was that crested wheatgrass was better adapted to the site, and the natural seed rain from the wolf plants from the previously failed seeding had produced seedlings that outcompeted the intermediate wheatgrass plants.

'Amur' intermediate wheatgrass was developed at the USDA Plant Material Center at Albuquerque, New Mexico, by J. A. Downs and T. F. Spaller.[26] They began the selection process with a plant from Manchuria that was labeled *Agropyron amurense* Drob but was subsequently identified as *Agropyron intermedium* (Host.) Beauv. by J. R. Swallen. Downs and Spaller noted variation in the plant material in terms of leaf pubescence that they attributed to "some introgression . . . with (*Agropyron) trichophorum* (Host) Nevski."[27] Despite this noted variation they described the cultivar as "leafy, vigorous-growing, but slow sod-forming type, uniform gray green."[28] We would agree with this description except for the "uniform" part. The stands Eckert and Evans established contained a mixture of green and gray plants. The 'Amur' that was seeded in northern Nevada was a highly variable plant material.

We visited Eckert's and Evans's old plots at Paradise Hill during the summer of 2001, 35–40 years after they had seeded crested and 'Amur' intermediate wheatgrass at the site. The old exclosure fences were completely gone, and the BLM had installed a stock water trough within the seeding site. The area had been burned in wildfires several times in the interim, and at the time of our visit was very heavily grazed. We identified the old seeding treatments by the yellow painted stakes that G. J. Klomp had brought down from Idaho when he transferred to Reno. Much to our surprise there was virtually no crested wheatgrass left in the now cheatgrass-dominated plots. Quite common, however, spreading outside the old plot rows, were the green and gray plants of 'Amur' intermediate wheatgrass. The adaptability of 'Amur' in the Great Basin is still open to question but certainly should not be automatically rejected.

This brings us to one of the great paradoxes of crested wheatgrass. To the best of our knowledge, all of the forms of crested wheatgrass that have been planted in the Great Basin are non-self-invasive plants. A. C. Hull and G. J. Klomp found that crested wheatgrass seeded in the middle of a badly abused area of rangeland or abandoned cropland with the disturbance or grazing pressure relaxed spread outward from the seeding and eventually occupied the ecological void.[29] Crested wheatgrass seeded in healthy stands of native vegetation, however, did not seem to spread. Hull and Klomp cited numerous examples of sites where initially poor stands of crested wheatgrass thickened with new seedling establishment over time.

The original stand of crested wheatgrass that Dr. Joe Robinson seeded at Arthur in the Ruby Mountains remained a sharp-edged, angular stand for years until it was invaded by big sagebrush. Generally, the drier the site the more likely that crested wheatgrass will not move out of the original drill rows. There are numerous examples of crested wheatgrass stands depleted by excessive grazing that thickened with establishment of new plants when grazing was removed. Obviously, crested wheatgrass is capable of producing highly viable seeds, and seedlings can establish in depleted stands. The question is why don't these seeds disperse widely, allowing the wheatgrass to act as a self-invasive species?

Eckert and Evans produced data that may be relevant to this question. They sowed two rows of crested and two rows of intermediate wheatgrass. In the fall, after the plants were mature and the seed heads were ripe, rodents always harvested intermediate wheatgrass seed heads but never seemed to touch the crested wheatgrass seeds. Granivorous rodents bury caches of various numbers of seeds at shallow depths in the soil surface, a practice known as scatter hoarding. The same rodent or others, or even different species of rodents, may return to these caches and retrieve the seeds for consumption or recaching. Sometimes the rodents return to graze on the seedlings emerging from the scatter-hoard caches.[30] Cheatgrass seedling emergence from such caches is common on Great Basin ranges, but we have never seen crested wheatgrass seedlings growing from a scatter-hoard cache.

We conducted preference trials with captive rodents using seed obtained from the ARS plant-breeding project at Logan, Utah. The study included seeds of all the grasses that remain *Agropyron* species under the new classification. The rodents preferred the seeds of some species over others, but they always refused the seeds of *Agropyron desertorum*. The seeds (caryopses) of many species of grasses have silicon-tipped minispines on the lemma and palea that cover the true seed. Crested wheatgrass seeds have such mini-appendages, but they are not as pronounced as those on several of the other species of *Agropyron*. The reason why rodents do not prefer crested wheatgrass seeds is unknown, but the fact that they avoid the seeds means that crested wheatgrass lacks a rodent-enhanced intermediate-range seed-dispersal system. This may be a major factor in its failure to be a vigorous self-invasive species.

The ARS grass-breeding program has released several new cultivars or registered germ plasm of grasses from the Tribe Triticeae in recent years. Among them are 'Newhy,' an advanced-generation hybrid between quackgrass and bluebunch wheatgrass, and the recently registered germ plasm RS-H hybrid wheatgrass.[31] This new plant material seems better adapted to areas of the Great Basin with higher precipitation than to the bulk of the big sagebrush and salt desert rangelands where cheatgrass is such a problem in northern Nevada. There is nothing inherently wrong with breeding and selecting plant material for higher-potential environments. The potential returns are much greater and the agricultural lands of such environments tend to be in private ownership rather than publicly owned rangelands, making acceptance of such new introduced plant material much more likely. During the last quarter of the twentieth century, however, the cheatgrass problem in the central Great Basin dramatically shifted from the upper two-thirds of the big sagebrush zone to the lower portion of the zone and portions of the salt desert ecosystem.

Plant breeders have responded to environmentalists' call for cultivars of native grasses to be used exclusively in revegetating Great Basin rangelands. 'Whitmar,' a bluebunch wheatgrass that has been available for many years in the seed trade, is an awnless cultivar developed from a population collected near Colton, Whitman County, Washington, and released in 1946 by the Soil Conservation Service.[32] An awned cultivar named 'Goldar' was released in 1989.[33] Neither cultivar was very successful in the central Great Basin. Melvin Meyers, who ranches in western Lassen County, California, and is an excellent grassland farmer, established a stand of 'Whitmar' on a mountain big sagebrush/western juniper site and had just one question afterward: "What do I do with it now? Only the black grass bug will eat the stuff!" Despite being awnless, 'Whitmar' plants were never highly preferred by livestock.

T. A. Jones, S. R. Larson, and their associates determined that the genetic diversity of 'Goldar' and 'Whitmar' represents only a tiny portion of the variability in the entire species.[34] They developed and released P-7 bluebunch wheatgrass, a multiple-origin polycross generated by intermating twenty-three open-pollinated native-site collections and the two cultivars Whitmar and Goldar. Their intention was to provide genetic diversity within a single germ plasm for semiarid to mesic sites where bluebunch wheatgrass was an original component of the vegetation. Apparently, they made no selections in the generations subsequent to the open crossing. We recognize that this plant material was released as germ plasm and not as a cultivar, but it leaves an important question unanswered. Do any of the twenty-five accessions that entered into the polycross have the inherent potential to compete with cheatgrass as seedlings? If only one of the twenty-five sources had some inherent potential to establish in the face of cheatgrass competition, then only 4 percent of the expensive certified seed planted would have any chance of successful establishment. This approach to plant breeding assumes that excessive grazing prevents bluebunch wheatgrass plants from reestablishing in native plant communities. If this were true, the species would return if livestock were removed or excluded. Numerous exclosure studies have shown that this is not the case. Bluebunch wheatgrass remains a minor component of plant communities on millions of acres of degraded rangelands because cheatgrass closes plant communities to the establishment of perennial grass seedlings.

The easiest way to select successfully among individuals in a segregating population is to base the process on some simply inherited, easily measured characteristic. Seedling competition, in contrast, is a complex of highly environmentally influenced, complexly inherited characteristics. One often expressed theory explaining the success of cheatgrass seedlings is the seeds' well-demonstrated capacity to germinate under very cold temperature regimes.[35] We have demonstrated, with help from plant breeders, that variation for potential germination at very cold temperatures existed in a segregation population of wheatgrass resulting from quack grass × bluebunch wheatgrass hybridization.[36]

During the late 1960s and the 1970s, environmentalists expressed concern that crested wheatgrass "monocultures" were harmful to wildlife habitat. Several attempts were made to scalp paths through established seedings in order to establish shrub seedlings.[37] Direct seeding of antelope bitterbrush along with crested wheatgrass has long been considered impossible, but it can be done by reducing the grass seed in the mixture. Four-wing saltbush has been successfully established in seeding mixtures with crested wheatgrass. Including sagebrush in seeding mixtures is more difficult. The seed and seedbed ecology of sagebrush species is very complex. The seeds of big sagebrush often establish best when they are broadcast and then pressed against the surface of the seedbed, but good examples of successful artificial seeding of big sagebrush are more difficult to find.[38]

Most crested wheatgrass seedings are invaded by big sagebrush.[39] Eventually, this encroachment results in a mixture of introduced perennial grasses and native shrubs that may support a diverse population of native wildlife.[40] On many sites the invasion by big sagebrush will continue until the herbaceous vegetation is almost entirely suppressed regardless of whether or not the site is grazed. This underscores the competitive advantage shrubs have in the Great Basin environment and the importance of recurrent wildfires in maintaining the ratio between herbaceous and woody plants. Crested wheatgrass seedings not invaded by big sagebrush do exist on specific sites. Some of these may be sites that were true desert grasslands before livestock was introduced.

Broadleaf species adapted to the same soils and climatic zone as crested wheatgrass are rare. Alfalfa, a legume, is one of them and is attractive to a variety of wildlife species. It has long been proposed as a species for seeding with crested wheatgrass.[41] On the northern Great Plains in environments with biologically significant summer precipitation, alfalfa and crested wheatgrass plantings have been successful. Dick Holland was very successful including alfalfa and sainfoin in the seeding mixture with wheatgrasses in the Long Valley–Peterson Mountain wildfire burn north of Reno. At various times cultivars of flax, penstemon, and small burnet have been used and extensively seeded with wheatgrasses in the Great Basin. Generally, the greater the amount and favorable distribution of moisture at a given site, the greater the chance these species will establish.

Recent proposals that native perennial forbs are the answer to the cheatgrass problem on Great Basin rangelands have no scientific basis. The same applies to the numerous "patent medicines" being offered to solve revegetation problems on areas infested with cheatgrass. These range from garlic powder and dried sagebrush juice to various microbiological additives. Perhaps soil microbiology holds the key to suppressing cheatgrass, but evaluation of such approaches to the problem must include experimental design, data collection, and statistical analysis. We tested a micronutrient proprietary product advertised to enhance seed germination and seedling enhancement and found nothing to support that claim. We treated eight native grasses (bluebunch wheatgrass, Idaho fescue, big bluegrass, thickspike wheatgrass, western wheatgrass, squirreltail, Indian ricegrass, and needle-and-thread grass) and one introduced ('Hycrest' crested wheatgrass) perennial grass with the proprietary product Germ-N-8 and compared seed germination, initial sprouting, and subsequent establishment of each species with its untreated counterpart. Bluebunch wheatgrass and needle-and-thread grass did seem to benefit initially from the product, but data on establishment showed no benefit to any of the treated species. In fact, thickspike wheatgrass and 'Hycrest' had significantly more establishment from seeds that were not treated with the product.

The perennial grass portion of big sagebrush/bunchgrass communities usually consists of one dominant species growing together with the short-lived native perennials squirreltail and Sandberg bluegrass. These short-lived species, which are much more ephemeral than the dominant bunchgrass, are the natural secondary successional species in big sagebrush/bunchgrass communities. From the earliest days of artificial seeding in the big sagebrush/bunchgrass environment, some scientists have suggested that crested wheatgrass needs a companion species to fill the niche occupied by squirreltail and Sandberg bluegrass. The latter was often considered a weed that competed with the long-lived perennial grasses in sagebrush/bunchgrass communities in the northern portion of the Intermountain Area.

"Biological diversity" is now the watchword, the key to solving the cheatgrass problem, and seeding mixtures proposed for use on rangelands almost always include some minimum number of species. Often the mixture includes such extremes as Indian ricegrass and intermediate wheatgrass. If one of these two is adapted to the site, the other is certainly not. This seems a waste of resources in all phases of the restoration process. Even if all the species should become established, they would create management problems if the goal of the seeding was to provide a stable forage base for livestock grazing.

The concentration on the Tribe Triticeae in restoration efforts has led to the development of numerous grass cultivars adapted to the portions of the big sagebrush/bunchgrass ecosystem with the greatest potential to support plant growth. Most of these areas were originally dominated by bluebunch wheatgrass or Idaho fescue. The dominant native perennial grasses in the central Great Basin are members of Hitchcock's Tribe Agrostideae, including species of needlegrass and Indian ricegrass.[42] Thurber's needlegrass fills a role in the Great Basin much like that filled by bluebunch wheatgrass in the Pacific Northwest. Robert Blank is the only scientist who has ever obtained a reasonable amount of germination from seeds of Thurber's needlegrass, and he did it by exposing the seeds to sagebrush smoke.[43] Virtually no seed of this perennial grass is available in the native plant seed trade. In 1999, thousands of acres of burned Wyoming big sagebrush/Thurber's needlegrass potential communities were seeded with 'Secar' Snake River wheatgrass because no Thurber's needlegrass was available and the public land agencies felt they had to use a native species. 'Secar' was originally released as a bluebunch wheatgrass but was found to be a different species. The chances of it establishing in Wyoming big sagebrush sites in northern Nevada are near zero at best.

Another common misconception in rangeland seedings to suppress cheatgrass is that if a little seed is good, more is much better. For the last 60 years the standard rate for seeding crested wheatgrass has been 7 pounds per acre with the rows spaced 12 inches apart; this works out to two seeds per inch of drill row, or twenty-four seeds per square foot. A fully stocked stand of crested wheatgrass in the 10–14-inch precipitation zone of the Great Basin is one plant per square foot.[44] Seedling germination is controlled by both the inherent physiological potential of the seed and the physical and biological potential of the seedbed to support germination. We discussed this concept in chapter 5 along with safe sites for germination and the concept of seedbeds controlling successional stages. Many years ago, the pioneer seedbed ecologist J. L. Harper used the analogy of a thimble and a pile of seeds.[45] If the potential of the seedbed to support germination is the capacity of the thimble per unit of seedbed, then a single thimbleful is sufficient. Crested wheatgrass seedbeds may have the potential for more than one of the twenty-four seeds planted per foot of row to germinate, but various perturbations, including intraspecific competition, will eventually thin the stand down to one plant per foot of row.

The overseeding syndrome took a strange twist during the massive seeding operations after the 1999 wildfires in northern Nevada. Someone took the opposite tack and decided that if ultra-low rates of crested wheatgrass were seeded, the resulting stands would suppress cheatgrass and allow the native perennial plants to establish and reoccupy the site. Apparently, no one had tested this idea in a valid experiment, or had even tried it. Nevertheless, public land management agencies did it in a failed seeding effort that reportedly cost $42 million.

Even worse, the same mistakes are made again and again. In 1948, for example, A. C. Hull Jr. suggested that the openers of a drill might peel back enough cheatgrass to allow crested wheatgrass to become established.[46] He was wrong; nearly complete control of cheatgrass in and between the rows is essential to give perennials a chance for establishment. Hull has contributed more to research on range seeding than any other individual (John Vallentine's bibliography of range management literature includes thirty-six first-author citations for Hull under "seeding"), so he is entitled to make a mistake now and then.[47] Fifty years later, during the winter of 1999–2000, a Reno newspaper quoted the person in charge of the massive seeding of the areas burned in the 1999 wildfires in Nevada (a former student of ours) as suggesting that the action of drill openers would peel back the cheatgrass and allow native perennial species to establish.

Numerous unproductive ideas have sidetracked research in range seeding technology over the years. The wholesale chase after a magic bullet in the form of pelleted seeds is the prime example. The pellets, created out of clay, diatomaceous earth, or some other carrier mixed with a nitrogen fertilizer, were supposed to create a portable safe site for germination around the seed. Seedbed preparation and drilling would not be necessary; the seed pellets could be broadcast from the air. Unfortunately, when the pelleted seeds were tested in designed experiments, not even marginal stands grew.[48] The pellets created an osmotic gradient that drew moisture from the seed to the coating and actually inhibited germination.

Stories about public land management agencies seeding crested wheatgrass over a million-acre area in Nevada imply that all of the big sagebrush/bunchgrass potential rangelands were seeded. In fact, less than 5 percent of Nevada's big sagebrush/bunchgrass potential rangeland was seeded to crested wheatgrass. Only the small portion of the total area that could be tilled and drilled was actually converted from cheatgrass to perennial grass. Cheatgrass on rocky sites and steep slopes was safe. Thousands of these non-tillable sites are broadcast-seeded after wildfires, but the chance of perennial grass seedling establishment is practically zero. If we are truly going to attack cheatgrass on all fronts, we need the right tool: plant materials that can be broadcast and will self-establish, successfully compete with cheatgrass, and produce seeds that repeat the process until cheatgrass is biologically suppressed. In other words, we need a highly competitive self-invasive species

Fig. 12.2. 'Immigrant' forage kochia is an extremely important addition to the plant material available to suppress cheatgrass. It is the first plant material of this type available that can be established in the salt desert sites and in the big sagebrush zone.

such as cheatgrass. Plant breeders are working toward this goal through hybridization of such perennial weeds as quackgrass with bluebunch wheatgrass in carefully controlled selection processes that evaluate the "weediness" character of segregating populations. The first perennial plant material that became available to range managers and more or less had these self-invasive characteristics was forage kochia.

'Immigrant' forage kochia (*Kochia prostrata*), native to the arid and semiarid regions of central Eurasia, is a long-lived semi-evergreen shrub that averages 1–3 feet in height (fig. 12.2). Forage kochia is another member of the goosefoot family, Chenopodiaceae, which contains many valuable arid rangeland species, including four-wing saltbush, winterfat, and black greasewood. Known as the "Russian alfalfa," forage kochia is adapted to many environments ranging from sandy to clay soil types, various alkalinity levels, and precipitation levels ranging from 5 to 27 inches annually. These attributes led early range researchers like Wesley Keller, longtime chief of the Forage and Arid Pasture Branch of the ARS, and Perry Plummer to acquire seeds of different accessions of forage kochia from the gardens of the V. I. Williams Museum at the Timiryazev Academy in Moscow in 1959.[49]

In 1984, after years of research with various accessions of forage kochia, the Soil Conservation Service registered and released 'Immigrant' for use in wildland rehabilitation seeding efforts. Forage kochia has been seeded on more than 150,000 acres in ten western states in efforts to combat cheatgrass-fueled wildfires and provide forage for wildlife.[50] In the current political climate, it is unpopular to suggest using introduced plant species like forage kochia in rangeland rehabilitation efforts. Some individuals and interest groups oppose the seeding of any introduced plant species for reasons ranging from concern over the genetic integrity of the plant community to unfounded claims that such species as forage kochia invade native plant communities.[51]

CHAPTER 13

Cheatgrass and Nitrogen

Nitrogen fertilizer in various forms came into common usage in crop production in the United States after World War II. Yield increases in cereal grain production obtained through nitrogen enrichment were often spectacular. Range scientists throughout western North America began to try various methods of fertilization with an array of forms of nitrogen to enhance forage production. Generations of graduate students repeated the trials over the next 20 years. Their results were similar; generally, applying nitrogen fertilizer markedly enhanced the herbage production of grasses. Unfortunately, the cost of nitrogen fertilization exceeded the value of the increase in forage production. Soon after James Young started work with the ARS in Reno, Nevada, Burgess Kay and Raymond Evans published the results of a fertilizer trial conducted in northeastern California.[1] Kay and Evans applied 60 pounds of nitrogen per acre as ammonium sulfate to a marginal stand of intermediate wheatgrass that had considerable cheatgrass between the perennial grass plants. Their hypothesis was that nitrogen enrichment would enhance the competitive ability of the perennial grass and result in the disappearance of the invasive annual cheatgrass. The plant parameter they measured in the field was herbage cover. They got the opposite of the result they hoped for: nitrogen fertilization increased cheatgrass cover and decreased the cover of intermediate wheatgrass. In other words, nitrogen fertilization favored cheatgrass to the point that the intermediate wheatgrass died.

A. M. Wilson, in cooperation with Grant Harris and Dillard Gates, conducted a very similar experiment in eastern Washington with a stand of bluebunch wheatgrass infested with cheatgrass.[2] They fertilized both good and poor stands of bluebunch wheatgrass. When they applied 80 pounds per acre of nitrogen as ammonium sulfate annually for 4 consecutive years, cheatgrass became dominant on both stands. The yield of bluebunch wheat-grass decreased by 50 percent and cheatgrass herbage yields increased by 400–600 percent. The authors' explanation of this response was based on the comparative phenology of the growth of cheatgrass and perennial grasses that A. C. Hull had reported a couple of decades earlier.[3] Cheatgrass is capa-ble of completing its life cycle a month before bluebunch wheatgrass seed matures. Nitrogen fertilization accelerated the rate and amount of cheatgrass growth until there was no soil moisture left for the growth of bluebunch wheatgrass. The cheatgrass plants crowded right against the bluebunch wheatgrass plants, and each year the perennial wheatgrass plants got smaller. Nitrogen fertilization also increased the density of cheatgrass seedlings and greatly increased the number of tillers per plant. The investigators did not see a single seedling of bluebunch wheatgrass in the fertilized or unfertilized plots during the 4-year course of the experiment.

W. E. Martin, a soil scientist at the University of California at Davis, was a well-known pot tester; that is, he used the nitrogen-phosphorous-sulfur testing method made famous by Hans Jenny at the University of Califor-nia at Berkeley. Zero (control), one, two, or all three of the nutrients were added at specific rates to measured amounts of potted soil. Martin also var-ied the rates of enrichment. The pots were planted with seeds of a plant with seedling foliage expressive of nutrient deficiencies; the classic species was romaine lettuce. Dr. Martin worked with the California Soil and Vegetation Survey, an ambitious project to map the range and forest soils and plant communities of the entire state. His contribution was to determine through bioassay the nitrogen, phosphorous, and sulfur status of every soil that was mapped. In the process of this exhaustive undertaking, Martin learned a lot about mineral nutrition of rangeland soils.

In the Mediterranean climate of cis-montane California, annual grasses—many of them species of *Bromus* (rarely cheatgrass)—have almost completely replaced the native perennial grasses. Scientists have been unable to find an introduced perennial grass able to replace and biologically suppress the exotic annuals as crested wheatgrass suppresses cheatgrass in the Great Basin. Harding grass was thought to be a possibility, but it was only marginally successful even on the better annual rangeland sites. Martin tried to enhance the ability of Harding grass to compete with annual grasses by fertilizing with nitrogen. When he fertilized with 60 pounds per acre of nitrogen, the herbage production of the annual grasses increased 91 percent, but there was no increase in the yield of Harding grass.[4] In fact, the Harding grass did not respond with increased herbage growth until the very high rate of nitrogen enrichment of 240 pounds per acre was reached. Grant Harris had previously made the same observation that bluebunch wheatgrass did not respond to low rates of nitrogen but did have increased growth with very high rates.[5]

Raymond Evans brought the pot test technique to Nevada from Berkeley and substituted cheatgrass for lettuce. Cheatgrass has wonderfully expressive leaves, especially for nitrogen deficiencies. Besides the visual expression of nutrient deficiencies, the amount of herbage production was the variable usually measured in these tests. Overwintering rosettes of cheatgrass from fall germination typically develop a bright red color symptomatic of nitrogen stress. In greenhouse experiments, Evans and Eckert compared the shoot and root growth responses of cheatgrass and crested wheatgrass seedlings after nitrogen and phosphorus enrichment.[6] They found that the shoot growth of cheatgrass was (1) more rapid in the early part of the experiment, (2) accelerated earlier by nitrogen, and (3) increased more by low to intermediate levels of nitrogen than that of crested wheatgrass. Shoot growth of crested wheatgrass was increased by higher levels of nitrogen and later in the experiment than cheatgrass. Root growth of crested wheatgrass was greater than that of cheatgrass at all levels of nitrogen and was increased by high levels of nitrogen. Cheatgrass removed more nitrogen and phosphorous from the growth solution at low and intermediate levels than did crested wheatgrass, but the latter removed more nitrogen from the solution than cheatgrass did at the highest levels of nitrogen enrichment. The relation observed in the field—that annuals such as cheatgrass are more efficient than perennial grasses in utilizing low levels of nitrogen and that high levels of nitrogen enrichment are required for perennials to respond—was confirmed. Eckert and Evans concluded that "differences in shoot and root growth and nutrient uptake of the two species in response to nitrogen and phosphorus levels suggest some of the possibilities and limitations of manipulating growth of crested wheatgrass through fertilization to favor its establishment and growth in relation to environmental factors and competition from downy brome."[7] They also field-tested the use of nitrogen fertilization to aid in the establishment of crested wheatgrass seedlings in experiments north of Reno. If the amount and periodicity of precipitation were favorable, nitrogen fertilization greatly enhanced crested wheatgrass seedling establishment, but on average it was not beneficial.

Burgess Kay conducted a cheatgrass fertilization experiment for 11 years in northeastern California in an attempt to get herbage production in very dry years.[8] The site was a former big sagebrush/bluebunch wheatgrass community that had burned and converted to cheatgrass dominance with occasional tansy and tumble mustard plants. Precipitation at the experimental site, located near Likely in Modoc County, California, ranged from 4.92 to 17.4 inches during the course of the experiment. Cheatgrass herbage production on control plots varied from 0 to 1,100 pounds per acre during the study period (average 450 pounds per acre). On this particular soil, cheatgrass production was limited by both nitrogen and sulfur. Applying 30 pounds per acre of nitrogen with 40 pounds per acre of sulfur (applied as gypsum) at least doubled cheatgrass herbage production in the years with moderate or higher precipitation. In dry years, there was no increase in cheatgrass forage but the production of tansy and tumble mustard did increase. In the wettest year, the yield of cheatgrass herbage on the control plot was 970 pounds per acre. The addition of 30 pounds of nitrogen and 40 pounds of sulfur increased herbage yield to 3,030 pounds per acre. When the nitrogen rate was increased to 120 pounds per acre in this very wet year, herbage production of cheatgrass shot up to an astounding 6,050 pounds per acre. One potential side benefit of nitrogen fertilization of cheatgrass might be an increase in herbage production very early in the spring when forage is in the greatest demand on this type of rangeland and native perennial grasses are most easily damaged by excessive grazing. Kay found no increase in cheatgrass herbage production following nitrogen fertilization in March. Leaves of cheatgrass fertilized with nitrogen were dark green in the early spring while the leaves of rosettes on the control plots were red—an indication of stress. The only plant that increased herbage growth with nitrogen fertilization in the early spring was tansy mustard. Instead of providing early forage, however, fertilization with nitrogen accelerated the phenology of tansy mustard, which was flowering by the time cattle were turned out on the range; cattle do not prefer the herbage of this plant at this stage of growth. In wet years, high rates of nitrogen increased the nitrate content of mustard herbage to the point that there was a danger of poisoning. In all experiments where nitrogen was added, the crude protein content of cheatgrass herbage was increased.

As far as the stated purpose of Kay's 11-year experiment was concerned, it was a failure. Instead of making cheatgrass herbage production consistent from year to year, nitrogen application accentuated the differences resulting from different moisture regimes. In dry years, nitrogen fertilization did nothing to increase cheatgrass herbage production; in wet years, it resulted in huge increases in the already very high herbage production. The study did confirm that cheatgrass loves nitrogen enrichment as long as moisture is available during the spring growing season.

Much of what is known about the cycling of nitrogen in the semiarid ecosystems of the Great Basin is the result of research conducted by Neil West and his students and collaborators. Despite the relatively low rainfall in the Great Basin, the annual input of nitrogen in rainfall is not much less than that in more mesic areas. Great Basin precipitation contains more dust, which raises the nitrogen content per unit of precipitation.[9] Dust erosion and deposition are major factors in soil formation in the Great Basin.[10] Robert Blank, for example, reported a nitrate flux of 2.4 pounds per acre per year in dust deposition from a playa surface in northwestern Nevada.[11] West concluded that erosion and deposition in desert areas with 20 percent vegetative cover were about equal as far as nitrogen was concerned.[12] The prevailing winds result in deposition to the northeast of playas rather than it being equal in all directions, and deposition of nitrate in specific topoedaphic locations in the Great Basin may thus be significant.[13] The depth of wetting in most big sagebrush communities is less than a meter, which eliminates deep leaching as a means of removing nitrogen from these ecosystems.

West considered the biological part of the nitrogen cycle in Great Basin ecosystems to be more interesting than the physical aspects because it is so different from nitrogen cycles in other kinds of systems.[14] Great Basin deserts have relatively few plant species and low densities of legumes. The major nitrogen fixers of cold winter ecosystems in the Great Basin are cryptogamic crusts. Some of the blue-green algae component of the cryptogamic crust is responsible for the surprisingly high rates of nitrogen fixation. Even though considerable amounts of nitrogen are fixed in the cryptogamic crust, nearly all of that is lost by denitrification, which occurs in the wet cryptogamic crust when a carbon source and partially anaerobic microsites in the bottom of the decaying crust are simultaneously available.

What effect does cheatgrass invasion and dominance have on the nitrogen cycle? We have already established that cheatgrass is a tremendous user of nitrogen, but does it also influence the input portion of the cycle? Dick Eckert followed nitrate nitrogen accumulation on control, mechanical, and herbicidal fallows in the late 1960s and encountered the typical variation in precipitation that haunts field research in the Great Basin. His sites were located in Humboldt County, Nevada, in the Trap Butte–Orvada area. The average October–early spring precipitation for the sites he studied was 7–8 inches. In 1967–1968 the site received 4–5 inches of precipitation, and in 1968–1969 it received 11–12 inches of precipitation. Despite the precipitation differences, nitrate nitrogen accumulations in the fall and winter of 1967 and 1968 were similar. The untreated cheatgrass stands accumulated an average of 5 pounds of nitrate nitrogen per acre. The mechanically fallowed areas averaged 27 pounds per acre, and the atrazine fallow plots accumulated 43 pounds per acre of nitrate nitrogen.[15] The differences between the atrazine treatment and mechanical fallow were due to timing. The atrazine was applied in the fall and killed cheatgrass plants that germinated that fall and early the following spring, preventing uptake of nitrate nitrogen. The mechanical fallow was not applied until late the next spring, after the fall- and early spring–germinating cheatgrass had markedly reduced the nitrate made available through nitrification. The critical factor is not how much nitrogen is produced during the fallow year, but how much remains in the seedbed the seedling year, which is the spring following the fallow year.

Nitrate nitrogen is susceptible to leaching. Eckert's results depended on the winter precipitation. In the spring of 1968, after the dry winter of 1967–1968, the surface 6 inches of soil on the atrazine fallow contained 30 pounds of nitrate nitrogen per acre, the mechanical fallow contained 29 pounds, and the cheatgrass stand had 13 pounds. After the wet winter of

1968–1969, the surface 6 inches of the atrazine fallow contained 5 pounds of nitrate nitrogen per acre and the cheatgrass control had 2 pounds per acre (Eckert did not provide a figure for the mechanical fallow). Overwinter precipitation at the end of the fallow year is important because of the leaching of nitrate in wet winters. The limited depth of wetting makes these closed systems, so the nitrate will still be in the profile when the wheatgrass seedlings planted after the fallow year become sufficiently established to reach lower depths in the soil profile.

Why not skip the herbicidal or mechanical fallow and simply fertilize wheatgrass with nitrogen? First, the fallows reduce competition from cheatgrass during the seedling year. If the process is done properly, the seedbed should be nearly free of cheatgrass. Second, soils store moisture during the fallow that allows the crested wheatgrass seedlings to take advantage of the nitrate enrichment. Cheatgrass and seedlings of perennial grasses compete primarily for moisture, but available nitrate plays a significant role in this competition. In the early 1990s James Trent, who came to work for the ARS at Reno as a research technician, came up with a remarkable idea. Jim pointed out that it was well established that cheatgrass and other invasive annual grasses such as medusahead thrive on nitrogen enrichment. What would happen, he asked, if the nitrogen were taken away? This could be accomplished in two ways: (1) the nitrogen could be immobilized by providing an ideal carbon source for the growth of soil microorganisms, or (2) the nitrification process could be inhibited by applying a chemical that had been developed to prevent the overwinter loss of nitrate nitrogen from agronomic fields. Trent had worked with immobilizing nitrogen while conducting research for his master of science degree at the University of Oklahoma, so he was the ideal person to answer this question.

Nitrogen management in agronomic crops had been a big issue for many decades in the United States. Among others, Tilman had been making big waves in plant ecology with nitrogen manipulations in grasslands in Minnesota, and McLendon and Redente had studied how nitrogen enrichment and immobilization influenced succession in big sagebrush communities in Colorado.[16] After testing other substances such as straw and sawdust, Trent began a search for a carbon source that would function in semiarid to arid rangeland conditions in the Great Basin. Finely ground wheat straw and sawdust proved ineffective under desert conditions. We finally settled on sucrose as a carbon source. Only one supermarket in the Reno area would readily take the federal credit card for tax-free purchases, which meant that we were regularly wheeling grocery carts full of 50-pound bags of sugar up to the checkout counter. Curiosity finally overwhelmed a clerk. "My, you must be doing a lot of baking," she said. Our reply was, "No, we're running a still back in the hills." After that, the researchers got very strange looks at the checkout counter.

Soil scientist Robert Blank suggested including treatments enriched with nitrogen, so we designed an experiment with eight treatments: (1) control, no treatment applied; (2) enrichment with calcium nitrate as a nitrate nitrogen source; (3) enrichment with urea as a urea nitrogen source; (4) enrichment with ammonium sulfate as an ammonia nitrogen source; (5) nitrogen immobilization with carbon; (6) inhibition of nitrification using the chemical nitrapyrin, a chlorinated pyridine (in nitrification, ammonia is converted to nitrite, which is converted to nitrate);[17] (7) immobilization-inhibition combination with both sucrose and nitrapyrin; and (8) enrichment with ammonium sulfate combined with inhibition of nitrification with nitrapyrin. Dr. Blank suggested the eighth treatment as a means of identifying plants that could use ammonia directly as a nitrogen source without nitrification proceeding to a nitrate source. The eight treatments were arranged in a randomized block design with four replications. It was a mark of the efficacy of nitrogen in annual invasive grass communities that individual treatments could be identified without using a plot diagram. Each treatment developed a unique vegetative signature.

Trent arrived at the amount of carbon to add by determining the biomass of the microorganisms in the top 6 inches of surface soil, which was about 1.2 pounds of carbon per 100 square feet. We applied twice this amount three times each year. The applications were applied in thirds in order to increase the soil microbiological organisms without overpowering them with a highly soluble source of carbon. One-third of the sucrose, nitrogen fertilizer, and nitrapyrin was applied to the soil before the first fall rain occurred, at the rate of 30 pounds of nitrogen per acre. The second application was made in early December before the soil was solidly frozen. The last application was in February when the soil had thawed and the herbaceous vegetation was beginning active growth. The nitrapyrin was applied at the suggested rate for cropland, 2 pounds per acre. The site we selected for the first large-scale field experiments was in Lassen County, California, north of the small town of Litchfield in a pass leading from the Honey Lake Valley north toward the Shinn Peak Highlands.[18] We named the site "Noble" after the historical marker on Highway 299. The vegetation at the time of European contact was probably Lahontan low sagebrush with a Sandberg bluegrass understory and a rich component of annual and perennial forbs. The site had seen diverse uses over the years: freighters on the road to Ravendale and Modoc County had camped there; it was a construction camp while the railroad was being built through the pass; and for many years it was a shearing site for a very large range sheep industry that wintered in the deserts of northwestern Nevada and summered in the High Sierra and the Warner Mountains.

Robert Blank analyzed the soils at the site and the effect on the native vegetation of these concentrated human activities.[19] The base soil was a thick layer of high shrink-swell smectitic clay. The contact time topography consisted of a series of low mounds and interspaces. The mounds supported the Lahontan sagebrush and most of the perennial grasses, and were largely composed of eolian-deposited silt and sand particles that had eroded from the lake plains of the vast pluvial lakes that occupied the basins of the Great Basin during the Pleistocene. The interspaces between the mounds formed miniplayas and were nearly impervious to water.[20] The ranchers in the Honey Lake Valley used the area as a turnout site in the spring. Excessive grazing in the early spring over the course of many years destroyed the perennial grasses. Cheatgrass invaded the site, which eventually burned in repeated wildfires. Trains going through the pass probably contributed to the wildfires with overheated brakes. Once the wildfires had destroyed the perennial grasses and Lahontan sagebrush, winds eroded away the mounds, exposing the underlying clay beds. The clay surface is now covered with basalt cobble pulled from the underlying rock by the shrink-swell clays and expelled on the profile surface. The clay soils are very high in cation exchange capacity, but virtually all of their physical characteristics are undesirable for seedling growth for most plants. The clay fraction of the soil is so high that the soil contains only about 34 percent moisture at the permanent wilting point. The structure of the clays when dry is granular, and they are readily eroded by wind and water. By late summer, the clays have shrunk and the soil surface is a network of cracks that can reach 10 inches in width and extend to the depth of the soil profile. The summer cracks provide habitat for the Great Basin rattlesnakes and large desert tarantulas. The shrink-and-swell clays are the primary invasion site for medusahead in the northwestern margin of the Great Basin.[21]

The site was predominantly medusahead, although some cheatgrass and rattlesnake brome grew there as well. Cheatgrass, the first invasive annual grass on the site, shares many characteristics with medusahead. Both build large seed banks,[22] and effective weed control must involve either getting all the seeds in the seed bank to germinate so they are susceptible to control or creating conditions that will hold the seeds in dormancy. Medusahead and cheatgrass have similar dormancy characteristics as well in that dormancy is acquired in seedbeds, and available nitrate and gibberellin enrichment of the germination substrate influence the breakdown of the dormancy.[23] We measured not only the characteristics of each treatment (plant density, herbage yield, and seed head density) but also the germinability of seeds in the seed bank through bioassay.

We ran the experiment for 5 years, from 1990 through 1995. Precipitation was a major factor. The long-term average precipitation for the site was 12 inches, but that included 3 consecutive years when precipitation dropped below 5 inches and 1 year when precipitation shot up to 20 inches. That was quite a spike in precipitation considering that the total precipitation for the 3 previous years was 11.6 inches. As expected, the results obtained with the nitrogen immobilization, nitrification inhibition, and nitrogen enrichment experiments depended on the amount of precipitation received. In all experiments with fertilization, ammonium sulfate was the most effective means of enriching nitrogen. Calcium nitrate should have been the most effective, but in semiarid to arid situations solubility becomes a problem, and ammonium sulfate was the most soluble form of nitrogen fertilizer used. In dry years, there was no difference in density of medusahead plants on control and ammonium sulfate–fertilized plots. In years with moderate precipitation, there were about one-third more annual grass plants on the ammonium sulfate–fertilized plots than on the control plots. In wet years, the medusahead density with ammonium sulfate fertilization was more than double that of the controls. The relations were apparent for herbage production and the number of seed heads.

The influence of carbon enrichment on nitrogen immobilization was spectacular. The density of annual grass was reduced to less than 1 percent of that on the control plots. Surprisingly, this tremendous reduction in annual grass was maintained annually despite the wide swings in the amount of precipitation. The reduction in annual grasses was also apparent in herbage and number of seed heads produced. Inhibiting nitrification was almost as effective as immobilizing nitrogen. In the wet years, this treatment was not as effective as the carbon enrichment treatments. Combining the two contrasting treatments proved redundant. The ability of the carbon enrichment to immobilize nitrogen in very dry years suggested that the lack of nitrogen was influencing the seedling recruitment very early in the life cycle of the annual grasses. The nitrate ion is generally considered to have the most influence on germination of many types of dormant seeds.[24] Thomas Monaco determined in a greenhouse study that under low nitrogen availability, cheatgrass and medusahead seedlings had root and shoot growth equal to or better than that of bluebunch wheatgrass or squirreltail seedlings.[25]

We demonstrated the delicate balance between cheatgrass and medusahead and the fact that dominance can be shifted very early in the growth of the two species through the use of the herbicide diuron back in the 1970s.[26] If the herbicide was applied very early in seedling establishment to a seemingly monospecific medusahead community, a cheatgrass-dominated community resulted. We did not combine this treatment with nitrogen enrichment or immobilization, but it would be a very interesting experiment.

Seeds of medusahead and cheatgrass are initially not dormant at maturity but acquire dormancy in field seedbeds over the winter.[27] This dormancy can be broken by enriching the germination substrate with nitrate nitrogen. In the experiment conducted at the medusahead site, nitrogen enrichment with ammonium sulfate reduced the annual grass seed bank.[28] Conversely, immobilization with a carbon source increased germination in greenhouse bioassay samples enriched with potassium nitrate. We concluded from these results that nitrogen abundance affects the germination of medusahead and cheatgrass seeds from the seed bank under natural conditions. A major disappointment with this research location was our inability to get perennial grasses to establish on the clay-textured seedbeds. Only once in 30 years of trials did we succeed in establishing perennial grasses by broadcast seeding on these soils, the only viable option because of the rock cover.

Our second venture into manipulating nitrogen to establish seedlings of perennial species in a site infested with cheatgrass involved the native shrub antelope bitterbrush. Wildlife habitat managers have long considered antelope bitterbrush the key species for management of mule deer, but it is very difficult to establish artificially.[29] Competition from cheatgrass is one of the major factors limiting seedling establishment, and cheatgrass-fueled wildfires destroy many antelope bitterbrush stands. Forest Service scientists A. L. Hormay and E. C. Nord did much of the pioneering research on antelope bitterbrush in northern California.[30] Nord "gave" us the location for our nitrogen research after the Forest Service discontinued the bitterbrush research project. The site was at Doyle, California, on a state wildlife management area managed by the California Department of Fish and Game for wintering mule deer. Doyle is located at the eastern base of the Sierra Nevada about 40 miles north of Reno in the delta of Long Valley Creek where it once flowed into pluvial Lake Lahontan. The soils are largely derived from decomposed granite. The annual precipitation recorded by two long-term recoding stations located on opposite sides of Highway 395 averages 12 inches on the west and barely 10 inches on the east side. Obviously, the site is in the rain shadow of the Sierra Nevada. Nord, who had vast experience with the ecology of antelope bitterbrush, considered the site too dry for artificial seeding of the shrub.

We instigated our standard eight-treatment experiment and established the antelope bitterbrush by direct seeding. The cheatgrass stand at this site was not particularly vigorous, probably due to both the low precipitation and the nutritional characteristics of the soil. Nitrogen enrichment, especially with ammonium sulfate, greatly enhanced cheatgrass density and herbage production. Within one season, the plots treated with ammonium sulfate became dense turfs of cheatgrass that looked as if they had grown there forever. Just before the cheatgrass on the ammonium sulfate plots matured, black-tailed jackrabbits pounced on the herbage. The next spring, jackrabbit pellets covered the ammonium sulfate–treated plots, but we could not find a single pellet on any other treatment in the experiment. The antelope bitterbrush seedlings grew spectacularly in both the nitrogen-immobilization plots and the inhibition-of-nitrification plots. The normally very slow growing seedlings reached a height of 2–4 inches their first growing season. The seedlings in the nitrogen-reduced plots were 8–10 inches tall by the late summer of their seedling year.

We showed the experiments to a touring group of range and wildlife managers late in the summer. It had been a dry year at Doyle, with slightly under 7 inches of precipitation, and grasshoppers had been very destructive on the surrounding rangelands. The tour group gazed in disbelief at the huge antelope bitterbrush seedlings. One experienced wildlife manager accused us of growing the seedlings in the greenhouse and transplanting them to the field just before the tour. When word of the field experiments began to circulate among individuals interested in restoring antelope bitterbrush stands, we were deluged with requests for how much sugar to put in the drill with the seed. We explained that our experiments were designed to study nitrogen interaction with competition between cheatgrass and seedlings of native perennials and were not directly applicable as practical field seeding treatments. We also explained that nitrapyrin was a commercial product, but anyone interested in using it as a pesticide would have to approach the manufacturer about obtaining registration for that use on rangelands. Pesticide manufacturers have traditionally been very reluctant to invest money in obtaining clearance for products to enhance wildlife habitat. This is unfortunate because wildlife habitat managers are willing and able to invest much more money per acre in establishing antelope bitterbrush for mule deer than ranchers can afford to spend for browse for domestic livestock. Despite the worldwide sugar glut, sucrose is too expensive to use as a carbon source for nitrogen immobilization on rangelands.

Economic considerations are not the major problem with nitrogen immobilization or inhibition of nitrification. Both techniques merely interrupt the supply of nitrogen available for plant growth; sooner or later there will be a surplus of nitrogen available on the treated area and cheatgrass will come back. The positive aspect of the process is that the near total inhibition of cheatgrass growth allows the antelope bitterbrush seedlings to establish. Although the antelope bitterbrush seedlings established on the nitrogen immobilization plots were 2 feet tall by the end of the second growing season, they were surrounded by a dense stand of cheatgrass that was 16 inches tall. Once the cheatgrass comes back, wildfires that can destroy the antelope bitterbrush seedlings are a distinct possibility. Seeding a combination of perennial grasses and antelope bitterbrush may be a practical compromise that will biologically suppress surges of cheatgrass. Managers of antelope bit-terbrush ranges for mule deer view perennial grasses as an enemy to shrub seedling establishment, but this should not be the case.

Our next venture into cheatgrass nitrogen relations was in a very different environment. Late in the twentieth century, cheatgrass suddenly became quite common in salt desert environments. No one paid much attention until 1985 when large wildfires roared through Flanigan north of Pyramid Lake and the Jungo Flat area of Humboldt County, Nevada. Both fires were largely fueled by cheatgrass growing in salt desert communities. Much of the fire at Flanigan occurred at a unique big sagebrush community whose original Indian ricegrass understory had been replaced by cheatgrass. The big sagebrush extended down into that very arid environment apparently because it was growing on an extensive field of Lake Lahontan–age sand. We established an experimental site on the Flanigan burn and did considerable research with the seeding of Indian ricegrass as well as transplanting native shrubs. Part of the experimental exclosure eventually became a dense stand of cheatgrass.

When James Young started working in Nevada, there was a one-room schoolhouse in Flanigan. The town slipped into ghost town status over the years, but weather observations continued to be made at the site for a while. The average annual precipitation was about 5 inches. Our records indicate about the same average, with rare years that doubled the average and not-infrequent winters when there was virtually no precipitation.

Although Flanigan was not technically a ghost town, it did have a ghost. Homer was still alive at the time, but his ability to appear seemingly out of nowhere in the middle of the desert was nothing short of super-natural. Homer occasionally worked as a cowboy-mechanic at the Espil Sheep Company in the Smoke Creek Desert. He was quite a mechanic, too; his collection of tools included some rare and unusual implements. Homer had a 10-acre dry ranch about 5 miles across the sand dunes above our experimental site at Flanigan. He got around in a 30-year-old Oldsmo-bile sedan with a hydromatic transmission, a car ideally suited for Homer's environment. His dry ranch was really dry. There was no source of potable water anywhere on it. Washing and drinking water apparently were not big concerns with Homer. We never thought to ask him if he was home the day the wildfire roared down from Dry Valley and burned several thousand acres while the Bureau of Land Management fire crews frantically started a back fire off the Flanigan county road. Homer is dead now, and his real ghost is in residence at the bunkhouse at the Buffalo Meadows Ranch where he died. The bunkhouse residents generally don't mind having him around, but his harmonica playing in the middle of the night is unsettling. We missed a great opportunity in not conducting an oral history interview while Homer was still living. He was a true child of the desert who lived in the dunes by choice. When he pulled in his cabin, which was mounted on skids, the dune vegetation was probably still big sagebrush/Indian ricegrass. He lived to see the range fenced and the conversion to a cheatgrass understory, and he had a close encounter with a wildfire storm in an environment that apparently had never known wildfires before.

We established the standard eight-treatment experiment on the sand-textured soils at Flanigan with considerable qualms. How effective would the nitrogen treatments be at this extremely dry site? The annual grass suppression with nitrogen immobilization and inhibition of nitrification was even more spectacular than it had been at the first two experimental areas. The edges of these plots were absolutely straight (fig. 13.1). Treatment locations were randomized, and an ammonium sulfate enrichment plot happened to adjoin a carbon-to-immobilize-nitrogen plot. The ammonium sulfate plot had a thick stand of cheatgrass that was 16 inches tall; the adjacent plot was bare.[31] We showed the plots to a visiting Australian scientist, who raced around taking photographs. Then he slipped up to us and whispered, "Just between friends, can you tell me what herbicide you sprayed on the plots with no annual grasses?" We were not sure he was ever convinced that the results were entirely the result of nitrogen.

The only cheatgrass that occurred on the plots with reduced or unavailable nitrogen grew in scatter-hoard caches made by rodents, which make good use of cheatgrass seeds.[32] The cheatgrass plants growing in the caches appeared stunted and had reddish leaves symptomatic of nitrogen stress. Bob Blank compared the nitrogen content of these cheatgrass plants with that in plants collected in the control or ammonium sulfate-enriched treatments. The nitrogen content of the cheatgrass herbage in the treated areas was significantly lower than that in the plants growing from rodent caches. The important point is not that the plants were suffering from nitrogen deficiency, however, but that they grew while the seeds of cheatgrass in the

Fig. 13.1. Nitrogen enrichment-immobilization experiments established on Lahontan sands at Flanigan, Nevada. The bare areas are carbon enrichment or nitrification inhibition treatments. The areas with rank growth of cheatgrass are the nitrogen enrichment plots.

seed bank did not. This is an indication that nitrogen stress occurs during the initial phases of germination. We previously mentioned this in regard to the medusahead site where we first applied the nitrogen experiment.

Rodents continually move seeds from existing seed banks to new sites. The nitrogen-immobilization and inhibition-of-nitrification treatments were free of vegetation and easy to dig in, and they attracted rodents with cheatgrass seeds to cache. Since the cached seeds responded differently to the treatment than the seeds emerging from the seed bank did, we assumed that they germinated at a different time. We initiated the treatments before the first fall rain, so for the seeds in the caches to respond differently they must have germinated in the spring, well after our last treatment application. This suggests that temperature and moisture conditions in the seedbed interact with nitrogen in determining cheatgrass seedling establishment. Fall-germinated cheatgrass plants normally show symptoms of nitrogen deficiency during the cold winter months. This entangled web of ecological and plant physiological interactions offers both a great challenge and potentially significant results to biologists.

Although we had planned to seed the nitrogen experiment with Indian ricegrass, that proved to be unnecessary on the plots where the availability of nitrogen was reduced. A huge number of Indian ricegrass seedlings spontaneously emerged from seeds in the sand seed banks. Indian ricegrass is one of the few perennial grasses native to the Great Basin that builds large seed banks.[33] Not only did the seedlings establish in copious numbers, they were very vigorous in the treatments where the few cheatgrass seedlings that grew were starving for nitrogen. Again we were slow to recognize the obvious. It is relatively easy to pull an established Indian ricegrass plant from the sands and get at least the coarser portion of the root system along with it. At first impression the Indian ricegrass roots look diseased. They appear greatly thickened and are coated with persistent sand. In fact, the thickening of the roots is caused by colonies of a variety of microorganisms that surround the roots and form a microphytic association with the Indian ricegrass plants. These microphytic root sheaths play a role in the plant's moisture relations and mineral nutrition. Perhaps the most interesting of these functions is nitrogen fixation! Like antelope bitterbrush, Indian ricegrass has evolved a system to survive in low nitrogen levels.

How did so many Indian ricegrass seeds come from deep dormancy to germination in seedbeds low in nitrate? The deep dormancy of Indian ricegrass intrigued seed physiologists for much of the twentieth century. Robert Blank found the answer to part of the dormancy requirements. The lemma and palea of the Indian ricegrass caryopses are tightly wrapped around the embryo and endosperm, forming a highly silicon enriched and extremely durable covering. James Young tried unsuccessfully to cut cross sections of these seeds and then hired the histology laboratory in the botany department of a major university to attempt such sections. No one was successful. Dr. Blank simply dropped a teaspoon of seeds into a resin solution. After it hardened, he polished thin sections for microscopic examination.[34] Not only did he have perfect sections, he randomly caught these cross sections in every conceivable orientation of the caryopses as they settled down through the hardening resin. In one of the sections Blank noted a gap between the pericarp and the interior of the true seed and reasoned that this might be a barrier to hydraulic conductivity to the embryo. In earlier laboratory germination tests, Indian ricegrass seeds germinated far better on sand than on a conventional germination paper substrate. Dr. Blank placed dormant

Indian ricegrass seeds in sand substrates where he had precisely established a gradient of matric potentials and found that at a matric potential very close to field capacity (the amount of water the soil can hold against the forces of gravity), 80–90 percent of the Indian ricegrass seeds would germinate; germination on paper in Petri dishes averaged only 2 percent.[35] Blank determined that once the sandy soil at Flanigan was wet from a moisture event, it maintained the proper matric potential for germination of Indian ricegrass about 3 inches below the soil surface. The seed of Indian ricegrass is small, but it has long coleoptiles that can emerge from great depths in the sand. As useful as this information is, it still does not answer the question of why Indian ricegrass germinated so well in the reduced nitrogen treatments and not at all in any of the other treatments.

Dr. Blank's paper on substrate matric potential reported that nitrogen enrichment had no effect on the germination of dormant seeds of Indian ricegrass. When we found that nitrogen immobilization or inhibition of nitrification resulted in flushes of germination of Indian ricegrass, we could cite this information as evidence that nitrogen dynamics had nothing to do with the apparent breaking of dormancy of Indian ricegrass seeds in the seed bank. Perhaps cheatgrass interference in the scant moisture supply is sufficient to inhibit germination of Indian ricegrass. If this is so, cheatgrass's invasion of this arid environment represents the crossing of an ecological threshold of immense significance.

One more nitrogen interaction was apparent in the Flanigan experimental plots. Indian ricegrass seedlings appeared to be markedly more vigorous in the treatment where nitrapyrin and ammonium sulfate were applied together than in plots treated with nitrapyrin alone. Does the nitrogen immobilization or nitrification inhibition technique have a potential use in weed control for invasive annual grasses in range seeding programs? The possibility certainly exists, but such research is fraught with many perils. Researchers who approach the problem with experience and understanding of field seedbeds in the Great Basin have the best chance of success. Greenhouse trials where moisture levels are kept near field capacity for the duration of the experiment are hopelessly out of touch with reality. Cheatgrass closes seedbeds to the recruitment of seedlings of perennials because it competes successfully for moisture. Nitrogen interacts with this competition. It remains to be seen whether nitrogen immobilization or inhibiting nitrification can survive the vast increase in the scale of nitrogen dynamics outside the aridity of the Great Basin and still have promise as a weed control technique.

And by the way, if you are out there in the cold deserts chasing the will-of-the-wisp nitrate nitrogen through winter seedbeds, don't look over your shoulder. Homer may be standing there.

CHAPTER 14

Grazing Management

Cheatgrass was the most abundant and most widely distributed forage species on the Great Basin rangelands during the twentieth century and has continued to expand its distribution in the twenty-first. The spectacular and sudden invasion of portions of the salt desert mentioned earlier is an example. For most of this period, federal land managers who calculated the amount of forage available on public rangelands were bound by agency rules to exclude cheatgrass. Such exclusion efforts were a gross misapplication of science to natural resource management. Very grudgingly, the USDA relaxed the rules late in the century to allow some cheatgrass in the forage allotment calculations, but only for restricted portions of the grazing season. Federal policy has continued to define cheatgrass as an exotic alien weed that, based on Clementsian plant ecology, will disappear if the ranges are properly grazed.

In July 1965 the Bureau of Land Management convened a cheatgrass research conference in Vale, Oregon, to bring range managers together with federal and university scientists.[1] Art Sawyer, who had recently retired as superintendent of the Squaw Butte Experiment Station near Burns, Oregon, voiced a ringing challenge to the assembled group: "No responsible person manages for cheatgrass, but is it now time we manage the cheatgrass we have." Tall, angular, and ramrod straight, Sawyer was the image of a scientist–livestock husbandryman.

As the twentieth century drew to a close, the types of professionals involved in the management of public rangelands and the demographic pool from which these professionals sprang was changing. Wildlife managers, recreation planners, and endangered species specialists supplanted the traditional cowboy range manager. Up until the 1960s, professional range managers often were recruited from rural environments. Shifts in the demography of the nation ended the rural dominance in range professions, especially as the field opened to women. The new range managers have brought new ideas and skills that are in many ways beneficial, but they have also brought new perspectives on the range livestock industry. Regardless of their views on range livestock, almost all of them believe that cows do not eat cheatgrass. They must have learned this from the literature, formal instruction, and peer exposure during undergraduate and possibly graduate training; it certainly is not based on their personal observations. Communication is key to the transfer of science-based technology. Communication has to start with the players using and understanding the same vocabulary. Toward that goal, we offer a brief overview of ruminant animal nutrition and the annual forage cycle for the range livestock industry in the Great Basin.

Ruminants—a group that includes cattle, sheep, deer, and elk—have highly evolved and modified stomachs that serve as internal anaerobic fermentation vats. Microorganisms living in these modified stomachs break down the complex, often-lignified cellulose carbohydrates that characterize the herbage of range grasses, and the host animal absorbs the by-products of this digestion. Many of the products of microbial digestion in the rumen are compounds the ruminants are not capable of synthesizing for themselves. Horses have a poor substitute for a rumen in the form of an enlarged large intestine that forms a similar type of digestion system. In severe winters when free-roaming horses are near starvation, they will eat their own dung and can derive enough previously undigested nutrition from it to stay alive. This is a testimonial both to the adaptability of feral horses and to the initial inefficiency of the horse's digestive system. Jackrabbits, the most abundant herbivores on sagebrush/bunchgrass rangelands before domestic livestock arrived, routinely eat their own feces for the same reason.[2]

The need for all of these digestive adaptations underscores the inherent characteristics of the tough, difficult-to-digest forage of the Great Basin. Range managers are under increasing pressure not to graze these ranges because of the nature and scarcity of the herbage they produce. Raising our beef animals in feedlots and our chicken and swine in factories is much more efficient than range production. The herbage produced by Great Basin rangelands is not directly consumable by humans. When Europeans first arrived, the Great Basin environment supported an extremely sparse population of humans whose staple diet was grass seeds. Much of the concentrated rations fed to animals in feedlots, on the other hand, *can* be directly consumed by humans. In a world where a significant portion of the human population is starving for protein, the very quality of the herbage that rangelands produce makes them essential now and in the future.

Range animals must receive an adequate supply of carbohydrates, fats, protein, vitamins, and minerals in their diet; these nutrition requirements are together referred to as total digestible nutrients, or TDN. The requirements for these nutrients vary among species and according to the physiological status of individual animals. Reproduction and lactation increase the nutritional requirements of grazing animals. Likewise, the quality of forage species in terms of nutrient content and digestibility changes with their phenological stage. Cheatgrass is an excellent example of a plant whose forage production changes with these factors.

Animal preference is a very significant part of range livestock nutrition. We can determine animals' preference for different forages under specific experimental conditions, but we cannot determine how palatable the animals find these forages because we do not know how a forage tastes to the consuming animal. Comparative preference varies with the density of grazers and the nature of the forage being grazed as well. Many years ago, for instance, we noted that black-tailed jackrabbits did not graze some populations of fourwing saltbush; we called this strain "rabbit-proof." Jackrabbit populations are highly cyclical. When the populations were low to moderate, rabbit-proof lived up to its name. When population levels were high, though, the rabbit-proof fourwing saltbush was eaten down to the ground just like the preferred forms. When jackrabbit numbers are extremely high, both the rabbit-proof and the preferred forms of fourwing saltbush will have their crowns eaten below the soil surface.

The range sheep industry in the Great Basin today is a faint shadow of the huge industry that existed in the Intermountain Area during the first half of the twentieth century, although sheep are still present in biologically significant numbers. These Nevada nomads wintered on the salt deserts and spent summers in the high mountain ranges. Coming and going, the sheep bands used the sagebrush/bunchgrass ranges of the foothills in the spring and fall. The sheep were better adapted to the forage conditions of much of the Great Basin than the cattle were. They could eat the spiny shrubs of the salt desert winter ranges and had the marked advantage of being able to utilize snow as a source of water on the winter ranges. Cows must have water every day in order to utilize forage on the range. They can get only part of the daily water requirement from noncrusted snow. In plant communities where virtually any grazing favors woody species, the browsing characteristics of sheep help to balance community dynamics. The sheer magnitude of the early-twentieth-century range sheep industry as a stand renewal process left an impact that is still apparent in the ecology of Great Basin rangelands. Despite the importance of the range sheep industry in past, present, and perhaps the future of Great Basin rangelands, though, the domestic animals that characterize the present rangelands of the Great Basin are cattle and feral horses.

The most common cattle production system is a cow-and-calf operation. In this type of system, the cow gives birth to a calf that is weaned and sold the same year it is born. The calves are sent to areas with winter pasture such as the annual range of cis-montane California or the winter wheat belt of the southern Great Plains. Some calves are sent directly to feedlots to add growth before they are transferred to a finishing lot and fed a high-concentrate ration. Typically, calves are born in the spring, when the maximum digestible nutrient requirements for the cow and her nursing calf coincide with the most nutritious and digestible forage production on the ranges; unfortunately, this is not the period of most abundant herbage production. If calves are dropped in the fall, the ranch operation has to have extra feed available for the cows and calves during the winter in much greater amounts than the normal wintering ration for brood cows. This has to come from hay and concentrate supplements.

Alternative cattle production systems include keeping long yearlings or purchasing stocker cattle. With the long yearling system, the producer saves the weaner calves produced by his herd, feeds them through the winter, and pastures them on the range, and then sells them the following year. The advantage to the livestock producer is a much heavier animal that will bring a better price than a weaner calf. The downside of this equation is the amount and quality of feed necessary to feed the animal over the winter and give it the quality forage it requires the next spring and summer. In most years, degraded big sagebrush/bunchgrass ranges cannot provide sufficient forage to produce an average daily rate of gain on long yearling calves to make a profit. Ranchers at an old ranch at Halleck, northeast of Elko, Nevada, on the Humboldt River who continued to hold calves as long yearlings well after the practice lost general popularity told us that two factors made the long yearling system profitable for them. First, their grazing allotment from the BLM included extensive areas that had been seeded to crested wheatgrass; properly managed these crested wheatgrass pastures were capable of producing sufficient gain on the yearlings to make them profitable. Second, the ranch produced enough hay to overwinter the yearlings without reducing the brood cow herd.

Until the early 1950s, every ranch in northern Nevada not only held over weaner calves as long yearlings, but also held steers over a second winter and sold them as 2-year-old beefs. Prior to that, in the early 1900s, ranches marketed 5-year-old steers because consumers demanded 5-pound steaks for individual servings.[3] Sometimes these animals went into feedlots for finishing and sometimes they went directly to the slaughterhouse as grass-fat beef. The central Great Basin does not have rangelands capable of producing grass-fat beef, so 2-year-old steers spend their second year grazing on irrigated meadows. Early in the twentieth century this entire industry was threatened by the endemic disease red water.[4] The development of a red water vaccine made the continued production of 2-year-old steers possible. By the late 1950s, almost every ranch in the Great Basin had converted to cow-and-calf operations. Consumer demand for smaller cuts of meat well marbled with fat fueled the change.

Buying stocker calves has several inherent drawbacks in the Great Basin rangelands related to the environment and year-to-year variation in forage production. The stocker cattle market is flooded with animals each fall when calves are weaned. If the Great Basin rancher buys calves when they are cheapest, he must have the hay supply necessary to feed them through the winter. If the next spring is perfect for cheatgrass forage production, the rancher has abundant forage to feed the hungry animals. If there is virtually no cheatgrass forage production the next spring, the rancher is out of luck. Buying stocker cattle for the winter wheat belt areas or the annual ranges of California is still risky, but generally the chances of near-complete failure of forage production on these ranges is much less than for cheatgrass-dominated ranges in the Great Basin. Stocker cattle ranchers can hedge their bet by holding off purchase until weather patterns in the late fall give a better estimate of potential forage production. Great Basin ranchers, however, cannot stall all winter until they know if the following spring and summer are going to have good forage production.

Hay production from native irrigated meadows provides the bulk of the forage for wintering brood cows. This production stayed constant in Nevada for much of the twentieth century at not quite 1 ton per acre. After the 2-plus-year-old steers and then the long yearling production systems disappeared, ranchers made up for the reduced revenue by increasing the number of brood cows. Cows with calves do not graze the same way that steers do, but this subtle but massive change in stand renewal was seldom if ever considered in evaluating the ranges of the Great Basin during the twentieth century. Cows have to be bred in early summer to have calves in March and early April. Good livestock managers leave the bulls with the cows for two estrous cycles. This ensures that the calf crop is very close in age and that the weaner calves are a uniform product. Smart managers also pregnancy-test the cows in the fall. Along with absolute detection and elimination of reproductive diseases such as brucellosis, this level of management ensures a high-percentage calf crop and subsequent high-percentage weaner calf crop, which is the product—along with culled cows—that brings in cash for the operation. Exceptional livestock operators may have 95 percent calf crops. At the other end of the scale are operations that are lucky to have one-third of the cows drop a calf every year. If bulls are left with the cows for the entire year, especially on extensive ranges with diverse topography, some cows will drop calves at off seasons, even in the dead of winter. Winter calves born on Great Basin ranges occasionally survive, but their frost-damaged tails and ears mark them for life.

The need to assist animals having difficulty calving is another reason to time calving for a single season. It is especially important to assist first-calf heifers during the calving period. About 20 percent of the cow herd is culled annually either deliberately or through death losses, so a significant portion of the female calves in each year's calf crop have to be selected as replacement animals. The nutritional requirements of replacement heifers are different from those of aged cows. Good management practices allow the replacement heifers to be bred as 2-year-olds rather than as long yearlings. All of these constraints funnel cow-and-calf production toward calving in the early spring.

Every range-based ranching operation has a different infrastructure in terms of forage resources. This makes describing the forage cycle for a typical ranch very difficult. We will provide examples of some of the variables involved. The carrying capacity, or the number of grazing animals a given unit of rangeland can support, is calculated in terms of animal unit months (AUM). The AUM is based on the amount of air-dry forage required to support a 1,000-pound cow for a month. This is generally about 20 pounds per day, or 600 pounds per AUM, but it may be as high as 1,000 pounds.[5] The variation is due in part to wide variation in the weight and therefore nutritional requirements of a mature cow and in part to interpretation. Not all of the standing crop of forage is harvestable; the common rule is to take half and leave half so the perennial plants can maintain themselves. Readily available tables provide the usable amount of a standing forage crop for a given forage species so ranchers can estimate the number of acres of rangeland necessary to support an AUM.[6] A common misconception is that only scrub cows that mature at 600 pounds can survive on truly arid ranges in the Great Basin. In fact, the unhealthy appearance of many cows on such ranges is often a product of poor management rather than the environment. Cows bred as yearlings and kept on a starvation diet mature at a light weight. Proper livestock and range management can produce high-quality cattle. Cull cows bring in a significant portion of a ranch's annual income, and there is a marked difference in profit between the sale of fifty cows averaging 600 pounds each and fifty weighing 1,000 pounds each.

We start our description of the annual forage cycle with the dropping of calves in March. At that time, the brood cows are in fenced native hay meadow fields at the home ranch. If the meadows become exceptionally wet as the ground thaws and the ranch has the infrastructure, the cows are moved to a foothills site off the valley floodplain. Ideal sites have a pinyon-juniper woodland cover. In northeastern California and adjacent Oregon, the ranchers prize western juniper woodlands adjacent to the home ranch because of the thermal cover the woodlands provide for wintering cattle. Ranchers in northwestern Nevada do not have that option because they are northwest of the pinyon-juniper zone and east of the western juniper woodlands. At calving time, the cows are consuming hay produced the previous summer. The quality of hay produced on native hay meadows is marginal at best. In midwinter, when the cows need a full rumen to be warmed by the microbial breakdown of coarse herbage, the poor-quality hay from the meadows is a plus. At calving time, the cows crave green forage with a high carotene and protein content to meet the demands of lactation. Severe labor and capital constraints converge on the rancher at calving time. Ranchers need to get the cows off the hay meadows so the winter collection of cow chips can be broken up and distributed by harrowing or brushing; in northeastern California the cowboys call this "slickening the meadows." This job needs to be done before the first irrigation, which is often timed by the arrival of the spring runoff crest from melting snow in mountain watersheds. The rancher has been feeding hay since the previous fall. It has been a seven-day-a-week, rain, blizzard, 30-degrees-below-zero job, and ranchers and cowboys are sick of it. Depending on the previous year's hay crop, the quantity and quality of the hay left in the stacks may be declining at an alarming rate. Hay left in the stack represents money in the bank, a hedge against a poor production year in the current growing season. The labor-starved ranchers are caught between increasing demands for hay production activities such as cleaning ditches, repairing sprinklers, and maintaining power equipment and the need to move cattle and check and repair fences, water troughs, and springs. It all comes down to one thing: the cows and calves must be turned out on the range. The cows are more than willing to go, but is the range ready for them? In the days before extensive fencing, the cows themselves chose the turnout date. One morning the rancher would go out to feed and there would be no cows on the feed ground. Apparently, the cows could sense when there was enough forage to support grazing and left on their own accord.[7]

"Spring" is a relative term in the West. During a discussion between papers at the first cheatgrass symposium in 1965, a discussant pointed out that the participants had spent several days arguing that cheatgrass had greatly increased during the past decade (1950s) and that the spring turnout date was steadily being pushed back. The group was about equally divided on the merits of these two being cause and effect. Another participant suggested that lack of agreement on when "early spring grazing" occurred was a major problem. Some speakers had described "early spring grazing" as beginning in early March, and others as beginning in late May.[8] This was all for nothing. After a week of papers and general discussion concerning the cheatgrass problem, the Idaho state director for the BLM, William Mathews, concluded the discussion with the statement, "What we need is less and later grazing, no seeding, and no burning"—exactly the opposite consensus of the entire audience. Unfortunately, this type of leadership persists in many circles today.

Latitude, elevation, and tradition contribute to the wide variation in what is meant by "early spring grazing." During February in the big sagebrush/bunchgrass zone of western Nevada, cheatgrass plants growing at elevations of 4,500–6,000 feet are rosettes of leaves flat on the ground if the plants germinated in the fall or one- or two-leafed seedlings barely emerged from the seedbed if germination was delayed to the spring. At this same time, the highly ephemeral, short-lived native perennial grasses Sandberg bluegrass and squirreltail have new upright leaves available for grazing. Growth begins later on north-facing slopes than on south-facing slopes. The amount, periodicity, and nature of spring precipitation interact with the phenology of cheatgrass. In exceptionally dry years, the current year's crop of seedlings may be dead by March. During springs with substantial snowfall into April, phenology is delayed. Among all the extremes, cheatgrass comes across as a virtually nonexistent forage species in late February and early March in the environment we described.

In the salt deserts of the lower elevations, the phenology of cheatgrass in the early spring is considerably accelerated. Except for areas of sand-textured soils where Indian ricegrass and needle-and-thread grass occur, perennial grasses were never a major forage source on these ranges. The dominant native perennial grasses are squirreltail and Sandberg bluegrass. The invasion of cheatgrass into these environments greatly increased herbaceous forage production, but in dry years cheatgrass forage production may be nonexistent. Cattle wintering on Indian ricegrass on these salt desert ranges have difficulty obtaining enough digestible protein. In the fall and winter, they lick the fruits of chenopod shrubs from the soil surface and graze Indian ricegrass seed heads, which retain one form of seed that is very high in protein. Most of the shrubs in the salt desert are deciduous, so leaves are not available for browsing during the winter. In any case, the spinescent nature of many chenopod shrubs does not encourage browsing. The important exceptions are the half shrubs winterfat, Nuttall saltbush, and red molly. In the early spring, patches of green cheatgrass with erect leaves are an extremely important source of protein and carotene. During April, cheatgrass is an important forage source on many formerly big sagebrush/bunchgrass ranges, although it is often not abundant. Burgess L. Kay described the cows on sagebrush ranges in Modoc County, California, as hunting at a fast pace in order to consume enough cheatgrass to provide a diet on which they could survive during April. Platt and Jackman credited D. E. Richards of Prairie City, Oregon, with coining the description of cheatgrass in the early spring as being only "half-a-bite high" (fig. 14.1)[9]

Fig. 14.1. Half-a-bite cheatgrass forage in the early spring. The problem with turning cattle out to graze cheatgrass range in the early spring is the lack of forage growth. Remnant bunchgrass plants usually have greater leaf growth in the early spring than cheatgrass and are therefore disproportionately utilized.

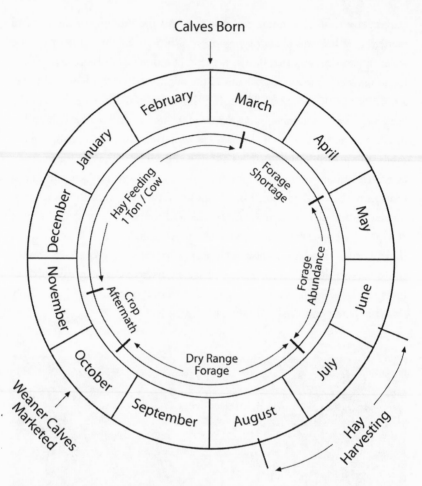

Fig. 14.2. Forage wheel showing forage sources, availability, and utilization during a yearly cycle on a typical Great Basin range livestock operation.

In their pioneer study on grazing cheatgrass, C. E. Fleming and associates started the cheatgrass grazing period on May 1.[10] Obviously, Fleming thought that cheatgrass grazing should start when the maximum growth of the annual grass was occurring. The ranges where Fleming was conducting grazing studies contained remnant stands of bluebunch wheatgrass and Idaho fescue. Holding off grazing until May 1 gave these perennial grasses at least a start toward restoring their carbohydrate reserves. Grazing them before April 15 will cause the perennials to disappear (for a scholarly discussion of the plant physiology aspects of this problem, see "Coping with Herbivory" by Caldwell et al.) (fig. 14.2).[11]

There are thus about 6 weeks in the spring when the demand for green forage is the highest of the entire year but forage alternatives are virtually nonexistent; if such forage does exist, it has the highest cost per AUM of the year. Very few ranchers were, or are, so self-defeating as to excessively graze Great Basin rangelands year after year in the early spring without giving the native perennial grasses a chance to recover. Scientific range management has tried to stop the practice for 80 years, but it still occurs. Take a ride around the central and northern Great Basin the first week in March on any year and observe how many cows are turned out on degraded formerly big sagebrush/bunchgrass rangelands "hunting cheatgrass for a living." Why? Virtually all federally owned big sagebrush/bunchgrass potential rangelands in the Great Basin are under some form of rest-rotation grazing management. With a three-pasture system, for example, a pasture is grazed in the early spring only once every 3 years. On other rangelands, however, livestock graze cheatgrass on degraded big sagebrush/bunchgrass ranges in the early spring because there is no alternative forage source. In fact, a clear alternative is available: introduced perennial crested wheatgrass (fig. 14.3).

Fig. 14.3. The introduced perennial grass crested wheatgrass provides an alternative forage during spring on degraded big sagebrush ranges and also biologically suppresses cheatgrass and decreases fuel loads that spread wildfires.

The history of crested wheatgrass introduction in North America and the Great Basin is discussed in chapter 6. The paper entitled "Some Growth Responses of Crested Wheatgrass Following Herbage Removal" by C. W. Cook and L. A. Stoddart contains valuable information on grazing crested wheatgrass in the early spring in the foothills of Utah. *Lambing on Crested Wheatgrass* by J. R. Gray and H. W. Springfield is also a good place to start.[12] Donald W. Hedrick published a case history of an actual crested wheatgrass seeding in eastern Oregon that became a source of early spring forage.[13] In the early 1950s, the Otley Brothers' ranch at Diamond in southeastern Oregon, in cooperation with the BLM, seeded about 1,900 acres of degraded formerly big sagebrush/bunchgrass rangelands to crested wheatgrass to supply early spring forage. Starting in 1958, cows and newly dropped calves were turned out on the crested wheatgrass from about mid-April until the end of May. By 1967 the ranch averaged 800 AUMs of grazing use at a stocking rate of 2.5 acres per AUM. Use of the crested wheatgrass in the early spring was credited with raising the formerly good calving percentage of 85 percent to an excellent 95 percent and with adding 40 pounds to the average calf weaning weight. One of the keys to the success of this operation was removing the cattle from the seeded pastures while there was still sufficient soil moisture for regrowth of the perennial grass. This dry herbage was available the next spring to be grazed along with new green crested wheatgrass leaves; if the spring was cold and the phenology of crested wheatgrass slowed down, the herbage provided an emergency ration until growth renewed on the perennial grass.

Why is it possible to harvest considerable forage from crested wheatgrass in the early spring year after year and not injure the stand while in the same environment such grazing practices will destroy native perennial grasses? Marten Caldwell and his many students have delved into the comparative morphology and physiology of crested and bluebunch wheatgrasses to answer this question. There is no single factor that makes crested wheatgrass resistant to herbivory and bluebunch wheatgrass susceptible. We suggested many years ago that the native perennial grasses of the Great Basin were perhaps more susceptible to early spring grazing because they had evolved in the absence of large herbivores.[14]

Given that an AUM of forage is most valuable in the early spring for Great Basin range livestock operations, what innovative options are available to alleviate the problems? Chapter 13 reviews the efforts of Burgess L. Kay to accelerate cheatgrass range readiness for grazing by fertilizing it with nitrogen.[15] Don Hyder and F. A. Sneva suggested the same thing at the Squaw Butte Experiment Station in Oregon.[16] Field studies and economic analyses of nitrogen fertilization of crested wheatgrass pastures in Utah revealed that in certain years, application of 25–30 pounds per acre of nitrogen as ammonium nitrate hastened spring range readiness by 11–13 days and was cheaper than buying hay for the same period.[17] Cheatgrass was not a factor at the time at the locations where these studies were conducted. Chapter 13 also notes that nitrogen fertilization had a negative effect on perennial grass stands if cheatgrass was present.

How important is eliminating the annual early spring forage shortage on Great Basin rangelands? The Vale project in southeastern Oregon was a landscape-scale attempt to restore degraded big sagebrush/bunchgrass rangelands through the extensive seeding of crested wheatgrass as well as water development and extensive fencing. An unbiased, science-based evaluation of the results of the Vale project emphasized that the seeding resulted in a tremendous improvement in the ecological condition of the unseeded portions of the native range, which constituted the majority of the landscape.[18] Much of this improvement may have been attributable to the availability of crested wheatgrass, which delayed turnout on degraded cheatgrass ranges. At the same time, we cannot dismiss the observations of numerous knowledgeable and very experienced observers who insisted at the first cheatgrass symposium that moving back the turnout date on degraded big sagebrush/bunchgrass ranges resulted in a tremendous increase in cheatgrass during the two decades immediately following World War II. Far too many qualified observers agreed with the observation to dismiss it out of hand. The fact that the tremendous spread and increase in dominance of cheatgrass corresponded with the period when the range turnout date was systematically being set back might have been a coincidence. We have frequently observed that application of Hormay rest-rotation grazing system results in nearly pure dominance by cheatgrass in low-elevation sites where few or no native perennial grasses exist. Under the three-pasture system, a pasture is grazed in the early spring only once every 3 years.

It has been suggested that the brutal level of range utilization from 1900 through the 1930s in Nevada suppressed cheatgrass. The pertinent question is, are we missing some important aspect of cheatgrass grazing management by dismissing the observations that cheatgrass increased as the range turnout date was set back? Unfortunately, the seeding of crested wheatgrass on public rangelands has virtually ceased because of the pressure from environmental organizations interested in improving the ecological condition of degraded rangelands. After all, why not seed native perennial grasses and accomplish the same thing as planting the introduced crested wheatgrass? As we discussed earlier, native perennial grasses lack seedling vigor and are susceptible to damage if grazed early in the spring. That is, even if the native perennial grasses could be established, they would lack the inherent potential to cope with herbivory in the early spring.[19]

Leaving annual grass herbage in the field is a standard practice in the cis-montane annual ranges in California.[20] It is hard for cattle to get enough bulk in their diet when they are grazing lush, rapidly growing, high-moisture-content annual grass seedlings. Mixing cured dry forage and new grass has dietary advantages. Ranchers can ensure some forage availability in the early spring on cheatgrass ranges when the combination of the amount and periodicity of precipitation and cold temperatures restricts cheatgrass growth.[21] This dry herbage acts as an insurance policy against future bad growing conditions, but it comes at the terrible risk of increased probability of wildfires. Platt and Jackman suggested that ranchers who depend on dry cheatgrass for future forage should keep an "agonizing clutch on a rabbit's foot."[22]

If you ask residents of the western Great Basin their favorite season of the year, the answer is often the long, glorious falls or perhaps the winter if the individual is a fan of winter sports; spring seldom receives a vote. Spring starts in February and makes an abrupt transition to summer after Memorial Day. The occasional beautiful, balmy spring days are very widely spaced among cold, windy, snowy, sometimes dusty days. Many a blooming daffodil bows its head under a cold skiff of icy snow. Through all this the cows are out there on the range chasing cheatgrass for half a bite of forage.

The period from May 15 to June 1—again depending on latitude, elevation, and the year—is milk-and-honey time for the cows and their calves on the range. Cheatgrass growth has accelerated, and it terminates with the development of flower stalks. The calves are growing rapidly and beginning to graze for themselves. In rare years, precipitation extends into June and the herbage production exceeds the amount required to support grazing animals (fig. 14.4). Platt and Jackman's fifty-page bulletin on cheatgrass in Oregon includes a sidebar titled, "You can have your cheat and eat it too."[23] The authors meant that if cheatgrass is grazed early during the rapid growth period and soil moisture is still available afterward, cheatgrass will respond with additional growth from tillers, resulting in much greater forage production than if the stand was not grazed. Platt and Jackman credited this finding to the research of C. E. Fleming in Nevada.

The most comprehensive grazing study on cheatgrass-dominated range was a cooperative project at Saylor Creek in southern Idaho involving the BLM and the Forest Service Intermountain Forest and Range Experiment Station. James O. Klemmedson and Robert B. Murray were the principal scientists, although several others participated in portions of the long-term study. A preliminary report of the project was given at the first cheatgrass symposium.[24] In one of the later papers published on the Saylor Creek experiments, Bob Murray and his associates discussed the potential vegetation of the site and provided an excellent phenological comparison of the major forage species in the pastures.[25] The experimental site was located at an elevation of 3,140 feet—that is, it was at a higher latitude but lower elevation than much of the Great Basin—and had an average annual precipitation of about 8.5 inches. Remnant vegetation indicated that a basin big sagebrush/Thurber's needlegrass community originally dominated the site. Crested wheatgrass (origin not specified) and Sandberg bluegrass both began growing in mid-February. Cheatgrass, squirreltail, and needle-and-thread grass began growth in mid-March. The authors mentioned fall germination of cheatgrass, so the early growth in the spring must refer to upright growth of the leaves from the overwintering rosette. The native perennials stream bank wheatgrass and basin wild ryegrass were the last species to initiate growth in the spring. Cheatgrass and Sandberg bluegrass were the first to initiate growth of flower stalks, in late April, and anthesis occurred about mid-May. Sandberg bluegrass was the first species to dry; cheatgrass was second, with maturity occurring from mid-June to early July. Crested and stream bank wheatgrass and needle-and-thread grass matured from mid-July to mid-August. The latest-maturing cheatgrass was flowering at about the same time as the earliest-maturing of the long-lived perennial grasses. This means that

Fig. 14.4. In some years, cheatgrass herbage production greatly exceeds the amount livestock can utilize. An abundant cheatgrass year in Emigrant Pass, Nevada, ended in a cheatgrass-fueled firestorm in 1964.

in dry years, cheatgrass would mature about a month earlier than the perennial grasses, and in wet years would mature earlier but might overlap with the earliest of the perennials to mature. Perhaps the most striking information in this phenology report for the Saylor Creek range is the indication that germination of cheatgrass and regrowth of the perennial grasses could be initiated by fall rains as early as September 1. Fall germination of cheatgrass in west-central Nevada happens only once about every 5 years, and then it happens in late fall.

The protein content of cheatgrass during the early part of the maximum growth period is usually above 10 percent.[26] Scientists began comparing the nutritive content of cheatgrass with that of other native or introduced plants in Montana during the 1930s.[27] Fleming and his early associates conducted proximate analysis and digestion trials on range forages a couple of decades before that, but their papers do not specifically mention cheatgrass.[28] Fleming later collected data on protein, ash, fat, fiber, and nitrogen-free extract for cheatgrass and native bunchgrasses throughout the grazing season.[29] The analysis of cheatgrass forage by Murray and associates reflects the improvement of analytical techniques in the 40 years following the initial work in Montana.[30] All of the nutritive value analyses report the same conclusion: the herbage of immature cheatgrass provides an adequate diet for grazing cattle, sheep, and horses. After reaching maturity, both cheatgrass and the native perennial grasses become deficient in protein. Ranchers have long known that even calves grazed on high-condition, big sagebrush/bluebunch wheatgrass communities gain less weight per day after the forage matures. In fact, the calves may actually start to lose weight. This is bad for a cow-and-calf operation, whose major profit comes from selling calves; the salable product remains steady or decreases in weight during the weaning and marketing period in mid-August to mid-October while the expense of maintaining the animals continues.

The grazing experiments at Saylor Creek used yearling steers. The highest average gain per day came in a year when cheatgrass production was not the greatest, but the combination of thin cheatgrass stands with a lot of bare openings and late rain resulted in exceptional summer growth of Russian thistle.[31] Given no choice, livestock instinctively graze species they do not highly prefer in order to balance their diet.

It is extremely important to distinguish between cheatgrass ranges that still have remnant native perennial grasses and shrubs such as antelope bitterbrush and ranges so badly degraded that there are no perennial grasses and repeated fires have eliminated all of the native shrubs. Yes, cows will still graze the dead dry cheatgrass after it matures if there are no perennial grasses, forbs, or browse species available to provide digestible protein. Yes, the dry herbage provides a diet deficient in protein and a protein supplement will be required to maintain optimum animal performance. Yes, cattle will use Russian thistle, barbwire Russian thistle, and perhaps annual kochia as protein sources when they are available on mature cheatgrass ranges. Oregon rangeland sages Platt and Jackman said, "Diet rule for livestock eat and get thin on dry grass."[32] It is important that they said "dry grass" rather than "dry cheatgrass." Once the native perennial grasses dry, they are just as deficient in protein as cheatgrass.

Cheatgrass maturity signals seed maturity. The fruits (caryopses) of cheatgrass—its "seeds"—are vicious. The end of the callus is needle sharp, and the flexible awn "works," or moves, when moistened. Platt and Jackman noted that "ripe cheat sends livestock to [an] eye, ear, nose and throat specialist."[33] Aven Nelson, the noted Rocky Mountain botanist, recognized the danger of grazing cheatgrass at seed ripe as far back as 1901.[34] The danger from the seeds is not restricted to grazing animals. Dogs with long hair and long ears such as cocker spaniels get cheatgrass seeds imbedded in their fur, ears, and feet. Serious complications can result if humans ingest or inhale the seeds. The ingestion of cheatgrass seeds by cattle can lead to lumpy jaw, a chronic infection of the head characterized by large swellings on the jaw. Cowboys doctor lumpy jaw by roping the animal, lancing the swelling with a pocket-knife, and dabbing the wound with a little KRS, a coal tar derivative that discourages flies. (Always use your own pocketknife if a cowboy invites you to share his lunch.) Eye, ear, and nose infections and physical injury from cheatgrass seeds take a toll in labor, medicine, and livestock performance, but they do not preclude grazing cheatgrass when it is mature.

In the big sagebrush/bunchgrass zone, the cheatgrass seeds disperse rather rapidly from the seed heads and the problem largely passes, leaving the cheatgrass safe to graze. The form of cheatgrass that invaded salt desert environments does not drop its seeds so readily. In large grazing units that vary in elevation, soils, and aspect, not all of the cheatgrass matures at the same time. Cows will avoid cheatgrass patches that are in full seed ripe and return to graze the forage after the seeds have dispersed. Cattle that spend the summers on high-elevation summer ranges do not have the problem at all. The discovery of high mountain ranges, even with the relatively short use period of July 15–October 15, revolutionized the range livestock industry in the far western United States during the nineteenth century.[35] These mountain ranges still exert an importance disproportionate to their acreage on the western range. The Forest Service's increasing reluctance to allow grazing on national forest land, where most of the high-elevation summer ranges are located, bodes ill for the livestock industry. Not every range livestock enterprise in the Great Basin has an infrastructure that includes a high mountain summer range. Some managers who do not have this asset have been very successful on cheatgrass ranges.

During the period when cheatgrass and the native perennial grasses are reaching their maximum annual herbage production and maturing, most ranchers are extremely busy making hay. Periodic severe winters, the extreme being the winter of 1889–1890, drove home the point during the formative years of range livestock production in the Great Basin: forage conserved as hay is necessary to get cows through the cold midwinter months.[36] The rule of thumb is 1 ton per brood cow being wintered. If the feeding period is November–March, hay constitutes 40 percent of the annual forage requirement. The equation is simple: if you want to winter one thousand cows, you need 1,000 tons of native hay, which requires 1,000-plus acres of hay meadow. Native hay meadows are no small-scale portion of the Great Basin agronomic landscape. In northern Nevada, there is no such thing as hay production from rain-fed fields. Hay has to be irrigated, and irrigated lands for native hay production constitute about 5 percent of the total landscape. Hay production for wintering range cows in the Great Basin has been described as a war in which ranchers spend half the year growing and putting the darn stuff in stacks and the other half of the year freezing to death while feeding it to bawling, ungrateful cows.[37] Capital and labor constraints make ranchers susceptible to losing this war. Ranchers attempt to solve the labor shortage by substituting increasingly technologically advanced and expensive mechanical equipment for labor. This is the Achilles' heel of ranching in the Great Basin. The native hay meadows provide a very significant portion of the total yearly forage supply for the range livestock industry in the Great Basin. In addition to the actual hay that is produced and conserved for winter feeding, the crop aftermath provides considerable forage in the form of regrowth after the hay crop is harvested and the many nooks and crannies in the hay meadows that are isolated by topography, sloughs, or willow growth and cannot be mowed for hay. One ton of hay for each brood cow is enough for two AUMs, but the feeding season runs from late November through March; thus, roughly two AUMs must come from crop aftermath for each brood cow.

Grazing the hay meadows fills the gap in the fall between gathering the cattle from the rangeland and feeding them hay. Their manure is recycled onto the meadows and provides nutrients for the subsequent year's hay production. The hay meadow ecosystem is a complex and poorly understood environment. The native hay meadows along the Humboldt River and its tributaries appear to be inefficient agricultural ecosystems, for example, but they are efficient filters that may improve water quality. The tailwater that returns to the river following irrigation of the meadows is often of better quality than it was when first diverted. The rhizomatous grasses and sedges that dominate native hay meadows form closed communities where it is difficult, but not impossible, to insert more desirable forage species. Alfalfa is not an option because the floodplain soils are not usually suitable for it. Nitrogen is usually the element that limits forage production. Nitrogen fertilization can often increase hay production, but careful economic analysis should determine its feasibility.

From late fall until early spring, livestock are fed hay every day until they are turned back out on the range. This completes the forage cycle for the year. New calves are dropped, and the range livestock production system begins again. Range managers assess the forage available during various periods of the annual forage cycle in widely diverse ways. Virtually every manager knows how many bales of hay are in the stack at the end of the haying season and can translate this information into the number of cows it will feed. Very few ranchers can assess range forage production, determine what proportion of the herbage is harvestable, and translate this into carrying capacity. Most assess range carrying capacity in one of two ways. They either rely on experience or depend on a public land management agent to tell them the stocking rate and timing of grazing. Ranchers have learned to value experience. If a certain grazing unit supported one hundred cow-and-calf pairs for 3 months last year, they think that is how it should be stocked this year. The problem with that approach is the great variability in forage production, especially if cheatgrass is a significant portion of the forage base. It is not possible to harvest the same number of AUMs of forage from cheatgrass range every year unless the range is grossly understocked, and even then there will be years when forage will be insufficient to support the grazing animals. If the forage base is primarily cheatgrass, ranchers need an alternative source of food to get cattle through the bad forage-production years. The alternatives are few. Supplemental forage can be purchased in the form of hay or irrigated pasture, or the cattle herd can be sold down to a level that the range will support. Both actions have drastic economic ramifications for the livestock production enterprise. Brilliant young economists such as Alan Torrell at New Mexico State University and Neal Rimby at the University of Idaho have come to the rescue, creating models that illustrate for ranchers the implications of such alternatives in livestock management. It behooves ranchers whose livelihood depends on grazing cheatgrass to learn about and understand such economic models.

Ranchers who depend solely on public land management agencies to determine the range's carrying capacity, ecological condition, and trend in ecological condition also endanger their operations. Public rangeland managers have consistently lacked a science-based, statistically precise sampling system to assess condition and trend on rangelands. Inadequate funding and lack of trained personnel have hindered efforts to develop one. If ranchers are to be held accountable for the ecological interactions of grazing animals with the environment, then they must empower themselves with the means to assess the nature, magnitude, and significance of such actions.[38] Many ranchers have turned to professional rangeland consultants to obtain range condition and trend information. If the rancher invests scarce capital merely to fight environmentalists and public land managers, he is wasting his resources. If he builds a database and uses the information to improve his range livestock production enterprise, he enhances both ecological and economic efficacy.

There is an inherent danger in the annual cycle of calves being born, cheatgrass turning green on the hillsides, a summer of putting up hay, and a long winter of feeding cows. Range livestock producers, who work under conditions of unmitigated stress, become mesmerized by the cycle and fail to consider new ideas. This seems to happen more with each successive generation. Grandfather did it this way, so that is the only way to do it. It is true that Gramps sold 2-year-old steers and now we sell weaner calves, but the cows still calve in the spring and we still work from dawn to dusk making hay. Breaking the cycle with innovative changes in management is close to heresy.

What changes in management are appropriate? Let us turn the question around and list the ecological and economic constraints on current range livestock operations. Ranchers are generally starved for capital and restricted by the availability and cost of labor. Besides servicing the debt on deeded land and the livestock, the major capital investment is for equipment to produce, harvest, store, and feed hay. The major labor constraints also center on hay production and feeding. From an animal nutrition standpoint, the critical lack of forage in the early spring and the poor quality of forage once the grasses are mature are major limiting factors. Management innovations depend on the physical and biological infrastructure of the particular ranching operation. When an old family ranch is sold and the purchaser is a corporate entity, large capital improvements such as new wells for irrigation, land leveling, new ground brought into production, and a big rock-and-pole arch over the front driveway to display the brand follow. These types of improvement may be justified in certain situations, but often they merely exacerbate the basic problems.

Keeping in mind that all Great Basin ranches differ in infrastructure, we will use a specific ranch to illustrate the type of management changes that can be made to accommodate the utilization of ranges where cheatgrass is a major forage species. The headquarters of the T Quarter Circle Ranch is located just west of Winnemucca, Nevada, on the meandering Humboldt River. Nancy and Frosty Tipton are its managers. The deeded land consists of native hay meadows along the Humboldt River and developed alfalfa hay fields on land that extends into Grass Valley south of Winnemucca. The T Quarter Circle Ranch has been in the same family since the 1860s.[39] The rangelands on which it has grazing privileges are extensive. The southern portion of the range is in the Central Pacific Railroad Grant, where grant sections alternate with National Resource sections administered by the BLM in a checkerboard pattern. Extending north from the Humboldt River the range encompasses portions of the Eugene Mountains, Blue Mountain, Krum Hills, Slumbering Hills (including Winnemucca Mountain), and the Bloody Run Mountains. North of the Krum Hills and between the Slumbering Hills on the northwest and the Bloody Run Mountains on the northeast, the T Quarter Circle range includes vast expanses of Silver State Valley, an area unique even by Nevada standards. A vast field of active, towering sand dunes marches up and over the Slumbering Hills from Desert Valley and flows across the valley and over the Bloody Run Mountains at Sand Pass. Portions of the sand fields are barren, wind-driven sand with tall dunes standing at the angle of repose. Other portions support excellent stands of Indian ricegrass and needle-and-thread grass. The alluvial fans spilling down from the mountains originally supported big sagebrush/Thurber's needlegrass communities, and the north-facing slopes on the higher ridges of the Bloody Run Mountains and deep canyons in the Slumbering Hills once supported bluebunch wheatgrass. The Eugene Mountains and Blue Mountain are the last extension of the single leaf pinyon–Utah juniper woodlands in the northwestern Great Basin. The margins of the rangeland touch railroads, highways, former military installations, and the town of Winnemucca. All of these human influences translate to wildfire ignition opportunities. Cheatgrass is the most frequent forage species occurring on the T Quarter Circle rangelands, and wildfires have been frequent. The BLM has seeded some of the burned areas in the Bloody Run Hills—as well as other areas within the district—to crested wheatgrass.

The Tiptons looked at the various plant communities and concluded that their land had great potential as winter and spring ranges but lacked high-elevation summer range. Unless they could purchase a BLM or Forest Service summer allotment, the historic range for the ranch had definite constraints for livestock production. Frosty Tipton is unique among ranchers, range managers, and scientists in his observational abilities. Bookshelves filled with an impressive collection of historical, botanical, geological, and ecological volumes pertaining to the environment of the Great Basin line the walls of the family room in the historic native-stone ranch house. Frosty has authored and coauthored scientific publications.[40] He was among the first to observe that cheatgrass was suddenly extending into salt desert environments. He was also among the first to grasp the fact that cheatgrass was enhancing the forage resources of truly arid sites, and as such constituted a potential new grazing resource.

The Tiptons had noted a steady increase in free-roaming horses on their rangeland, and they had detailed records proving that no free-roaming horses were present when Congress passed the Wild Free-Roaming Horses and Burros Act of 1971. The family asked the BLM to remove the illegal horses. The BLM agreed that the horses did not fall under the protection of the federal legislation but indicated that no funds were available to accomplish the roundup. The Tiptons went to federal court and obtained a court order forcing the BLM to remove the horses. When the roundup was completed, an astounding two thousand horses had been removed. These horses had been on the range year-round and represented more AUMs than the ranch was licensed to graze.

The Tiptons' next step in modernizing the management of the ranch was to go to a reputable consulting firm and commission a science-based, statistically precise sampling to assess the ecological condition and trend of the rangeland communities on their grazing allotment. The next management decision was the most radical. The Tiptons brought the cows in from the range in the summer when the cheatgrass and perennial grasses were mature and grazed the hay meadows instead of cutting hay. They made up for the drop in hay production by wintering the dry cows on the desert ranges. This was a heretical move. Every rancher's child has heard how the winter of 1889–1890 killed 90 percent of the cows on the range in northern Nevada. Certainly a winter equal to or worse than the winter of 1889–1890 is bound to come, but when? It did not come during the twentieth century. Do you manage the range for an event that has a return interval greater than 100 years, or do you manage for the "normal" year, which characteristically is drier than average?

Turning dry cows back on the range in the fall to graze mature dry cheatgrass is about as heretical as grazing hay meadows in the summer instead of cutting hay. Everyone knows that dry cheatgrass is deficient in protein. Frosty Tipton also knew that cows wintering on Indian ricegrass would travel far from water sources to graze dry seed heads of the one phenotype of cheatgrass whose seeds do not dehisce at maturity.[41] In fact, Frosty was the first to observe that the form of cheatgrass that grows in salt desert environments does not immediately drop its seeds, as the type found in the sagebrush–bunchgrass zone does. Cows turned out in the fall readily utilize the high protein seeds by licking them from the dry seed heads before consuming the cheatgrass herbage itself.

Grazing the native hay meadows along the Humboldt River in the summer has the additional advantage of biologically suppressing the invasive perennial weed tall whitetop.[42] We noted earlier that it is hard to introduce improved forage species into the hay meadows because the grasses and sedges that dominate them are sod-forming rhizomatous species that exclude other plants. The creeping-rooted tall whitetop is perfectly adapted to spread through and rapidly dominate the grasses and sedges if the meadow is cut for hay. Tall whitetop rapidly regrows aerial stems after the haying and spreads at the expense of the grasses and sedges. Tall whitetop dominance may actually increase hay production on the meadows because of its rank, semiwoody stem growth, but it greatly reduces the nutritional quality of what is already poor-quality hay. Cattle will not utilize tall whitetop in pure stands, but they will graze it at the stage when it is rapidly growing and before the flower stalks elongate. This grazing helps tip the competitive balance enough to give the grasses a chance to compete. This is another Frosty Tipton observation that has proven correct through experimentation.

The most important aspect of grazing management on cheatgrass ranges is not rainfall or temperature; it is observant range and livestock managers willing and able to react to an always-varying forage resource. Innovative management of the forage pulses of cheatgrass is going to be much more successful than managing cheatgrass-dominated ranges as perennial grass ranges, because cows do eat cheatgrass.

CHAPTER 15

Cheatgrass and Wildlife

Father and son head out for a hunting trip to central Nevada in early October. As they travel the road to their destination, the father recalls memories of making this very trip with his father 25 years earlier. The father describes the beauty of the land, the color of the trees, and the abundance of deer and sage grouse. The wide-eyed son asks his father endless questions, just as his father had asked the boy's grandfather. At first light the next morning, the two walk out for a day of sage grouse hunting. As they walk, the father notices how different the landscape appears. The land must have burned in recent years. Woody species such as sagebrush are gone except for a few islands here and there. The sage grouse and mule deer are not nearly as abundant as he remembered. As they rest on a knoll and look out over the land, the son asks his father where the sage grouse and mule deer are. The father looks at his son, picks up a handful of grass, and tells him, "This is cheatgrass son, it is changing many wildlife habitats in the west, and the plants and animals as well."

Joel Peterson, retired wildlife biologist with the Montana Fish and Game Department, wrote in 1971:

A race of people once lived on an isolated island somewhere in the Pacific Ocean. The land was extremely fertile, providing the essentials of life. Their mainstay was cattle and grain crops for food, while trees provided wood to make shelter. One day the island began to sink in the ocean, causing an onslaught of water which covered the trees, grass, and croplands. The island's people were forced to move to the high and barren mountains. Here there was little or no food or shelter, and no grass for their cattle. Eventually they perished.[1]

Peterson noted that this scenario occurs far too often. Land management practices, or their lack of, as well as exotic species have made modern range and wildlife management a complex field.

The invasion of cheatgrass has indeed contributed significantly to the transformation of millions of acres of wildlife habitats throughout the West. Cheatgrass outcompetes native plant seedlings for moisture, and the mature plants provide a fine-textured, early-maturing fuel that feeds wild-fires. Resource managers do not agree on vegetation characteristics of the western landscape before European settlement, but most authorities believe that herbaceous vegetation was a dominant part of the landscape (fig. 15.1). The domestic livestock that grazed the ranges year-round, year after year, in enormous numbers depleted the native perennial grasses.[2] Extensive archival research on the journals written by early explorers of the Great Basin and on old newspaper articles convinced Bob McQuivey, former chief of habitat of the Nevada Department of Wildlife, that Nevada had once been an ocean of sagebrush (fig. 15.2).[3] Had the vast majority of western rangelands been dominated by shrubby species like sagebrush, though, they could never have supported the hundreds of thousands of sheep, cattle, and horses that they were suddenly required to support in the late nineteenth century. The work of Forest Sneva at the Squaw Butte Experiment Station in eastern Oregon supports this conclusion.[4]

Perceptions of the diversity and richness of wildlife before European contact vary as well. The journals of early explorers like Jedediah Smith, John Work, and Peter Skene Ogden, who journeyed through the Great Basin during the second and third decades of the nineteenth century, describe the harsh conditions they encountered. They noted when wildlife was plentiful and when game was so scarce that they had to kill their own horses for food to survive.[5] In November 1832 near Alturas, California, John Work saw "few tracks but no sight of deer."[6] A century later, this same location

Fig. 15.1. A formerly big sagebrush/bunchgrass community converted to perennial grass following a wildfire. Most authorities agree that herbaceous vegetation dominated many habitats when Europeans arrived in the Great Basin.

had plentiful mule deer. "Crossing the road was a singular barrier, built by Indians, to pen in large hares when they hunt them, for there is no other game here," reported J. Goldsborough Bruff on September 25, 1849, while traveling near Soldiers Meadows north of Gerlach in northwestern Nevada.[7] Today, sage grouse, mule deer, and pronghorn antelope are common in this area. In 1826, at the present location of Malheur Lake in east-central Oregon, Ogden wrote, "I may say without exaggeration, man in this country is deprived of every comfort that can tend to make existence desirable. If I escape this year I trust I shall not be doomed to endure another."[8] Resource managers attempting to restore rangeland often focus their efforts on establishing plant species that are native, beneficial to wildlife, and represent *pris-*

tine conditions. As the journals of early explorers note, *pristine* and big game animals were not compatible.

The twentieth century brought conservation efforts to protect and enhance wildlife, especially game species. Pronghorns are an example of the success of these efforts. The species dropped from an estimated several million in the very early 1800s to 10,000 by the beginning of the twentieth century. By 1925, populations were estimated to be back up to 24,000 animals, and the next 50 years saw the return of the pronghorn to the western range with an estimated 1,500 percent increase (fig. 15.3).[9] Rocky Mountain elk were reported to have increased about 20 percent in the West during the 1970s and 1980s alone. Dave Mathis reported in his history of wildlife management in Nevada that elk were almost nonexistent in the state when the first explorers entered the areas that now provide elk habitat. Army surveyor and mapmaker Captain J. H. Simpson, for example, saw only two elk in 1859 in eastern Nevada near the present town of Ely.[10] By the end of the twentieth century, Rocky Mountain elk populations in Nevada were estimated at 6,400 and were still on the increase.[11] Most residents of elk

Fig. 15.2. The oceans of big sagebrush on the western ranges provided very little grazing for the vast numbers of livestock that were eventually introduced.

Fig. 15.3. Wildlife management efforts have greatly benefited native wildlife such as the pronghorn antelope.

country do not initially view cheatgrass as a threat to elk habitat; once they understand that cheatgrass can invade and alter virtually any habitat, however, they become very concerned. In a presentation to White Pine County commissioners, we described collections of cheatgrass seeds from sites at elevations of 3,400–10,000 feet. The highest collection was made at the top of Kalamazoo Creek in the Schell Creek Range in White Pine County. We reminded them that cheatgrass was originally present only next to railroads and roadsides and in other disturbed areas. Cheatgrass is invading and changing the wildlife habitats of eastern Nevada just as it has done in other parts of the western United States.

Mule deer populations have experienced boom-and-bust cycles since Europeans first arrived. At the turn of the twentieth century, taking a mule deer near the present city of Reno was front-page news; by the 1950s, mule deer populations were exploding not only in western Nevada but throughout the West.[12] The invasion of cheatgrass has had significant impacts on mule deer. As previously mentioned, the introduction of large numbers of domestic livestock onto western rangelands along with unregulated grazing practices depleted perennial grass species and increased the dominance of shrub species such as sagebrush and rabbitbrush. Inferring grass dominance compared to shrubs does not mean that an area that was a grassland became a shrubland, but rather that the relative densities of herbaceous and woody species changed. Great Basin plant communities that have sagebrush present today most likely had sagebrush prior to European contact as well.[13] Mule deer benefit when woody species such as sagebrush, antelope bitterbrush, and mountain mahogany are productive components of their habitat. These woody species provide desirable nutrition and cover for the deer, especially during the fall, winter, and early spring months. Grasses and forbs also play an important role in the health of mule deer herds as they provide much-needed protein and phosphorous.[14]

When cheatgrass moves in, wildfires that destroy shrubs follow. This scenario has played out all across the Intermountain West, which now experiences larger and more frequent wildfires than ever before. With each passing wildfire season, more and more shrub habitats are converted to cheatgrass dominance, and the fire cycle continues. Historical wildfires probably occurred at 60–110-year intervals; cheatgrass dominance has reduced the interfire interval to 5–10 years in many habitats. Woody species simply do not have enough time to recover in between wildfires.[15] The loss of the shrubs has important consequences for wildlife. Ken Gray, a wildlife biologist with the Nevada Department of Wildlife in northeastern Nevada, reported that the Independence mule deer herd in northeastern Nevada plummeted from 35,000 in the 1960s to under 10,000 in 2003; recurrent wildfires and loss of important shrub species on the winter ranges were the primary factors.[16] The Wyoming and basin big sagebrush habitats at lower elevations, which make up a large part of the mule deer winter ranges throughout the West, have converted from sagebrush/bunchgrass steppe vegetation to cheatgrass rangelands. By identifying the problem (fires fueled by cheatgrass) and taking advantage of mining mitigation funds and contributions from sportsmen, Gray has been very successful in preventing wildfires while at the same time providing critical forage for wildlife (fig. 15.4). Much of his success is due to the use of such nonnative species as crested wheatgrass and 'Immigrant' forage kochia. Crested wheatgrass provides the necessary perennial grass needed to suppress cheatgrass while forage kochia is a nutritious shrub that can compete with cheatgrass, reduce fuel loads, and resprout following wildfires.

Fig. 15.4. Range improvement practices that include the seeding of introduced species such as crested wheatgrass and 'Immigrant' forage kochia have been very beneficial to wildlife. The efforts of wildlife biologist Ken Gray and others at Dunphy Hills in northeastern Nevada provide just one example.

In the 1980s, wildlife biologists looked on crested wheatgrass as a forage species for livestock only. The removal of millions of acres of big sagebrush stands in order to seed crested wheatgrass angered many wildlife managers, but some have since seen the light and have begun adding nonnative forage species to their wildlife seed mixes. During much of his career as a wildlife biologist with the Nevada Department of Wildlife, Jim Jeffress viewed crested wheatgrass as beneficial only to ranchers and their livestock and fought efforts to plant the species.[17] Eventually, however, he realized that critical habitats were being converted to cheatgrass. He also realized that most of the winter forage available to his mule deer herds in the Santa Rosa Range of northern Nevada came from old crested wheatgrass seedings. Further, these seedings eventually returned to sagebrush as the longer fire interval allowed the shrubs to return; this did not happen on cheatgrass-dominated ranges. Another wildlife biologist with the Nevada Department of Wildlife assured us in the early 1990s that "crested wheatgrass has no benefit to wildlife." Less than 10 years later he wished that he had thousands of acres of crested wheatgrass rather than thousands of acres of cheatgrass. Ken Gray lost 778,000 acres—1,170 square miles, or the equivalent of a 4-mile-wide swath from Reno to Elko—in a northeastern Nevada hunting zone under his supervision in 1999–2001.[18]

Wildlife managers must deal with cheatgrass and protect the remaining shrub communities. Perennial grasses like crested wheatgrass are more effective at biologically suppressing cheatgrass than any other plant growth form (fig. 15.5). In 2002, Jim Jeffress found that some two-thirds of the 3 million acres of Wyoming sagebrush mapped in the BLM's Winnemucca District in the mid-1980s had been converted to cheatgrass dominance by cheatgrass-fueled wildfires.[19] The diminishing sagebrush/bunchgrass habitat and concomitant increase of monotypic cheatgrass communities is an ecological disaster for wildlife throughout the Great Basin. Active range improvement practices must be implemented to ensure better success of perennial grass establishment, decreased fire frequency and size, and the return of browse species such as sagebrush to these communities. Rather than trying to preserve

Fig. 15.5. Crested wheatgrass acts as an ecological Band-Aid because it reduces fire frequency and allows for the recruitment of shrubs. Here, mules browse an older crested wheatgrass seeding in early spring.

the remaining sagebrush habitats untouched, wildlife biologists and other resource managers should focus on developing methodologies to interact with the problem of cheatgrass-fueled wildfires. The hands-off protection approach has a very predictable end: continued fuel for wildfires and further loss of critical shrub communities.

The Charles Sheldon National Pronghorn Refuge in northwestern Nevada provides a good example of a narrow approach to range management. The livestock that were permitted to graze this half million acres of rangeland were permanently removed in 1991 in an attempt to improve rangeland health. The increase in fuel from herbaceous vegetation resulted in large, frequent wildfires that destroyed critical browse communities (fig. 15.6). Ten years after the Badger Mountain fire on the Sheldon Refuge, the shrub composition is 73 percent lower than preburn densities (USDA-ARS unpublished data). The odds favor this habitat burning again before it ever reaches preburn shrub densities. Perhaps these managed rangelands do resemble vegetative conditions prior to European contact, but important browse communities have been lost to the detriment of mule deer, sage grouse, pigmy rabbits, and other wildlife species that rely on them.

Sage grouse occur throughout the range of big sagebrush, which includes California, Colorado, Idaho, Montana, Nevada, North Dakota, Oregon, Utah, Washington, and Wyoming; they have been extirpated from Arizona, Kansas, Nebraska, New Mexico, Oklahoma, and British Columbia.[20] Although forbs and insects are critical for newborn chicks and hens from May through September, sage grouse rely on sagebrush in their diet throughout the year.[21] A 2004 petition seeking protection for sage grouse under the Threatened and Endangered Species Act was denied. Forty years ago, D. A. Klebenow estimated that some 700,000 sage grouse remained.[22] Some environmentalists would like to use declining sage grouse numbers as an excuse to eliminate livestock grazing and other activities on sagebrush rangelands. The Montana Mountains region in Humboldt County has the largest documented population of sage grouse in Nevada, and perhaps in North America. Capture-recapture studies indicate that this population numbers 8,000–12,000 birds. The single largest factor that threatens this population, according to Jim Jeffress, is the loss of sagebrush in intermediate and winter range habitats and conversion to cheatgrass-dominated landscapes.[23]

Fig. 15.6. A critical mountain brush community that burned in 1994 and again in 1997 on the Charles Sheldon National Antelope Refuge in northern Nevada. The return of critical browse species is a very slow process, as this photo taken in 2007 shows.

Sage grouse numbers prior to European contact are unknown. Clait Braun, a wildlife biologist with the Colorado Division of Wildlife, estimated precontact populations at 2 million birds.[24] The journals of early explorers in the Great Basin, on the other hand, do not mention large sage grouse populations other than in a few localities such as the eastern slope of the northern Sierra Nevada in Honey Lake Valley of northeastern California. Trappers and settlers who traversed Honey Lake Valley in the 1850s saw plentiful waterfowl, pronghorn, and jackrabbits, but no sage grouse.[25] Perhaps the scarcity of sage grouse was related to the scarcity of mule deer. Mule deer and sage grouse occupy the same sagebrush-dominated habitats. The increase in sagebrush density that followed excessive grazing of native bunchgrasses may have benefited both species.

When early resource managers removed thousands of acres of sagebrush in the effort to seed crested wheatgrass for additional livestock forage and to suppress invasive exotic weeds such as halogeton, they destroyed vital sage grouse wintering, nesting, and strutting grounds (leks). The impact of sagebrush removal on leks probably varied depending on the site, but generally it was disastrous. Sage grouse are believed to use the same lek year after year, and the removal of sagebrush at a lek can significantly reduce or even eliminate sage grouse attendance.[26] More than 80 percent of sage grouse nests reported in various states were directly under sagebrush shrubs.[27] Accumulating snow sometimes forces sage grouse down to lower elevations. Typically, the lower-elevation habitats are Wyoming big sagebrush, but some of those habitats were seeded to crested wheatgrass. Range management wildlife conflicts of this sort resulted in sage grouse committees, workshops, and regulations. In 1969, the Sixth Biennial Western States Sage Grouse Workshop published *Guidelines for Habitat Protection in Sage Grouse Range*.[28] The pamphlet stated the specific needs of sage grouse, noted that sage grouse were not adjusting to the habitat changes, and predicted that sagebrush removal practices might endanger the future of this native grouse. The guidelines recommended no vegetation control programs on sage grouse ranges. If land management agencies decided that sagebrush control was necessary, the guidelines stated, the state's wildlife agency should be contacted and notified of the proposed control 2 years prior to its implementation. This would give the state wildlife departments enough time to map the areas sage grouse were using at the proposed site and evaluate the effects the proposed control method might have on the sage grouse population. After the proposal, notification, and mapping, the cooperative agencies (state and federal) would visit the site and formulate a comprehensive multiple-use management plan. All sagebrush control project plans should include long-term quantitative measurements of vegetation before and after the control method, its effects on the wildlife habitat, and a report on whether the objectives of the project were accomplished. No control work should be considered in areas with less than 20 percent live sagebrush crown cover or within a 2-mile radius of strutting grounds, nesting habitats, or other special-use areas. Further, the guidelines suggested no control in areas known to be important sage grouse wintering habitats within the previous 10 years or along streams, meadows, or secondary drainages; and also suggested retaining a 100-yard-wide strip of sagebrush on each edge of the meadow or drainage. The guidelines were a recipe for cheatgrass invasion and expansion. They virtually eliminated active range improvement practices because they were designed to preserve an environment that was already degraded while doing nothing about the cause of the degradation. The degraded environments were dynamic, not stable, so the degradation processes continued in a downward spiral. Pioneers called sage grouse "fool's hens." The adjective is more appropriate for management strategies that stop habitat restoration attempts and ensure the decline of sage grouse. A few professional wildlife managers voiced their concern and insisted that range improvement practices had to be implemented in some areas.[29]

Various state wildlife agencies supported the guidelines, adding to the livestock-wildlife conflicts of that time. Rather than helping the sage grouse, however, the guidelines actually ensured the degradation of many sagebrush habitats. Sagebrush ultimately invaded many meadows, decreasing their size and quality (fig. 15.7). Livestock continued to graze the herbaceous vegetation in the meadows in the absence of any range improvement practices to increase the forage base. As herbaceous vegetation decreased, sagebrush densities increased, ultimately ending any hope for perennial grass recruitment in these closed communities. The communities were not closed to cheatgrass, though, and without the presence of perennial grasses to suppress its growth, cheatgrass invaded the sagebrush-dominated habitats. With virtually millions of acres of sagebrush/bunchgrass rangelands being grazed

Fig. 15.7. A mountain meadow being invaded by big sagebrush and western juniper. If the invasion continues, this critical riparian habitat will be lost.

without herbaceous vegetation being added, sagebrush continued to close out the recruitment of new perennial grass seedlings while cheatgrass moved into the grass niche. Cheatgrass provided a fine-textured, early-maturing fuel that increased the chance, rate, spread, and season of wildfires, which ultimately destroyed more sage grouse habitat than any range improvement practice. As Jim Jeffress observed, former crested wheatgrass seedings have now returned to sagebrush communities (fig. 15.8). Cheatgrass-fueled wildfires have burned millions of acres of important browse communities and closed them to natural succession. The loss of sagebrush habitats to range improvement practices was minor in scale compared with those lost to cheatgrass-fueled wildfires and cheatgrass-dominated rangelands.

When Bruce Welch, a retired Forest Service plant physiologist, delivered a lecture to resource managers on the nutritional value of sagebrush, he was astounded to have one of the resource managers stand up and comment that Bruce had made sagebrush sound so good and it was "too bad nothing eats it." Bruce was speechless for a moment, but quickly named the ten or twelve animal species that he had personally watched eat sagebrush; that number has now increased to more than fifteen species.[30] If resource managers do not understand the importance of sagebrush as forage for browsers, what hope can there be for game animals such as pronghorn, mule deer, Rocky Mountain elk, cottontail rabbit, and sage grouse; and other wildlife species such as black-tailed jackrabbit, pigmy rabbit, dark-eyed junco, Uinta ground squirrel, and white-crowned sparrow? Besides browse, sagebrush provides nesting, escape, and thermal cover, and its removal can immediately reduce populations of sagebrush-nesting species.[31] While sagebrush-obligate species are forced to look elsewhere for sagebrush habitats where they can survive, wildlife species such as the white-tailed jackrabbit, horned lark, and western meadowlark that prefer more open environments may benefit from the removal. Long-term loss of the shrubby species on which animals such as sage grouse depend results in increasingly limited survival. Increasing the diversity of woody, herbaceous, and edge effect vegetation, on the other hand, increases bird species diversity and wildlife species diversity in general.

Plant-animal interactions have assumed increasing importance in community ecology as our understanding of how animals interact with plants as herbivores, pollinators, and handlers of seed has grown. Granivorous

Fig. 15.8. An old crested wheatgrass seeding that has been reinvaded by big sagebrush. The habitat provides good sage grouse habitat.

rodents, for example, play a very active and important role in the harvesting and dispersal of many rangeland plant species.[32] The granivorous rodents benefit because the seed is a nutritious food source, and the plant benefits because its seed is dispersed, and sometimes germination and establishment are enhanced. Some seeds are larder-cached; that is, harvested and buried deep inside the rodent's burrow. These seeds are too deep to germinate. Granivorous rodents also scatter-hoard cache, harvesting the seed and burying it in shallow depressions on the soil surface. Seeds in the scatter-hoard caches that are not consumed germinate the following spring and are an important element in the recruitment of many plant species (fig. 15.9). Excessively large rodent populations have been reported to reduce or even eliminate certain plant species.[33] Kangaroo rats—rodents with external cheek pouches and the ability to harvest large quantities of seed in a very short time—can have dramatic effects on plant cover and composition. We have spent many hours in the field observing the interactions of rodents with antelope bitterbrush (an important western browse species) and Indian ricegrass (an important winter forage perennial grass species)

Fig. 15.9. Antelope bitterbrush seedling resulting from a rodent cache. Granivorous rodents are very important in the dispersal of antelope bitterbrush seeds and other plant species in the Great Basin.

seed in Nevada and California. The rodents' activities likewise affect the growth and distribution of exotic and invasive species. Native vegetative species are eliminated around rodent burrows while Russian thistle and cheatgrass are common.

Granivorous rodents prefer cheatgrass seeds over those of many other plants.[34] They readily cache the seeds, and in very dry years the only cheatgrass seed production may be from plants growing from these caches. The rodents' preference for cheatgrass has long-term effects on other rangeland species, both resident and migratory. Rodents are important prey species for raptors, snakes, and other predators. The catastrophic wildfires that frequently occur in cheatgrass-dominated habitats can reduce rodent populations, and this reduction can ramify all the way up the food chain to dominant predators such as golden eagles. The immediate decrease in rodent populations may be felt in the short term, but the significant loss of black-tailed jackrabbits, eagles' staple food, may be long term and devastating to that particular eagle population. The transformation that results from cheatgrass invasion and conversion thus often goes far beyond our immediate diagnosis concerning plant and animal interactions.

Not all species suffer from the conversion to cheatgrass, of course. Jim Yoakum, formerly a wildlife biologist with the BLM and now retired, has been following pronghorn populations very closely since the 1950s.[35] Within the last 10 years Yoakum has noticed pronghorns expanding into new areas, including crested wheatgrass and wildfire-burned habitats converted to cheatgrass. Pronghorns prefer habitats with less than 50 percent vegetative cover. Shrubs should make up 5–30 percent of the covered area, depending on site potential, but the habitat must also be rich in forbs (10–50 percent) and grasses (5–10 percent). Sagebrush is a very important browse species as well as providing cover.

Elk benefit when perennial grasses flourish. They are primarily grazers, though they seek important browse species like sagebrush, serviceberry, mountain mahogany, and antelope bitterbrush during certain seasons. When the first Europeans arrived, elk were present on the Great Plains in large numbers, commonly in association with bison. Subsequent hunting pressure may have forced the elk into the mountains.[36] Today, elk are very abundant in eastern Nevada where once they were very scarce. In some areas their numbers have risen well above their allotted management levels; in fact, elk increased from an estimated 1,965 animals in 1990 to more than 7,000 in 2003. At the same time, mule deer decreased from an estimated 202,000 to 109,000. Pinyon-juniper encroachment, which has been a problem for resource managers in eastern Nevada, has apparently played a role in these reversals. In some areas pinyon and juniper trees have crowded out herbaceous and browse species as well as sucking up the water from natural springs and meadows (fig. 15.10). Pinyon and juniper have invaded an estimated 18 million acres of rangeland in various successional stages, three times the amount present before Europeans arrived.[37] When these habitats burn, they usually return to perennial grass dominance and productive shrub communities. With cheatgrass a possible invader, resource managers are going to have to pay special attention (fig. 15.11). At present, many of the eastern Nevada habitats in danger of cheatgrass invasion receive measurable summer precipitation that benefits perennial grasses; if this should change, cheatgrass could be the beneficiary.

Fig. 15.10. Encroaching pinyon-juniper woodlands have drastically reduced browsing, grazing, and water resources for wildlife, feral horses, and domestic livestock.

Fig. 15.11. Sage grouse in one of the last remaining big sagebrush islands in northwestern Nevada. Without active range improvement practices this will be a graveyard for the species in the future.

One of the few wildlife species that finds cheatgrass a good home is the chukar partridge (fig. 15.12), an important upland game bird in the West.[38] The chukar, which originated in northern India, Pakistan, and Afghanistan, is one of the red-legged partridges of the genus *Alectoris,* which includes more than twenty species and subspecies. Glenn Christensen, formerly a Nevada Department of Wildlife biologist, has assembled a great deal of information on the chukar. The first known introduction into North America occurred in 1893 when W. O. Blaisdell of Illinois brought in five pairs from Karachi, Pakistan. Many successful introductions followed, including one in Nevada in the 1930s. By 1947 there was a two-day chukar hunting season in Nevada, and by 1951 chukars were plentiful and were becoming very popular game birds. The arid western landscapes closely resemble southwestern and Central Asia in climate and vegetation. Both areas are arid to semiarid, and the vegetation is primarily a grass-forb understory with short brush; the grass in the Asian understory is cheatgrass. Chukars eat a variety of forbs, grasses, leaves, insects, and seeds, but the one consistent dietary component year-round is cheatgrass leaves and seeds, even in good years when other plant species are abundant.

Wildlife managers are well aware of the danger of cheatgrass but less aware of the danger posed by a less desirable exotic invasive weed that has no value as forage. Medusahead originated in Central Asia and was first identified in North America in 1887.[39] It is a highly competitive plant that replaces cheatgrass on sites characterized by clay soils. The very long awns bear reversed barbs that injure animals. Medusahead accumulates silica that makes it unpalatable to grazing animals, including important game species such as mule deer, pronghorn, and elk. The high silica content of mature plants preserves them, and the plants tend to accumulate, forming litter that chokes out other plants and creates severe wildfire danger. The presence of medusahead on some rangelands has had a dramatic influence on chukar and some other wildlife species, and it has been spreading at an alarming rate.[40] In Idaho alone medusahead went from a few scattered plants to the dominant plant on 750,000 acres within a period of 15 years. The sudden invasion caught range managers and management agencies off guard. There were no plans in place to deal with medusahead, and the ranges have suffered as a consequence.

Range managers need to decrease sagebrush control methods detrimental to wildlife while at the same time implementing range improvement projects

Fig. 15.12. The chukar partridge has become a very popular upland game bird in the Inter-mountain West. Chukars are associated with cheatgrass-infested rangelands in rough terrain.

for future wildlife populations. They must incorporate techniques to manage invasive exotics, increase edge effect, and actively manage the land for a foreseeable community. Passive management is not an option. Sitting and arguing about it at the table, as the participants at the first cheatgrass symposium in 1965 did, can lead to uncontrolled results such as the invasion of exotic weeds. Now, more than 40 years later, resource managers and wildlife biologists are trying to find ways to get perennial grass species established to suppress cheatgrass, decrease fire intervals, and ultimately save and allow for the return of important browse species like sagebrush. The problem, though, is that cheatgrass is either dominating or has a foothold in many habitats, which for the most part was not the case in the 1950s during the sagebrush removal and crested wheatgrass seeding era. One of the many problems with the current approach to seeding perennial grasses in an effort to eventually restore sagebrush rangelands comes back to the use of native versus nonnative plant material. Opponents of the use of nonnative species in rangeland seeding efforts argue on, apparently oblivious to the fact that more and more critical habitats are burning and converting to cheatgrass. Emotions have no role in the battle for the western ranges. Opponents' passion to shoot down any range improvement practice that prescribes the use of nonnative plant material because it does not represent the "pristine" environment or will benefit livestock only is leading to disaster. We suggest that readers remember the so-called pristine environments that early settlers traversed as well as the experiences of wildlife biologists like Jim Jeffress.

CHAPTER 16

Wildfire on the Range

The central issue in any discussion of the ecological ramifications of cheat-grass invasion and dominance on Great Basin rangelands is the role cheat-grass plays in the stand renewal process—the process by which a given plant community perpetuates itself on a specific site. Much of what appears in the literature concerning wildfires fueled by cheatgrass is observational and not based on measured results from designed experiments. Such observations are valuable, but it is difficult to interpret them with scientific precision.

Plants do not live forever. Sooner or later they die and new seedlings have to be recruited either to perpetuate the preexisting community or to introduce a new community to the site. Essentially, the stand renewal concept is based on the premise that the species composition and dominance of a given com-munity are highly influenced by how the tenure of the previous assemblage of plants on the site was terminated. Many years ago in a paper authored with Jack Major, we recognized burning as the stand renewal process for big sage-brush communities.[1] Big sagebrush stands tend to consist of plants of roughly the same age, within 10–15 years, indicating that the previous community on the site was catastrophically destroyed. The decade-plus variation in age is a reflection of sagebrush's slow recolonization due to limited seed dispersal.

Historically, the Great Basin rangelands have experienced eleven kinds of renewal processes.

1. A dynamic—in relation to climate—fire interval that occurred on specific sites before European contact (including aboriginal uses of wildfires)

2. Vertebrate and invertebrate herbivory

3. Promiscuous burning introduced largely, but by no means entirely, by herdsmen

4. Attempts at absolute control of wildfires on rangelands

5. Substitution of mechanical control by plowing for wildfires as a stand renewal process for big sagebrush/bunchgrass plant communities

6. Substitution of herbicide applications for wildfire for sagebrush/bunchgrass community renewal

7. Wildfires fueled by cheatgrass

8. Grazing management introduced as a substitute for mechanical or herbicidal stand renewal in sagebrush communities

9. To a limited extent, wildfires burning in native grasslands that were restored through grazing management

10. Deliberate use of prescribed burning to renew stands

11. Fires burning in annual grassland communities

It is within this array of stand renewal processes that the current burning of Great Basin rangelands operates. Let us examine each of them more closely. If you searched long enough in the Great Basin, you probably could find one spot where all eleven types of stand renewal by wildfires are within the field of view. You will need very good vision to view the precontact example because it probably will be on a very steep slope at high elevation, with a cliff that served as a firebreak located below the stand. We are not suggesting that anyone put in a lot of time searching for such a spot; our point is to emphasize the diverse array of stand renewal processes and therefore successional plant communities that can occur in big sagebrush/bunchgrass potential communities in the Great Basin.

Fire Intervals before European Contact. The first stand renewal process that occurred in the Great Basin is the most difficult to identify and quantify. Estimates of the interval between wildfires in sagebrush/bunchgrass communities before European contact are highly variable and often controversial. Only recently have comprehensive studies tried to reconstruct the post-Pleistocene vegetation of the Great Basin. Analysis of fossil pollen indicates that five-needle pine species (subalpine tree species such as limber and bristlecone pines) were growing near the maximum levels (4,380 feet elevation) of the pluvial lakes of the Great Basin during the glacial maxima of the Pleistocene. Currently, five-needle pines are found at elevations from 8,000 to 10,000 feet in the Great Basin. Apparently, many low-elevation refugia existed during the pluvial lake periods. Perhaps the geological diversity associated with fault-block mountain ranges contributed to these plant refugia, which somehow existed before such diverse species as the prototype salt desert shrubs or Utah juniper.[2]

As the climate became warmer and drier during the early Holocene (10,000–8000 BP [years before the present]), the great pluvial lake basins became barren, wind-eroded lake plains that were gradually colonized by evolving chenopod species. The subalpine forest of five-needle pines retreated up the mountains to refugia on ranges that extended near or above 10,000 feet in elevation. Lower-elevation mixed coniferous forest, with species such as ponderosa and Jeffrey pine and Douglas-fir, rapidly disappeared in the early Holocene. From the upper reaches of the salt deserts to the highest mountain ranges, there was a lot of area for expansion of big sagebrush/bunchgrass communities.[3]

The area dominated by sagebrush communities was gradually reduced as pinyon-juniper woodlands invaded the Great Basin from the southeast.[4] The lower edge of the pinyon-juniper woodlands was about 1,500 feet higher than it had been during the mid-Holocene (8000–4000 bp).[5] Following the mid-Holocene, precipitation increased while temperatures remained warm;[6] grasses became abundant, and with them came wildfires.

The Neoglacial period (3000–2500 BP) that followed the mid-Holocene was characterized by much cooler and wetter conditions.[7] The woodlands of the Great Basin probably occupied an area similar to their distribution today. The late Holocene (2500–140 BP) was a period of changing climatic patterns.[8] Severe droughts and major fires decreased woodlands and grasslands and increased dominance by sagebrush. In the central Great Basin, extensive hill slope erosion accompanied by deposition of alluvial fans in drainage channels occurred during the first two-thirds of this drought period.[9] The warming trend known as the Medieval Climatic Anomaly occurred between 1500 and 1100 BP, bringing increased precipitation, a return of grasses, and expansion of the distribution of woodlands downslope and northward.[10]

During the sixteenth century, a severe drought, possibly accompanied by large fires, caused extensive mortality of pinyon and juniper woodlands across the American Southwest.[11] Since the end of the Little Ice Age (700–150 BP) there has been a general warming trend in the Great Basin. Climate conditions during the twentieth century were similar to those in the period immediately following the Neoglacial period.[12] It is evident that both moisture abundance and drought can increase the occurrence and extent of wildfires in the Great Basin. This paradox is apparently a result of the continuity of the vegetation and the nature of the fuels. Wetter climatic conditions favor the growth of herbaceous vegetation, especially perennial grasses, which supply fuels to carry fire among shrubs. Extreme drought results in dead and dying woody vegetation that provides a high-energy concentrated fuel source with very low moisture content. Under extreme drought conditions, on a hot day with the wind intensity and consistency sufficiently high, a continuous fuel source is not necessary for the fire to spread.

These climatic extremes would make the prehistoric record of wildfires in the Great Basin very difficult to reconstruct even if we had excellent tools to aid in the reconstruction, which we do not. Various events leave behind clues, though. Several species of conifers develop fire scars called "cat faces" when successive wildfires burning through the wooded community as ground fires char the base of the tree sufficiently to damage the cambium on a portion of the trunk. The damage is not sufficient to kill the tree, and the remaining cambium eventually grows over the edges of the scar, leaving a characteristic growth pattern in the annual growth rings. Dendrochronology techniques make it possible to determine the apparent age of the fire by counting the rings in a cross section cut from the trunk or in a well-placed increment boring of the trunk.[13] The fire scar technique applies only to communities on the margins of the sagebrush grasslands where woodlands intermingle. It is generally of little use in sagebrush/bunchgrass areas because there are no trees to develop fire scars. Forest ecologists John and Carol Thilenius told us about a field trip in the southern Cascade Mountains of Oregon where they escaped a sudden rain shower by comfortably stepping inside the cat face on a very large incense cedar tree. It was like standing inside a living museum wrapped in several centuries of fire history.

Remarkable prehistoric chronologies of wildfires have been developed for ponderosa pine woodlands using the fire scar technique, although several factors must be considered in interpreting such chronologies. The fire must be a ground fire. A crown fire would kill the tree, which would eventually fall and decay, leaving no living fire-scarred tree to be dated. To be repeatedly scarred by wildfires and not killed, the tree has to grow in a favorable topoedaphic situation that allows ground fires that are not too intense. These requirements are so constraining that the fire scar technique probably underestimates the actual frequency of wildfires. Small, low-intensity fires probably would not leave a scar in these specific tree habitats.

Likewise, the specific habitat in which the fire-scarred trees occur influences the dating. Wayne Burkhardt, for example, used western juniper trees growing in the Owyhee Mountains of southwestern Idaho to estimate the frequency of wildfires. The interfire interval he calculated was so short (less than 5 years) that anyone visiting these woodlands would want to be ready to jump up on a stump to let the fire burn past.[14] Young and Evans used fire scars on western juniper trees growing in low sagebrush communities in northeastern California and estimated the interval between fires to be up to a century.[15] Although both studies took place in western juniper woodlands, the sagebrush component was different. Burkhardt sampled western juniper/big sagebrush communities while Young and Evans sampled western juniper/low sagebrush communities. Our site in northeastern California has extensive western juniper/mountain big sagebrush communities immediately adjacent to and surrounding the low sagebrush communities. There are no fire-scarred trees in these communities. The western junipers on the low sagebrush sites were more than four centuries old. The trees on the mountain big sagebrush sites were less than a century old at the time Young and Evans did the study.

The noted archaeologist P. J. Mehringer developed many novel methods for estimating dates for archaeology sites, including the use of the fossil pollen record.[16] Mehringer had the insight that smoke from wildfires is a charcoal aerosol (solid suspended in air) and as such has deposition characteristics similar to pollen. He used that fact to develop a wildfire chronology for sediment samples obtained from Diamond Craters in eastern Oregon. Diamond Craters has an exceptionally high rate of deposition, which provided Mehringer with yards of sediment with which to examine the post-Pleistocene record. Finding sites in the Great Basin with similar or even fractional rates of deposition and profile preservation may be difficult but is probably not impossible. Young and Evans excavated a peat spring mound at Walti Hot Springs in central Nevada, currently a salt desert, and traced the fossil pollen record back to spruce woodlands (USDA-ARS, Reno, unpublished research). The spring mound was peat interspersed with layers of fine sediment eroded from the Grass Valley playa.[17] Young and Evans were not familiar with Mehringer's work with subaerially deposited charcoal at the time and did not think to examine the sequence for evidence of wildfires. Bob Blank used radiocarbon dating of charcoal and volcanic tephra in a mountain meadow soil profile from the eastern Sierra Nevada to develop a fire chronology, but again this was a site of rather rapid rates of deposition, and the coarse charcoal from forest fires was not necessarily subaerially deposited.[18]

The lack of a technique to determine precisely the interval between prehistoric wildfires in sagebrush/bunchgrass communities leaves us with an abundance of estimates based on opinions rather than hard data. The opinions can be roughly divided between those wishing to return wildfires to the environment through the application of prescribed burning and those who wish to continue a policy of absolute fire suppression on range and forest lands. The burners are sure that wildfires were frequent in precontact time, and the nonburners are equally sure that wildfires were rare events.

Young and Evans inadvertently contributed to estimates of fire intervals in big sagebrush/bunchgrass communities through their research on low rabbitbrush invasion and dominance following wildfires.[19] This study indicated that low rabbitbrush was the seral dominant of big sagebrush/Thurber's needlegrass communities for roughly 15 years following stand renewal in wildfires, before it was replaced by big sagebrush. Henry Wright, who along with Art Bailey was a leading authority on rangeland burning, picked up on this research and pointed out the widespread distribution of species and subspecies of rabbitbrush in big sagebrush communities. If the rabbitbrush species generally followed the successional dynamics Young and Evans suggested, Wright said, then the interval between wildfires under precontact conditions must have been greater than 15 years or early explorers would have reported on the prevalence of rabbitbrush/bunchgrass communities rather than sagebrush grasslands. Steven Whisenant, who studied the intervals between wildfires in sagebrush/bunchgrass vegetation on the Snake River Plains of Idaho, suggested that the interfire interval there was 60–110 years prior to the arrival of Europeans.[20] Whisenant studied both high-condition sagebrush/bunchgrass sites with minimum amounts of cheatgrass and sites of the same apparent ecological potential dominated by cheatgrass. The *amount* of fine fuels actually increased in the healthy communities, but the *continuity* of fine fuels dramatically increased with cheatgrass dominance (fig. 16.1). The interval between wildfires in sagebrush/bunchgrass communities before contact with Europeans remains an intriguing puzzle. Quite a few people believe the answer will solve many of the current management problems of rangeland sites with the potential to support big sagebrush/bunchgrass communities. Considering the demonstrated variation in short-, moderate-, and long-term climatic conditions in the American West during the Holocene, this quest is literally looking for the source of smoke in a maze of mirrors.

Vertebrate and Invertebrate Herbivory. Obviously, herbivory was a stand renewal process in big sagebrush/bunchgrass communities in the Great Basin prior to European contact. It must not have been a major factor, however, because the population density of large herbivores was extremely sparse.[21] Pronghorns used big sagebrush browse at all seasons of the year, but no one has ever reported sagebrush stands killed by excessive pronghorn browsing. As discussed in chapter 15, mule deer population densities at contact time were so low that their impact on sagebrush stand renewal had to be minimal. Bighorn sheep were probably restricted to the more rugged, mountainous portions of the Great Basin where escape topography and cover existed. Rocky Mountain elk were absent from the far western Great Basin, and the American bison had withdrawn from that region before Europeans arrived.[22] Vertebrate herbivory when Europeans came on the scene was confined to small mammals, with jackrabbits being the most numerous and voracious grazers. Peak jackrabbit populations can decimate herbaceous vegetation in limited areas.[23] Some forms of big sagebrush are virtually immune to jackrabbit herbivory. Do jackrabbits constitute a stand renewal process in the Great Basin? Certainly they are a stand renewal factor for herbaceous vegetation, but not for established big sagebrush plants.

Sage grouse and pigmy rabbits are sagebrush obligates in the Great Basin, especially during the winter months. They hardly can be considered stand renewal factors, although if they should be given endangered species status

Fig. 16.1. Cheatgrass provides the continuity of fuels that increases the rate and spread of wildfires in degraded big sagebrush/bunchgrass communities.

they could become factors, radically changing land management policies. Even though the federal government has not yet listed the sage grouse as an endangered species, the possibility that it may soon do so has already changed rangeland management policies. These policy changes will be reflected in changing stand renewal and in the species composition and dominance of future plant communities. Predicting what these changes in stand renewal will be is very difficult, but the potential consequences of such policy changes should be thoroughly examined before they are implemented.

Insect herbivory was a form of stand renewal in big sagebrush stands in the Great Basin before European contact and continues to play a role in these communities. The sagebrush defoliator (more often known by its scientific name, *Aroga websteri*), a moth belonging to the family Gelechiidae, defoliates large blocks of big sagebrush in some years.[24] Severe infestations can kill a large percentage of the older big sagebrush plants in a community, but the younger, more vigorous sagebrush plants usually recover. The resulting stand renewal occurs without the woody species giving up total dominance of the site, as happens with burning. The sagebrush defoliator has a marked interaction with wildfires. The feeding larvae produce webs that cover the shrub canopy and collect dry sagebrush leaves and assorted wind-deposited litter. This creates an extremely flammable mixture that when combined with abundant cheatgrass herbage results in wildfires that are impossible to suppress. This was the case in the 1973 Hallelujah Junction wildfire north of Reno and the huge fires in northeastern Nevada in 2006, where more than 1 million acres burned in Elko County alone.

In chapter 7 we discussed the abundant secondary compounds in the leaves of sagebrush species that protect portions of populations from excessive herbivory by large vertebrates and perhaps invertebrates. Whether or not the abundant secondary compounds that limit herbivory for the browse of most woody sagebrush species are metabolic by-products that evolved as anti-herbivore protection is unknown. Some forms of big sagebrush are highly desirable winter browse for mule deer.[25] Whether such preferential grazing could constitute a stand renewal process is not known either. Changes in the relative abundance of preferred and nonpreferred forms of big sagebrush resulting from the huge populations of mule deer that peaked on Great Basin ranges in the 1950s may have been the most significant modification in sagebrush/bunchgrass ranges during the twentieth century.

The influence of secondary plant compounds on domestic livestock will be crucial in the rise of cheatgrass on Great Basin rangelands. Cattle and horses generally prefer perennial grasses to big sagebrush, and the shrubs increase in density and dominance as a result. Domestic sheep prefer black sagebrush to many other plants, yet there is no evidence of entire communities destroyed by excessive grazing. In the Churchill Canyon area of western Nevada, for example, a large range sheep operation had winter headquarters just north of the Buckskin Mountains for 100 years starting in the late nineteenth century. Most of the black sagebrush/desert needle-grass communities in the Buckskin Mountains are in excellent condition. A holding pasture close to the winter headquarters was used for sick or injured animals and as a gathering point for shearing, and the black sagebrush plants in this pasture were excessively browsed annually. These plants are dwarfed, even though the sheep operation closed a couple of decades ago, but the community is still dominated by black sagebrush. This suggests that herbivory is not a major stand renewal process in the woody portion of sagebrush/grass communities.

Promiscuous Burning. Promiscuous burning has become uncommon, but people do still deliberately set fires. Passive or unintentional promiscuous burning caused by such agents as improperly discarded cigarettes or vehicle exhaust has increased as the human population has expanded into the Great Basin. Range livestock husbandry seems to foster promiscuous burning behavior. When David Griffiths visited the northwestern Great Basin in 1899, he saw numerous fires in the mountains supposedly set by sheep herders in the hope of enhancing early spring forage production during the next grazing season.[26] Scientists may not give them credit for it, but sheep herders and ranchers have long known that fire reduces competition from woody species, increases the production of nutritious herbage, and promotes nutrient cycling and availability. It also has a blackbody effect on early spring growth. If these interacting factors do not always result in increased forage production and harvesting, they often create the appearance of such increases. If the rancher adds more grazing animals to take advantage of the apparent increase in forage production on burned rangelands that were already fully or excessively stocked, degradation of the forage resources is inevitable. This is the subject of the classic paper by G. D. Pickford published in 1932.[27] J. F. Pechanec and A. C. Hull went to considerable lengths to dispel the idea that burning cheatgrass-infested communities would produce feed for lambs earlier the following spring.[28]

Those not experienced with camping in a big sagebrush environment may not realize that sagebrush was a necessary and widely used fuel for much of the nineteenth and twentieth centuries in the Great Basin.[29] It was even harvested on a moderate scale to power steam engines used in the mining and milling of ore. Cowboys, sheep herders, and prospectors preferred to camp where there was a ready supply of pinyon or, even better, juniper or mountain mahogany wood; at lower elevations, especially in northwestern Nevada, sagebrush was the fuel of last resort. Camps were usually near a water source, which usually encompassed at least a small meadow, and the largest big sagebrush plants tended to occur where the shrub invaded former meadows. This suggests a disproportionate harvesting pattern that may have been significant as a stand renewal process for these specific communities. Escaped campfires probably contributed to a disproportionate distribution of wildfires.

Absolute Control of Wildfires. The early-twentieth-century conservation movement in the United States was built on a program of stamping out forest fires—promiscuous, accidental, and natural. Choking smoke followed by acres of charred and blackened landscape, not to mention inconsequential losses of human life in forest fires, provided the dramatic images and fodder for ringing literary compositions that galvanized the American public into embracing the concept of conservation of natural resources.

After the Forest Service was established in 1905, the federal government set aside certain areas as national forests.[30] Nevada had few forests that could qualify, but a group of cattlemen with large ranch holdings petitioned the U.S. Department of Agriculture to establish a national forest on the higher mountain ranges in Nevada. Their motives were based not on conservation but on the hope that their ownership of commensurate property and their long history of using the mountain ranges would establish them as the sole users of the national forest ranges, excluding tramp sheep ranchers.[31] The Nevada cattle ranchers did succeed in excluding the landless sheep operations, but they found themselves subject to severe regulations designed to eliminate wildfires as well.

The ecological consequences of attempting to exclude fire as a stand renewal process in American forests, woodlands, and rangelands were not understood for much of the twentieth century. Early Forest Service silviculture bulletins often included a photograph of an old-growth ponderosa pine tree with a pronounced fire scar accompanied by a caption noting how destructive wildfires were in pine woodlands.[32] The government's policy of excluding wildfires during the twentieth century led to a greater expansion of woody vegetation in the Great Basin than had occurred since the Neoglacial period 5,000 years ago.[33]

University professors and natural resource scientists such as Harold Biswell at the University of California School of Forestry, R. R. Humphrey in the Southwest, Harold Weaver for ponderosa pine woodlands, and Rexford Daubenmire in the Pacific Northwest defied conventional wisdom to preach that fire was a natural part of the environment.[34] It remained for knowledgeable social scientists to communicate the ecological pitfalls of fire exclusion in natural resource management to a wider spectrum of the general public. Stephen J. Pyne, a leading authority on the history of fire on wildlands in North America, has published numerous books on wildfires based on fifteen summers he spent serving on fire suppression crews.[35]

For the first third of the twentieth century, the non–Forest Service federal lands in Nevada were technically under the administration of the Land Office of the Department of the Interior. The Land Office offered the land to homesteaders, but no one wanted the dry, dusty ranges where rain-fed crop production was impossible and irrigation water was fully (often excessively) appropriated. The Taylor Grazing Act in 1934 brought these lands under the control of the Grazing Service of the Department of the Interior, and President Franklin D. Roosevelt issued an executive order closing the lands to homesteading. The initial control of these lands by the Grazing Service was rather tenuous. The Battle Mountain Grazing District was not established in Nevada until 1954, long after the Grazing Service had become the Bureau of Land Management.

The tenure of the Forest Service and the Grazing Service during the first 60 years of the twentieth century coincided with a dramatic decrease in the number of acres of sagebrush/bunchgrass rangelands burned in the Great Basin. The two agencies deserve partial credit (or blame) for this reduction in wildfires, which was achieved through campaigns to stop illegal promiscuous burning and to prevent wildfires. In the case of big sagebrush communities, the reduction in wildfires was more biological than political-regulatory. There simply was not enough herbaceous vegetation on these shrub-dominated, overgrazed rangelands to support wildfire ignition and spread (fig. 16.2). Even cheatgrass was biologically suppressed by excessive grazing. Vast acres of big sagebrush communities, lacking an understory of herbaceous vegetation, were caught in an ecological time warp as big sagebrush plants maintained a viselike dominance. The sagebrush roots mined the barren interspaces between the shrub mounds for nutrients and water. An overwhelming annual seed rain ensured an abundant crop of sagebrush seedlings to occupy any opening in the community.

Mechanical Removal of Sagebrush. Dense big sagebrush stands were monumental barriers to the development of irrigated farms in the early days of reclamation projects in the Intermountain Area. Homesteaders had to laboriously grub out each sagebrush plant. Federal researchers financed by efforts to improve the productivity of the western ranges during World War II developed mechanical equipment to plow and seed degraded big sagebrush

Fig. 16.2. Big sagebrush community with virtually no herbaceous vegetation in the understory. It is very difficult to get this vegetation to ignite and the flames to spread from shrub to shrub.

stands.[36] As a substitute for burning as a stand renewal process, custom-constructed plows pulled by large track-laying tractors destroyed degraded big sagebrush stands that lacked sufficient herbaceous understory to burn. Plowing at least partially incorporated the woody biomass from the big sagebrush plants into the seedbed. Apparently, no scientist has ever investigated the relative influence of this process on nutrient cycling compared with burning. About 1 million of Nevada's 19 million acres of big sagebrush were plowed and seeded to crested wheatgrass, probably more than were seeded in any other western state. The seedings were largely successful because there was virtually no cheatgrass at that time to compete with the seedlings. The success of these seedings tremendously enhanced the forage supply on Nevada rangelands. Estimates in the early 1960s suggested that crested wheatgrass seedings furnished 25 percent of the forage base for the state's range livestock industry.[37] Increasing the forage base and using the crested wheatgrass seedings in the early spring, when native bunchgrass ranges were most susceptible to damage from grazing, reduced grazing pressure on the native herbaceous perennials, but it also reduced the grazing pressure on cheatgrass.

Federal agencies offered seedings to grazing permittees as bait to gain acceptance for reductions in the number of permitted animals or changes in the season of use of the ranges. This also favored cheatgrass. In hindsight, the cause and effect is obvious, but at the time it was lost in the euphoric belief that adding to the forage base and grazing management would return native perennial grasses to dominance in big sagebrush/bunchgrass communities.

Application of Herbicides. Toward the middle of the twentieth century, as previously mentioned, Don Hyder and A. C. Hull proposed substituting 2,4-D for wildfires as a stand renewal process in big sagebrush communities because public land management agencies were completely committed at the time to a policy of excluding wildfires. The herbicide experimenters were careful to spray only big sagebrush communities that still contained sufficient native perennial grasses to occupy the environmental potential released by killing the overstory shrub. During the 2004 science panel hearings on the status of sage grouse populations, J. W. Connelly estimated that one quarter of the total distribution of big sagebrush had been treated with 2,4-D;[38] that is a gross exaggeration. Perhaps 0.1 percent of the big sagebrush rangeland was treated with 2,4-D, and by 2004 it was impossible to identify these sites by their appearance. All had either reverted back to sagebrush dominance (a testimony to sagebrush reproductive potential) or converted to cheatgrass dominance.

The stand renewal process is relevant to the current status of cheatgrass and wildfires in several ways. Both Hyder and Hull stressed that the herbicide application treatment to renew big sagebrush stands would work only if sufficient native perennial grasses were present. The people who actually applied the herbicide did not always follow this guideline, though, or did not accurately assess the perennial grasses. If sufficient perennial grasses were not present, the herbicide treatment often resulted in conversion to cheatgrass dominance. Such a conversion also resulted if the site was not rested from grazing following the herbicide treatment to allow perennial grasses to increase or if subsequent grazing management or lack of management allowed excessive grazing to destroy the released perennial grasses. The herbicide treatments did create a transitory period of perennial grass dominance followed by the recruitment of sagebrush seedlings, resulting in a temporary community with a mixture of shrubs and grasses.

Superficially, mechanical and herbicidal substitutes for recurring wildfires to renew big sagebrush/bunchgrass plant communities appear to be identical processes, but there are important differences. The mechanical control and seeding process can be applied only to certain topographic areas and rock-free sites, while no such restrictions apply to aerial herbicide applications. The plowing action incorporates a portion of the woody biomass into the seedbed, but the herbicide leaves dead woody skeletons of the big sagebrush standing.

Wildfires Fueled by Cheatgrass. Wildfire can be reintroduced to degraded big sagebrush communities, with cheatgrass providing the fuel to spread the fire from shrub to shrub. Intentional exclusion of wildfires from sagebrush/bunchgrass communities by fire suppression and unintentional exclusion through removal of herbaceous fuel by excessive grazing coupled to produce an overabundance of sagebrush on millions of acres by the mid-twentieth century.[39] If bunchgrass was still present in sufficient quantities, burning followed by grazing management restored the grazing resource. If the native bunchgrasses were largely gone and cheatgrass provided the fuel for spreading fire among big sagebrush plants, the results of the fire were certain to be bad unless the burn was artificially seeded to perennial grasses.

In the late 1990s, the Nevada Section of the Wildlife Society put together a working group to develop a white paper on the Society's "Fire Statement." Wildlife officials were not pleased with the end product because they viewed it as "pro-fire." The paper stated that fire is beneficial in the mountain brush communities but not in the lower sagebrush and salt desert shrub communities. Even though mountain brush communities make up less than 7 percent of the Nevada landscape, most state wildlife managers would not support burning. Platt and Jackman struggled with the ethics of burning degraded big sagebrush stands that had converted to cheatgrass.[40] Because big sagebrush plants in degraded communities can live for decades, succession appears to be frozen and the lack of forage perpetual. If the degraded communities are burned, the community will convert to cheatgrass and sprouting shrubs. In years with above average precipitation, these burned areas produce a large amount of forage. Once a cheatgrass plant is established, no matter how diminutive, a seed bank is established.[41] Even during dry years, when cheatgrass seems to disappear from the sagebrush stands, the seed bank is intact. Each year with sufficient moisture, cheatgrass enlarges its distribution and density in the stands until conditions are right to support a wildfire. Platt and Jackman recognized this process at a time when it was not common knowledge among the scientific community, and certainly not among professional land managers. Platt and Jackman also recognized that while cheatgrass invasion results in a huge short-term increase in forage availability, once the cheatgrass is established, the site cannot be returned to the original big sagebrush/bunchgrass community. This is the most difficult concept to impart to the general public concerning the management of sagebrush/bunchgrass rangelands. The problem is not that Wyoming big sagebrush is burned in wildfires, it is that the communities did not burn while there was still a perennial grass understory to occupy the ecological potential (space, moisture, nutrients) released by the destruction of the nonsprouting big sagebrush.

Millions of acres of Wyoming big sagebrush/Thurber's needlegrass communities in northern Nevada have been converted to cheatgrass. The only way to restore those communities is through weed control and artificial seeding.[42] If there is enough woody biomass in the degraded community when it burns, the intensity of the fire will be sufficient to destroy a lot of the cheatgrass seed bank. This destruction will allow perennial grass seedlings to establish without weed control as long as the site is seeded the season immediately following the fire.

Grazing Management. Rest-rotation grazing management—usually using three pastures—as a substitute for brush and weed control followed by artificial seeding as stand renewal processes in sagebrush/bunchgrass plant communities has been nearly universally applied since the mid-1960s. Several factors combined to end the golden age of range improvement (mechanical and herbicidal brush control and artificial seeding). These included the passage of the National Environmental Policy Act (NEPA), which requires environmental impact studies before range improvement projects are implemented, and growing opposition to range improvement practices from vocal environmental and wildlife professional manager groups. Seeding of crested wheatgrass in particular was viewed as benefiting only ranchers and degrading the natural environment. There was also an issue of cost. Federal land management agencies had to obtain appropriations to pay for range improvements, and politicians were sensitive to adverse publicity about seeding crested wheatgrass. Rest-rotation was the answer to the conflict.

Rest-rotation grazing was successful in restoring the grass portion of the community at higher elevations in the Great Basin where remnant stands of native perennial grasses remained in the understory. On a much larger scale, at lower elevations, the application of this grazing system ensured dominance by cheatgrass. Most important to this discussion, this stand renewal process resulted in the accumulation of huge amounts of cheatgrass fuel during the years the pastures were rested and during the period of deferred grazing until after seed ripe. This was a recipe for massive wildfires. A. L. Hormay did excellent research developing the rest-rotation system, but he was not working in pure stands of cheatgrass with no remnant native perennial grass component. It is unfortunate that Hormay does not get the credit he deserves, because the grazing system works wonderfully when applied to the correct site. Wayne Burkhardt was the first to recognize that rest-rotation grazing management was interacting with wildfires to create a new frequency of stand renewal in the Intermountain Area. Ranchers were among the first to realize that rest-rotation grazing management was enhancing wildfires. Professional range managers eventually joined ranchers in grasping the magnitude of the problem but were powerless against established agency policies. Most scientist-educators, with the exception of Burkhardt, were the last to understand the relationship. This probably reflects on the relative frequency of field observations made by ranchers, range managers, and most wildland scientists.

Rest-rotation grazing on rangelands managed by the federal government is probably the most extensive application of a new stand renewal process to grazing lands ever undertaken in the world on a long time scale. In retrospect, the selling of this program as an answer to all the problems of range management in the Intermountain Area was one of the great con jobs of all time. Government employees who wanted rapid promotion rabidly supported rest-rotation grazing. It did not hurt to have a photograph of Gus Hormay on one's desk.

The results of the rest-rotation grazing management system have not yet been evaluated scientifically on a landscape scale. One of the primary reasons for dropping range improvement practices in favor of grazing management systems was to sidestep the legal requirements for environmental impact statements under NEPA. Eventually the courts required such compliance, but it was done as a blanket statement. The original implementation of regulations for environmental impact statements (EISs) limited such documents to 150 pages. By 2005, 35 years after the law was enacted, the average EIS was 570 pages long.[43] During that time, millions of acres of rangelands in the Intermountain Area were added to the ecological downward spiral of cheatgrass and recurrent wildfires.

The most serious consequence of implementing grazing management as the standard tool for sagebrush rangelands is the failure to artificially seed areas burned in wildfires. During the last decades of the twentieth century, millions of acres of burned big sagebrush on publicly owned land were left untouched after wildfires. Opposition from vocal environmentalists to a seed mixture that included crested wheatgrass was a major reason. Environmentalists tended to view introduced crested wheatgrass as a tool the range livestock industry was using to exploit public rangelands. If a native species has no chance of competing with cheatgrass, they thought, it is a waste of money to include it in postwildfire seeding mixtures.

The Deliberate Use of Prescribed Burning to Renew Stands. When Europeans arrived in the Great Basin, wildfires in native perennial grass communities may have been the most frequent and extensive form of stand renewal in the sagebrush/bunchgrass environments. Today, wildfires are becoming more frequent at higher-elevation sites under improved grazing management. One of the consequences of such wildfires is the suppression of browse species on summer ranges. This has a serious impact on mule deer populations, especially when coupled with cheatgrass-fueled wildfires that decimate winter ranges at lower elevations.

Fires Burning in Annual Grassland Communities. The current, but definitely not the last, act in the fire relations of big sagebrush/bunchgrass-cheatgrass communities is repeated fires burning in cheatgrass-dominated landscapes. This is probably the least-studied aspect of the continuum from precontact to present conditions, but it is rapidly becoming the most extensive type of landscape. After the 1999 firestorms in northern Nevada, we surveyed portions of the 1.6 million acres that burned. We commonly noted that fires that burned pure cheatgrass stands had not destroyed the litter on the soil surface and had left a near continuous layer of cheatgrass caryopses (fig. 16.3).

Fig. 16.3. Seedbed of cheatgrass-dominated community that burned in the 1999 firestorms that struck northern Nevada. The litter is largely composed of cheatgrass caryopses that are virtually untouched by the rapidly moving fire.

As portions of the area of rangelands currently dominated by cheatgrass are converted to dominance by other exotic invasive species, the wildfire characteristics of the communities will also change. Medusahead constitutes an even worse hazard for wildfires than cheatgrass.[44] Medusahead invasion results in abundant herbaceous fuel and ensures that sites will burn, destroying the low sagebrush and converting to perpetual medusahead dominance. The only way to biologically suppress medusahead is to encourage a good stand of perennial grasses.

The abundance of cheatgrass herbage obviously influences its fuel characteristics. If the year is favorable for cheatgrass growth but the site does not burn, the dried herbage adds to the next year's fuel load, as happened in northern Nevada in 1984–1985 and 1998–1999. If a critical mass of cheatgrass herbage is available, the chance of ignition and the rate with which wildfires spread increase. The presence of a critical mass of cheatgrass herbage also extends the wildfire season earlier and later in the calendar year. Cheatgrass invades not because of wildfires per se, but because of the relative lack of competition from degraded populations of herbaceous perennials, especially native perennial grasses. It also invades high ecological condition native plant communities because there is a niche available for a competitive annual grass that native plant species do not fill. Excessive, improperly timed, and annually repeated grazing enlarges the adaptive niche available to cheatgrass. Cheatgrass dominates because of this degradation and the destruction of sagebrush by wildfires. It may inhibit the establishment of sagebrush seedlings, but cheatgrass does not kill established sagebrush plants through competition. Wildfires are a necessary step in the transformation from shrub steppe to annual grass domination. It is easy to fall into the trap of believing that all of the ecological problems associated with cheatgrass in the Great Basin would be solved if wildfires were eliminated. Fire as a stand renewal process is an inherent part of the Great Basin ecology, however, especially in the big sagebrush/bunchgrass zone.

The native perennial grasses are not immune to wildfires. From the earliest days of range research in the Great Basin, researchers have understood that wildfires injure the perennial grass portion of plant communities and that some period is required for the perennial grasses to recover. Henry Wright used a portable combustion chamber to test the effect of burning on native grasses in southern Idaho.[45] He compared three stages of phenology (June, July, and August), perennial grass plant size, and two soil maximum temperatures (200 and 400°F). The short-lived perennial grasses squirreltail and Sandberg bluegrass were largely immune to the effects of burning. Needle-and-thread grass and Thurber's needlegrass were damaged by June and July burns. Before cheatgrass became widely established, wildfires historically occurred in August and September, after the dominant perennial grasses were mature. Extensive wildfires in June and July in the sagebrush/bunchgrass zone are almost entirely fueled by cheatgrass, at least in terms of the fine fuel that provides the continuity for fire spread. Wright also found that larger needlegrass plants were more susceptible to damage in wildfires than smaller plants, and the more intense the fire (higher soil temperature), the more the grasses were damaged. Sandberg bluegrass and squirreltail were not severely injured by the burning because even in June these ephemeral species had already produced seeds and were largely dormant. Wright considered bluebunch wheatgrass to be less susceptible to damage by wildfires because of its stem structure versus the leafy nature of the needlegrasses.

Bunchgrasses with dense vegetative culms are severely damaged by wildfires because their dense culms will burn for 2–3 hours after a fire passes. Wright recorded temperatures in bunchgrasses as high as 1,000°F 45 minutes after the fire front had passed.

Studies by Robert Blank and his colleagues have shown an increase in the levels of water-soluble nutrients in soil seedbeds following wildfires in sagebrush/bunchgrass communities.[46] The magnitude of these increases depends on a complex interaction between the fire temperature, the burn time, and the plant species being combusted. Blank suggested that changes in nutrient content of the seedbed are dictated by the mosaic of shrubs, perennial bunchgrasses, and annual grasses such as cheatgrass that are present at the time of the burn. If the community that burns has a relative abundance of woody species, postburn succession will be stimulated by an abundance of nitrogen. If the community that burns is dominated by cheatgrass, there will be a tiny spike in nutrient availability the season following the burn. Cheatgrass can respond to slight increases in available nitrogen while perennial grasses generally require a much higher rate of enrichment before responding. The higher the nonsprouting woody sagebrush's contribution to the species composition of the preburn community, the greater will be the space available for colonization in the postburn community.

Blank's research offers a fresh approach to the understanding of wildfire as a stand renewal process in former big sagebrush/bunchgrass plant communities currently occupied by cheatgrass. Escape from the maze of smoky mirrors that hinders the interpretation of past wildfires as chronologies lie literally at our feet in the chemical records etched in the genesis of soil profiles.

CHAPTER 17

Conclusions

Cheatgrass is still a paradox. This exotic alien is in firm control of plant succession in Great Basin plant communities. Cheatgrass truncates the natural progression toward native perennial grass and shrub dominance in a seemingly endless act of dominance. At the same time, cheatgrass appears to be only a single step in a progression of invasive and exotic species. As a species, cheatgrass is the ultimate high-stakes gambler. It hones biological efficiency to the maximum by producing millions of stable duplicates of successful genotypes through self-pollination. Mix in abundant phenotypic variability, and cheatgrass in a given spot can suck a seedbed dry, excluding seedlings of native species. At the same time, cheatgrass apparently catches the faintest breath of change blowing across the desert and responds with new genotypes through hybridization. If we do not intervene to restore cheatgrass-dominated areas to perennial dominance, the next stage in dominance of the weed communities may be even less desirable for the environment.

Fire and ice are the natural allies of cheatgrass. Wildfires fueled by cheatgrass destroy competing native woody vegetation. Cheatgrass cannot compete with established native vegetation, but stand renewal through burning in wildfires brings competition down to the seedling level, where cheatgrass is always going to win. Moisture is nearly completely out of phase with temperatures that permit plant growth in the Great Basin. Cheatgrass has the inherent ability to germinate at freezing temperatures and to grow roots that explore soil profiles throughout the depth of wetting during the winter months.

Despite strident vocal protest to the contrary, cheatgrass has become a major forage species in the Great Basin. Domestic livestock and wildlife species depend on it for significant portions of their diet, but the great variability in cheatgrass herbage production from one year to the next makes that a risky venture. Grazing domestic livestock on cheatgrass rather than allowing it to build fuel loads that result in catastrophic wildfires that consume critical habitats seems a logical thing to do, but few livestock grazing operations in the Great Basin can perform such a task. Cow-calf ranches simply cannot adjust stocking rates to meet extremes in cheatgrass herbage production. Nor is it yet possible to predict when a superabundant cheatgrass year is going to occur.

The invasion of cheatgrass on salt desert rangelands has increased the annual production of herbaceous forage on these winter ranges. Dry cheatgrass is only a spark away from disaster during the summer and fall. Ignition not only destroys the potential winter forage, it also closes federally owned lands to livestock grazing for 2 years.

More research is necessary to break the cycle of cheatgrass invasion and wildfires, but this is not solely a research problem. Society as a whole has to reach to a consensus on several vital issues before a scientifically valid cheatgrass policy can be crafted and implemented.

Consensus issue 1: Grazing livestock on rangelands. Will domestic livestock be allowed to graze on federal lands in the Great Basin? Closing public lands to grazing would have a devastating effect on deeded land property values and on the economy of small rural towns, and it would ensure an increase in cheatgrass and, concomitantly, more catastrophic wildfires. More than 60 years ago, A. C. Hull reported that two decades of trying to return native grasses to areas completely dominated by cheatgrass through grazing management had resulted in uniformly dismal failures.

Consensus issue 2. Plant material for restoring cheatgrass-infested rangelands. Richard Eckert pointed out 50 years ago that degraded big sagebrush/cheatgrass stands in the Great Basin would produce ten times more harvestable forage if they were converted to crested wheatgrass. In dry years, it would be closer to one hundred times more. Crested wheatgrass is available in the early spring and persists when native perennial grasses are most harmed by grazing. Grazed crested wheatgrass stands are virtually immune to wildfires. Crested wheatgrass excludes cheatgrass in dry years and suppresses it in wet years. Those who oppose crested wheatgrass because it is an alien species, and thus must be bad for the environment, overlook the fact that wheat, barley, oats, rye, and alfalfa, the cornerstones of American agriculture, are also alien to North America. Unlike cheatgrass, crested wheatgrass is not a self-invasive species. Others have complained that crested wheatgrass was seeded solely to benefit ranchers. Certainly it did that, but while benefiting ranchers it also biologically suppressed cheatgrass, prevented accelerated erosion, and broke the short return interval of wildfires.

Consensus issue 3. Integrated weed control for cheatgrass control and perennial seeding. R. L. Piemeisel clearly demonstrated 70 years ago that controlling cheatgrass without seeding a replacement produces a bumper crop of herbaceous weeds such as tumble mustard and Russian thistle that represent lower successional stages. Effective control of cheatgrass requires near complete eradication, otherwise perennial seedlings will not establish.

If the American public will not accept the use of herbicides on rangelands, researchers had better start thinking of innovative alternatives. Classic biological control with insects has not generally been applied to grasses, although the biological control of cheatgrass with plant pathogens has been researched for use in winter wheat fields in the Pacific Northwest. The development of such "outside-the-box" thinking will be enhanced by the comprehensive understanding of the physiological and ecological knowledge of cheatgrass that already exists.

More than 40 years ago, James Young was thoroughly castigated by an editor of an article on Great Basin potential plant communities with the question, "Who weeps for the plant communities that existed at contact with Europeans in the Great Basin?" Successful suppression of cheatgrass on Great Basin rangelands requires a passion for the environment, the knowledge base to understand the complex ecosystem, and the political savvy to reach consensuses that will work scientifically and practically. There is no time to waste.

Common and Scientific Names of Plants Mentioned in the Text

PLANTS

Alfalfa, *Medicago sativa*
Annual kochia, *Kochia scoparia*
Bailey's greasewood, *Sarcobatus baileyii*
Barbed goatgrass, *Aegilops triuncialis*
Barley, *Hordeum vulgare*
Basin wildrye grass, *Leymus cinereus*
Black greasewood, *Sarcobatus vermiculatus*
Bristlecone pine, *Pinus longaeva*
Budsage, *Artemisia spinescens*
Bulbous bluegrass, *Poa bulbosa*
Canadian thistle, *Circium arvense*
Cheatgrass, *Bromus tectorum*
Chilean chess, *Bromus trinii*
Common mullein, *Verbasum thapsus*
Desert needlegrass, *Achnatherum speciosum*
Dyer's woad, *Isatis tinctoria*
Filaree, *Erodium ciccutarium*
Forage kochia, 'Immigrant,' *Kochia prostrata*

Fourwing saltbush, *Atriplex canescens*
Fremont cottonwood, *Populus fremontii*
Halogeton, *Halogeton glomeratus*
Hoary chess, *Bromus commutatus*
Idaho fescue, *Festuca idahoensis*
Indian ricegrass, *Achnatherum hymenoides*
Lamb's-quarters, *Chenopodium atrovirons*
Medusahead, *Taeniatherum caput-medusae* ssp. *asperum*
Needle-and-thread grass, *Hesperostipa comata*
Poverty brome, *Bromus sterilis*
Rattlesnake chess, *Bromus briziformis*
Red brome, *Bromus rubens*
Sandberg bluegrass, *Poa sandbergii*
Six-weeks fescue, *Vulpia octoflora*
Slender oats, *Avena barbata*
Smooth brome, *Bromus mango*
Tansy mustard, *Descurania pinnata*
Thurber's needlegrass, *Achnatherum thurberianum*
Tumble mustard, *Sisymbrium altissimum*
Utah juniper, *Juniperus osteosperma*
Western juniper, *Juniperus occidentalis*
Whitebark pine, *Pinus albicaulis*
Wild oats, *Avena fatua*
Yellow pepperweed, *Lepidium perfoliatum*

Notes

1 : THE MANY FACES OF CHEATGRASS

1. J. A. Young and A. B. Sparks, *Cattle in the Cold Desert* (Reno: University of Nevada Press, 2002).

2. Ibid.

3. Ibid.

4. Ibid.

5. J. A. Young, "Hay making: The mechanical revolution on the western range," *Western Historical Quarterly* 14 (1983): 311–326. The narrative is a composite based on this article and many interviews conducted in the Great Basin; interviews on file at Nevada Historical Society, Reno.

6. K. Platt and E. R. Jackman, *The Cheatgrass Problem in Oregon*, Oregon Agricultural Experiment Station Bulletin 688, Corvallis, 1946.

7. Gloria Griffen Cline, *Exploring the Great Basin* (Norman: University of Oklahoma Press, 1963).

8. J. A. Young, "Chukar partridge," *Rangelands* 3 (1981): 166–168.

9. F. L. Emmerich, F. H. Tipton, and J. A. Young, "Cheatgrass: Changing perspectives and management strategies," *Rangelands* 15, no. 9 (1983): 37–40.

10. S. B. Doten, *Fifty-five Years of Agricultural Research on Ranch and Range*, Bulletin 163, Nevada Agricultural Experiment Station, University of Nevada, Reno, 1942.

11. J. A. Young, "Range research in the far western United States, the first generation," *Journal of Range Management* 53 (2000): 2–11.

12. J. A. Young, Fay Allen, and C. D. Clements, "The Wooton Plan: Division of Nevada rangelands based on water," in *Rangeland Management and Water Resources*, American Water Resources Association, Special Conference, Reno, Nev., 1998, pp. 293–298.

13. H. R. Kylie, G. H. Hieronymus, and A. G. Hall, CCC *Forestry* (Washington, D.C.: USDA Forest Service, 1937).

14. J. A. Young and R. A. Evans, "Erosion and deposition of fine sediments from playas," *Journal of Arid Environments* 10 (1986): 103–115.

15. Details on the wildfire are from various issues of the *Humboldt Star*, which reported on the subsequent events for more than a year.

2 : DEVELOPING A PERSPECTIVE OF THE ENVIRONMENT

1. Gloria G. Cline, *Exploring the Great Basin* (Norman: University of Oklahoma Press, 1963); Maurice S. Sullivan, ed., *The Travels of Jedediah S. Smith* (Santa Ana, Calif.: Fine Arts Press, 1960).

2. Samuel G. Houghton, *A Trace of Desert Waters* (Glendale, Calif.: Arthur H. Clark, 1976).

3. I. C. Russell, *Geological History of Lake Lahontan: A Quarternary Lake of Northwestern Nevada* (Washington, D.C.: U.S. Geological Survey, 1885).

4. W. D. Billings, "The shadscale vegetation zone of Nevada and eastern California in relation to climate and soils," *American Midland Naturalist* 42 (1949): 87–109.

5. Russell, *Geological History of Lake Lahontan.*

6. J. A. Young and A. B. Sparks, *Cattle in the Cold Desert* (Reno: University of Nevada Press, 2002).

3 : PREADAPTATION OF CHEATGRASS FOR THE GREAT BASIN

1. Hans Helbaek, "Plant collecting, dry farming, and irrigation agriculture in prehistoric Deh Luran," appendix A in *Prehistory and Human Ecology of the Deh Luran Plain,* ed. Frank Hole, Kent Flannery, and James Neely, pp. 383–426, Memoirs of the Museum of Anthropology (Ann Arbor: University of Michigan, 1969).

2. Vladimir Kostivkovsky and J. A. Young, "Invasive exotic rangeland weeds: A glimpse at their native habitat," *Rangelands* 22 (2000): 3–6.

3. A. G. Babaev, *Desert Problems and Desertification in Central Asia* (Berlin: Springer-Verlag, 1999).

4. J. A. Young and A. B. Sparks, *Cattle in the Cold Desert* (Reno: University of Nevada Press, 2002).

5. Kostivkovsky and Young, "Invasive exotic rangeland weeds."

6. A. S. Hitchcock, *Manual of the Grasses of the United States,* USDA Miscellaneous Publication 200 (Washington, D.C.: USDA, 1951).

7. J. O. Klemmedson and J. G. Smith, "Cheatgrass (*Bromus tectorum* L.)," *Botanical Review* (April–June 1964): 226–262.

8. Samuel Botsford Buckley, "Plants and insects from Georgia and Texas," *Proceedings of the Academy of Science of Philadelphia* (1862):98.

9. Helbaek, "Plant collecting, dry farming, and irrigation agriculture in prehistoric Deh Luran."

10. F. H. Hillman, *Clover Seeds and Their Impurities,* Nevada Agricultural Experiment Station Bulletin 47, Reno, 1900.

11. R. L. Piemeisel, *Changes in Weedy Plant Cover on Cleared Sagebrush Land and Their Probable Causes,* Technical Bulletin 654, Washington, D.C., 1938.

12. George W. Hendry and M. K. Bellue, "The plant content of adobe bricks," *California Historical Society Quarterly* 4 (1925): 110–127.

13. J. C. Frémont, *Report of the Exploring Expedition to the Rocky Mountains in the Year 1842 and to Oregon and North California in the Years 1843–44* (Washington, D.C.: Gales and Seaton Printers, 1845).

14. W. W. Robbins, "*Alien Plants Growing without Cultivation in California,* Bulletin 637 (Berkeley: University of California, 1940).

15. S. B. Parish, "The immigrant plants of southern California," *Southern California Academy of Science Bulletin* 20 (1920): 3–30.

16. E. G. Beckwith, *Report of Exploration for a Route for the Pacific Railroad of the Line of the Forty-first Parallel of North Latitude,* vol. 2, U.S. House of Representatives Executive Document 91, 33rd Congress, 2nd session, 1854 (Washington, D.C.: Government Printing Office, 1854).

17. Sereno Watson, "Botany," in *U.S. Geological Exploration of the Fortieth Parallel,* vol. 5 (Washington, D.C.: Government Printing Office, 1871).

18. Young and Sparks, *Cattle in the Cold Desert.*

19. I. C. Russell, *Geological History of Lake Lahontan: A Quaternary Lake of Northwestern Nevada* (Washington, D.C.: U.S. Geological Survey, 1885).

20. *Carson Morning Appeal,* December 4, 1886.

21. David Griffiths, *Forage Conditions on the Northern Border of the Great Basin,* Bulletin 15, USDA Bureau of Plant Industry, Washington, D.C., 1902.

22. F. Lamson-Scribner was one of the foremost agrostologists in the United States during the nineteenth century. See J. A. Young, "Range research in the far western United States: The first generation," *Journal of Range Management* 53 (2000): 2–11.

23. Young and Sparks, *Cattle in the Cold Desert.*

24. Ibid.

25. R. N. Mack, "Invasion of *Bromus tectorum* into western North America: An ecological chronicle," *Agro-Ecosystems* 7 (1981): 145–165.

26. W. D. Billings, "*Bromus tectorum*: A biotic cause of ecosystem impoverishment in the Great Basin," in *The Earth in Transition: Patterns and Processes of Biotic Impoverishment,* ed. G. M. Woodwell, 131–214 (New York: Cambridge University Press, 1989).

27. Ibid.

28. J. A. Young, "The public responses to the catastrophic spread of Russian thistle (1880) and halogeton (1945)," *Agricultural History* 62 (1988): 122–130.

29. J. A. Young, R. A. Evans, and B. L. Kay, "Germination of dyer's woad seeds," *Weed Science* 19 (1971): 76–78.

30. D. L. Yensen, "The 1900 invasion of alien plants in southern Idaho," *Great Basin Naturalist* 41 (1981): 176–183.

31. Young, "The public responses to the catastrophic spread of Russian thistle (1880) and halogeton (1945)."

32. Yensen, "The 1900 invasion of alien plants in southern Idaho."

33. Griffiths, *Forage Conditions on the Northern Border of the Great Basin.*

34. N. E. West, ed., *Temperate Deserts and Semi-deserts* (Amsterdam: Elsevier Scientific, 1982).

35. F. A. Sneva, "Grasses return following sagebrush control in eastern Oregon," *Journal of Range Management* 25 (1972): 174–178.

36. T. H. Kearney, L. J. Briggs, and H. T. Shantz, "Indicator significance of vegetation in Toole Valley, Utah," *Agricultural Research* 1 (1914): 365–417.

37. G. D. Pickford, "The influence of continued heavy grazing and promiscuous burning on spring-fall ranges in Utah," *Ecology* 13 (1932): 159–171.

38. Ibid.

39. G. Bennion, "Re-grassing the range," *National Wool Grower* 14, no. 6 (1924): 19–21.

40. Pickford, "The influence of continued heavy grazing and promiscuous burning on spring-fall ranges in Utah."

41. C. L. Forsling, "Fire and range improvement problems," *National Wool Grower* 24, no. 7 (1924): 14.

42. Yensen, "The 1900 invasion of alien plants in southern Idaho." Yensen cited two sources of information: O. R. Hick, cited as Idaho pioneer, personal communication; and J. E. Stablein, regional grazer, typed statement, in Works Progress Administration, History of Grazing, Manuscript files, Utah State University, Logan, 1940.

4 : SCIENTIFIC PERCEPTIONS OF CHEATGRASS

1. Related to James A. Young by the late Warren Whitman, professor of range ecology at North Dakota State University, 1961.

2. J. E. Weaver and F. E. Clements, *Plant Ecology* (New York: McGraw-Hill, 1929).

3. Ibid.

4. H. L. Shantz, "Plant succession on abandoned roads in eastern Colorado," *Journal of Ecology* 5 (1918): 1942.

5. H. C. Cowles, "The ecological relations of the vegetation on the sand dunes of Lake Michigan," *Botanical Gazette* 27 (1899): 95–116, 167–202, 281–308, 361–391.

6. F. E. Clements, *Plant Succession,* Publication 242 (Washington, D.C.: Carnegie Institute, 1916).

7. Weaver and Clements, *Plant Ecology,* p. 421.

8. J. A. Young, "Range research in the far western United States: The first generation," *Journal of Range Management* 53 (2000): 2–11.

9. W. R. Chapline, "Range research in the United States," *Herbage Review* 5 (1937): 1–13.

10. W. D. Rowley, *U.S. Forest Service Grazing and Rangelands* (College Station: Texas A&M University Press, 1985).

11. G. S. Strickler and W. B. Hall, "The Standley allotment: A history of range recovery," Research Paper 315, USDA Forest Service, Portland, Ore., 1980.

12. W. M. Keck, "Great Basin Station," Research Paper 118, USDA Forest Service, Ogden, Utah, 1972.

13. A. W. Sampson, *Plant Succession in Relation to Range Management,* USDA Bulletin 791 (Washington, D.C.: USDA, 1919), p. 3.

14. Ibid., p. 46.

15. Ibid.

16. For example, J. A. Young, R. A. Evans, and J. Major, "Alien plants in the Great Basin," *Journal of Range Management* 25 (1972): 194–201; D. L. Yensen, "The 1900 invasion of alien plants in southern Idaho," *Great Basin Naturalist* 41 (1981): 176–183; R. N. Mack, "Invasion of *Bromus tectorum* L. into western North America: An ecological chronicle," *Agro-Ecosystems* 7 (1981): 145–165.

17. G. H. Glover and W. W. Robbins, *Colorado Plants Injurious to Livestock,* Colorado Agricultural Experiment Station Bulletin 211, Fort Collins, 1915.

18. For example, D. Griffiths, *Forage Conditions on the Northern Border of the Great Basin,* Bulletin 15, USDA Bureau of Plant Industry (Washington, D.C.: USDA, 1902).

19. Mack, "Invasion of *Bromus tectorum* L. into western North America," and numerous citations; for example, C. V. Piper and R. K. Beattie, *The Flora of Southeastern Washington and Adjacent Idaho* (Pullman, Wash.: Allen Press, 1914).

20. G. D. Pickford, "The influence of continued heavy grazing and the promiscuous burning on spring-fall ranges in Utah," *Ecology* 13 (1932): 159–171.

21. J. F. Pechanec and G. Stewart, *Grazing Spring-Fall Sheep Ranges in Southern Idaho,* Agriculture Circular 808 (Washington, D.C.: USDA, 1949).

22. Pickford, "The influence of continued heavy grazing and the promiscuous burning on spring-fall ranges in Utah."

23. George Stewart and A. E. Young, "The hazard of basing permanent grazing capacity on *Bromus tectorum,*" *Agronomy Journal* 31 (1939): 1002–1015.

24. Ibid., p. 1002.

25. Ibid., p. 1003.

26. Ibid., p. 1004.

27. Ibid., p. 1005.

28. L. C. Hurtt, "Downy brome (cheatgrass) range for horses," Applied Forestry Notes, Northern Rocky Mountain Forest and Range Experiment Station, Missoula, Mt., 1939.

29. Stewart and Young, "The hazard of basing permanent grazing capacity on *Bromus tectorum*," p. 1004.

30. Ibid., p. 1007.

31. C. E. Fleming, M. A. Shipley, and M. R. Miller, *Bronco Grass* (Bromus tectorum) *on Nevada Ranges*, Bulletin 159, Nevada Agricultural Experiment Station, Reno, 1942.

32. C. E. Fleming, "Blanket system of handling sheep on the Madison National Forest," *National Wool Grower* 5 (1915): 7–10.

33. Fleming et al., *Bronco Grass*.

5 : SERAL CONTINUUM: THE FIRST STEP

1. J. A. Young, "Tumbleweed," *Scientific American* 264, no. 3 (1991): 82–85.

2. L. H. Dewey, *Russian Thistle*, USDA Bulletin 15 (Washington, D.C.: USDA, 1892).

3. Ibid.

4. Ibid.

5. A. Wallace, A. Rhods, and R. R. N. Raynor, "Germination behavior of *Salsola* influenced by temperature, moisture, depths of planting and gamma radiation," *Agronomy Journal* 60 (1968): 76–78.

6. J. A. Young and R. A. Evans, "Germination and establishment of *Salsola* in relation to seedbed environment. I. Temperature, after ripening, and moisture relations of *Salsola* seeds as determined by laboratory studies," *Agronomy Journal* 64 (1972): 214–218.

7. J. A. Young, "The response to the catastrophic spread of Russian thistle (1880) and halogeton (1945)," *Agricultural History* 62 (1988): 122–130.

8. Observation of Frosty Tipton, T Quarter Circle Ranch, Winnemucca, Nevada, July 15, 1980.

9. J. C. Beatley, "Russian thistle (*Salsola*) species in the western United States," *Journal of Range Management* 26 (1973): 225–226.

10. J. A. Young and R. A. Evans, "Barbwire Russian thistle seed germination," *Range Management* 32 (1979): 390–394.

11. J. K. McAdoo, C. C. Evans, B. A. Roundy, J. A. Young, and R. A. Evans, "Influence of heteromyid rodents on *Oryzopsis hymenoides* seeds in Lahontan sands," *Journal of Range Management* 36 (1983): 61–64.

12. J. A. Young, P. C. Martinelli, R. E. Eckert Jr., and R. A. Evans, *Halogeton*, Miscellaneous Publication 1553 (Washington, D.C.: USDA Agricultural Research Service, 1999).

13. Ibid.

14. R. W. Pemberton, "The distribution of halogeton in North America," *Journal of Range Management* 39 (1986): 281–283.

15. M. C. Williams, "Biochemical analysis, germination, and production of black and brown seed of *Halogeton glomeratus*," *Journal of Plant Physiology* 35 (1960): 500–505.

16. Young et al., *Halogeton.*

17. R. E. Eckert Jr. and F. E. Kinsinger, "Effect of *Halogeton glomeratus* leachate on chemical and physical characteristics of soils," *Ecology* 41 (1960): 785–790.

18. G. W. Hendry, "The adobe brick as a historical source," *Agricultural History* 5, no. 3 (1931): 110–127.

19. J. C. Frémont, *Report of the Exploring Expedition to the Rock Mountains in the Year 1842, and to Oregon and North California in the Years 1843–44* (Washington, D.C.: Gales and Seaton Printers, 1845).

20. J. A. Young, R. A. Evans, and B. L. Kay, "Dispersal and germination of broadleaf filaree," *Agronomy Journal* 67 (1975): 54–57.

6 : SERAL CONTINUUM: THE INTERMEDIATE STEP

1. J. A. Young and R. A. Evans, "Invasion of medusahead into the Great Basin," *Weed Science* 18 (1970): 89–97.

2. J. A. Young and R. A. Evans, "Erosion and deposition of fine sediments from playas," *Journal of Arid Environments* 10 (1986): 103–115.

3. W. W. Robbins, *Alien Plants Growing without Cultivation in California,* Bulletin 637, California Agricultural Experiment Station, Berkeley, 1940.

4. J. C. Hickman, ed., "Tansy mustard," in *The Jepson Manual* (Berkeley: University of California Press, 1993), pp. 413–416.

5. J. A. Young and R. A. Evans, "Mucilaginous seed costs," *Weed Science* 21 (1973): 52–54.

6. E. J. Salisbury, *The Reproductive Capacity of Plants* (London: Bells, 1942).

7. J. L. Harper, J. T. Williams, and G. R. Sager, "The behavior of seeds in soil. I. The heterogeneity of soil surfaces and its role in determining the establishment of plants," *Journal of Ecology* 539 (1965): 273–286.

8. N. Collis-George and J. E. Sands, "The control of seed germination by moisture as a soil physical property," *Australian Journal of Agricultural Research* 10 (1959): 628–636.

9. J. A. Young and R. A. Evans, "Germination of Great Basin wildrye seeds collected from native stands," *Agronomy Journal* 73 (1981): 917–920.

10. J. A. Young, R. A. Evans, R. O. Gifford, and R. E. Eckert Jr., "Germination characteristics of three species of Cruciferae," *Weed Science* 18 (1970): 41–47.

11. Ibid.

12. Ibid.

13. Ibid.

1. R. L. Piemeisel, *Changes in Weedy Plant Cover on Cleared Sagebrush Land and Their Probable Causes,* USDA Bulletin 654 (Washington, D.C.: USDA, 1951).

2. J. A. Young, "Snake River country—a rangeland heritage," *Rangelands* 8 (1986): 199–201.

3. J. A. Young and C. D. Clements, "Range research: The second generation," *Journal of Range Management* 54 (2001): 115–121.

4. R. Daubenmire, *Steppe Vegetation of Washington,* Technical Bulletin 62, Washington Agricultural Experiment Station, Pullman, 1970, p. 80.

5. Ibid., p. 81.

6. A. Leopold, "Cheatgrass takes over," *Land* 1 (1940): 310–313.

7. S. A. Warg, "Life history and economic studies on *Bromus tectorum,*" master's thesis, University of Montana, 1938.

8. E. W. Tisdale, "The grass of the southern interior of British Columbia," *Ecology* 27 (1947): 159–167.

9. A. C. Hull and J. F. Pechanec, "Cheatgrass—a challenge to range research," *Journal of Forestry* 45 (1947): 555–524.

10. R. Daubenmire, "Plant succession due to overgrazing in the *Agropyron* bunchgrass prairie of southeastern Washington," *Ecology* 21 (1940): 55–65.

11. J. A. Young and D. E. Palmquist, "Plant age/size distribution in black sagebrush: Effects on community structure," *Great Basin Naturalist* 52 (1992): 313–320.

12. J. L. Charley and N. E. West, "Micro patterns of nitrogen mineralization activity in soils of some shrub dominated semi-desert ecosystems of Utah," *Soil Biology and Biochemistry* 9 (1977): 357–365.

13. J. A. Young, R. A. Evans, and J. Major, "Alien plants in the Great Basin," *Journal of Range Management* 25 (1972): 194–201.

14. N. E. West, "Western intermountain steppe," in *Temperate Desert and Semideserts,* vol. 5 of *Ecosystems of the World,* ed. N. E. West (Amsterdam: Elsevier, 1983), pp. 351–374.

15. R. N. Mack and J. N. Thompson, "Evolution of steppe with few large, hoofed mammals," *American Naturalist* 119 (1982): 757–773.

16. J. W. Burkhardt, *Herbivory in the Intermountain West,* University of Idaho Forest and Range Experiment Station Bulletin 58, Moscow, 1996.

17. N. E. West and M. A. Hassan, "Population dynamics after wildfires in sagebrush grasslands," *Journal of Range Management* 31 (1985): 283–289.

18. J. A. Young and H. Mayeux, "Seed ecology of woody species of *Artemisia* and *Chrysothamnus,*" in *Compositae: Biology and Utilization,* vol. 2, ed. P. D. S. Caligari and D. J. N. Hinds (Kew, U.K.: Royal Botanic Gardens, 1996), pp. 93–104.

19. J. A. Young and C. D. Clements, "Cheatgrass control and seeding," *Rangelands* 22, no. 4 (2000): 3–7.

20. R. A. Evans, R. E. Eckert, B. L. Kay, and J. A. Young, "Downy brome control with soil active herbicides," *Weed Science* 17 (1969): 166–169.

21. J. O. Klemmedson and J. G. Smith, "Cheatgrass (*Bromus tectorum* L.)," *Botanical Review* (April–June 1964): 226–262.

22. L. W. Kreitlow and A. T. Bleak, "*Podosporriella verticillata*, a soil borne pathogen of some western Gramineae," *Phytopathology* 54 (1964): 353–357.

23. J. A. Young, R. A. Evans, and R. E. Eckert, "Population dynamics of downy brome," *Weed Science* 17 (1968): 20–26.

24. Ibid.

25. Ibid.

26. S. A. Warg, "Life history and economic studies on *Bromus tectorum*."

27. J. A. Young and R. A. Evans, "Erosion and deposition of fine sediments from playas," *Journal of Arid Environments* 10 (1979): 103–115.

28. N. I. Newman, "Factors controlling the germination date of winter annuals," *Journal of Ecology* 52 (1963): 391–404.

29. J. L. Harper, J. T. Williams, and G. R. Sanger, "The behavior of seeds in the soil. I. The heterogeneity of the soil surface and its role in determining the establishment of plants," *Journal of Ecology* 539 (1965): 273–286.

30. A. M. Mayer and Poljakoff-Mayer, *The Germination of Seeds* (Oxford: Pergamon Press, 1965).

31. R. A. Evans and J. A. Young, "Enhancing germination of dormant seeds of downy brome," *Weed Science* 23 (1975): 354–357; J. A. Young and R. A. Evans, "Germinability of seed reserves in a big sagebrush community," *Weed Science* 23 (1975): 358–364.

32. R. A. Evans and J. A. Young, "Plant litter and establishment of alien annual species in rangeland communities," *Weed Science* 18 (1970): 697–703; R. A. Evans and J. A. Young, "Microsite requirements for establishment of annual rangeland weeds," *Weed Science* 20 (1972): 350–356.

8 : THE COMPETITIVE NATURE OF CHEATGRASS

1. J. O. Klemmedson and J. G. Smith, "Cheatgrass (*Bromus tectorum*)," *Botanical Review* 30 (1964): 226–262.

2. J. H. Robertson and C. K. Pearse, "Artificial reseeding and the closed community," *Northwest Science* 19 (1946): 58–68.

3. B. Davis, "Joseph H. Robertson—range scientist pioneer," *Rangelands* 11 (1989): 210–212.

4. Robertson and Pearse, "Artificial reseeding and the closed community," 58.

5. J. H. Robertson, "A quantitative study of the true prairie vegetation after three years of extreme drought," *Ecological Monographs* 9 (1939): 431–492.

6. Robertson and Pearse, "Artificial reseeding and the closed community," 58.

7. Ibid., 60.

8. J. A. Young and D. McKenzie, "Rangeland drill," *Rangelands* 4 (1982): 108–113.

9. S. A. Warg, "Life history and economic studies of *Bromus tectorum*," master's thesis, Montana State University, 1938.

10. L. C. Hurtt, "Cheatgrass ranges for horses," *Montana Farmer* 25, no. 11 (1938): 2; L. C. Hurtt, "Downy brome (cheatgrass) range for horses," Applied Forestry Notes, USDA Forest Service, Northern Rocky Mountain Forest and Range Experiment Station, Missoula, Mt., 1939.

11. J. A. Young, "Range research in the far western United States: The first generation," *Journal of Range Management* 53 (2000): 1–14.

12. A. C. Hull Jr. and J. F. Pechanec, "Cheatgrass—a challenge to range research," *Journal of Forestry* 45 (1947): 554–564.

13. R. Daubenmire, "Plant succession due to overgrazing in the *Agropyron* bunchgrass prairie of southeastern Washington," *Ecology* 21 (1940): 55–64.

14. G. D. Piemeisel, *Changes in Weed Plant Cover on Cleared Sagebrush Land and Their Probable Causes,* USDA Technical Bulletin 654 (Washington, D.C.: USDA, 1938).

15. Ibid.

16. L. E. Spence, "Root studies of important range plants of the Boise River watershed," *Journal of Forestry* 35 (1937): 747–754.

17. G. W. Craddock and C. K. Pearse, *Surface Run-off and Erosion on Granitic Mountain Soils of Idaho as Influenced by Range Cover, Soil Disturbance, Slope, and Precipitation Intensity,* USDA Circular 428 (Washington, D.C.: USDA, 1938).

18. See G. Stewart and A. E. Young, "The hazard of basing permanent grazing capacity on *Bromus tectorum*," *Agronomy Journal* 31 (1939): 1002–1015.

19. C. E. Fleming, M. A Shipley, and M. R. Miller, *Bronco Grass* (Bromus tectorum) *on Nevada Ranges,* Bulletin 159, Nevada Agricultural Experiment Station, Reno, 1942.

20. J. A. Young, R. A. Evans, R. E. Eckert Jr., and B. L. Kay, "Cheatgrass," *Rangelands* 9 (1987): 266–270.

21. L. A. Sharp, K. Sanders, and N. Rimbey, "Variability of crested wheatgrass production over 35 years," *Rangelands* 14 (1992): 153–168.

22. Piemeisel, *Changes in Weed Plant Cover on Cleared Sagebrush Land and Their Probable Causes.*

23. J. A. Young, R. A. Evans, and R. E. Eckert, "Population dynamics of downy brome," *Weed Science* 17 (1968): 20–26.

24. W. T. Hinds, "Energy and carbon balance in cheatgrass: An essay in autecology," *Ecological Monographs* 45 (1975): 367–388.

25. J. F. Pechanec and A. C. Hull Jr., "Cheatgrass fires reduce next year's early spring forage," *National Wool Grower* 35, no. 4 (1945): 13.

26. G. Stewart and A. C. Hull, "Cheatgrass (*Bromus tectorum* L.)—an ecologic intruder in southern Idaho," *Ecology* 30 (1939): 58–72.

27. L. C. Hulbert, "Ecological studies of *Bromus tectorum* and other annual bromegrasses," *Ecological Monographs* 24 (1955): 181–213.

28. L. E. Spence, "Root studies of important range plants of the Boise River watershed," *Journal of Forestry* 35 (1937): 747–754.

29. E. W. Tisdale, "The grasslands of the southern interior of British Columbia," *Ecology* 28 (1947): 346–382, 350.

30. H. C. Hanson, "Ecology of grasslands, II," *Botanical Review* 16 (1950): 283–360, 301.

31. L. C. Hulbert, "Ecological studies of *Bromus tectorum* and other annual bromegrasses."

32. Hulbert used the methods of J. D. Sayre and V. H. Morris, "The lithium method of measuring the extent of corn root systems," *Plant Physiology* 15 (1940): 761–764.

33. Robertson and Pearse, "Artificial reseeding and the closed community," 64.

34. R. A. Evans, "Effects of different densities of downy brome (*Bromus tectorum*) on growth and survival of crested wheatgrass (*Agropyron desertorum*) in the greenhouse," *Weeds* 9 (1961): 216–223.

35. A. C. Hull, "Competition and water requirements of cheatgrass and wheatgrass in the greenhouse," *Journal of Range Management* 16 (1963): 199–204.

36. W. S. Chepil, "Germination of weed seeds. I. Longevity, periodicity of germination, and vitality of seeds in cultivated soil," *Scientific Agriculture* 26 (1946): 307–346.

37. Hulbert, "Ecological studies of *Bromus tectorum* and other annual bromegrasses."

38. D. W. Finnerty and D. L. Klingman, "Life cycles and control studies of some weed bromegrasses," *Weeds* 10 (1962): 40–47.

39. G. A. Harris, "Some competitive relationships between *Agropyron spicatum* and *Bromus tectorum*," *Ecological Monographs* 37 (1967): 89–111.

40. J. A. Young and R. A. Evans, *Temperature Profiles for Germination of Cool Season Range Grasses*, Bulletin ARR-W-27, USDA Agricultural Research Service, Oakland, Calif., 1982.

41. J. A. Young, R. A. Evans, and B. A. Roundy, "Quantity and germinability of *Oryzopsis hymenoides* seeds in Lahontan sands," *Journal of Range Management* 36 (1983): 61–64.

42. M. A. Hassan and N. E. West, "Dynamics of soil seed pools in burned and unburned sagebrush semi-deserts," *Ecology* 67 (1986): 269–272.

43. G. A. Harris, "Some competitive relationships between *Agropyron spicatum* and *Bromus tectorum.*"

44. G. Harris and A. M. Wilson, "Competition for moisture among seedlings of annual and perennial grasses as influenced by root elongation at low temperatures," *Ecology* 51 (1970): 530–534.

9 : GENETIC VARIATION AND BREEDING SYSTEMS

1. H. G. Baker, "The evolution of weeds," *Annual Review of Ecology and Systematics* 5 (1974): 1–24, 1.

2. E. J. Salisbury, *Weeds and Aliens* (London: Collins, 1961).

3. Baker, "The evolution of weeds," p. 304.

4. R. W. Allard, "Genetic systems associated with colonizing ability in predominantly self-pollinated species," in *The Genetics of Colonizing Species,* ed. H. G. Baker and G. L. Stebbins (New York: Academic Press, 1965), pp. 49–75.

5. G. L. Stebbins, "Self fertilization and population variability in higher plants," *American Naturalist* 91 (1957): 337–338.

6. G. L. Stebbins, *Variation and Evolution in Plants* (New York: Columbia University Press, 1950), p. 166.

7. M. J. McKone, "Reproductive biology of several bromegrasses (*Bromus*): Breeding system, pattern of fruit maturation, and seed set," *American Journal of Botany* 72 (1960): 1334–1339.

8. Stebbins, *Variation and Evolution in Plants.*

9. J. R. Harlan, "Cleistogamy and chasmogamy in *Bromus carinatus* Hook. and Arn," *American Journal of Botany* 32, no. 3 (1945): 142–148; H. G. Baker, "Self-compatibility and establishment after "long-distance" dispersal," *Evolution* 9 (1955): 347–348.

10. Stebbins, "Self fertilization and population variability in higher plants."

11. Ibid., 346.

12. Ibid.

13. Ibid.

14. Baker, "The evolution of weeds," p. 75.

15. W. T. Hinds, "Energy and carbon balances in cheatgrass: An essay in autecology," *Ecological Monographs* 45 (1975): 367–388.

16. C. A. Suneson, "An evolutionary plant breeding method," *Agronony Journal* 48 (1956): 188–190.

17. Baker, "The evolution of weeds."

18. S. J. Novak, R. N. Mack, and D. E. Soltis, "Genetic variation in *Bromus tectorum* (Poaceae) population differentiation in its North American range," *American Journal of Botany* 78 (1991): 1150–1161.

19. M. Nei, "Estimating average heterozygosity and genetic distance from a small number of individuals," *Genetics* 89 (1978): 583–590.

20. L. C. Hulbert, "Ecological studies of *Bromus tectorum* and other annual bromegrasses," *Ecological Monographs* 25 (1955): 181–213.

21. C. P. Mallory-Smith, P. Hendrickson, and G. Mueller-Warrent, "Cross-resistance of primisulfuron-resistant *Bromus tectorum* L. (downy brome) to sulfosul-furon," *Weed Science* 47 (1999): 256–257.

22. R. A. Evans and J. A. Young, "Plant litter and the establishment of alien annual species in rangeland communities," *Weed Science* 18 (1970): 697–703.

23. R. R. Blank, F. Allen, and J. A. Young, "Extractable anions in soils following wildfire in a sagebrush-grass community," *Soil Science Society of America Journal* 58 (1994): 564–570.

24. J. A. Young and R. A. Evans, "Responses of weed populations to human manipulations of the natural environment," *Weed Science* 19 (1976): 186–190.

25. Ibid.

26. W. D. Billings, "The plant associations of the Carson Desert, western Nevada," *Butler University Botany Studies* 8 (1945): 89–123.

27. J. A. Young and F. H. Tipton, "Invasion of cheatgrass into arid environments of the Lahontan Basin," General Technical Report, USDA Forest Service, Ogden, Utah, 1990, pp. 37–40.

28. A. A. Agrawal, "Phenotypic plasticity in the interactions and evolution of species," *Science* 294 (2001): 321–326.

29. D. Elbert, C. Haag, M. Kirpatrick, M. Riek, J. W. Hottinger, and V. Pajunen, "A selective advantage to immigrant genes in *Daphnia* metapopulations," *Science* 295 (2002): 485–487.

10 : CONTROL OF CHEATGRASS AND SEEDING PRIOR TO HERBICIDES

1. J. A. Young and D. McKenzie, "Rangeland drill," *Rangelands* 4 (1982): 108–113.

2. J. A. Young, "Range research in the far western United States: The first genera-tion," *Journal of Range Management* 53 (2000): 2–11.

3. J. A. Young, Fay Allen, and C. D. Clements. "The Wooton Plan—division of Nevada rangelands based on water," in Proceedings of the Specialty Conference on Rangeland Management and Water Resources, American Water Resources Associa-tion, Reno, Nev., May 27–29, 1998, pp. 293–298.

4. J. A. Young, P. C. Martinelli, R. E. Eckert Jr., and R. A. Evans, *Halogeton*, USDA Miscellaneous Publication 1553 (Washington, D.C.: USDA Agricultural Research Service, 1999).

5. G. Bennion, "Re-grassing the range," *National Wool Grower* 14, no. 6 (1924): 19–21; quoted from p. 19.

6. J. A. Young and A. B. Sparks, *Cattle in the Cold Desert* (Reno: University of Nevada Press, 2002).

7. J. A. Young, "Range research in the far western United States: The first generation."

8. J. F. Pechanec, A. P. Plummer, J. H. Robertson, and A. C. Hull, "Eradication of big sagebrush (*Artemisia tridentata*)," Research Paper 10, USDA Forest Service, Ogden, Utah, 1944.

9. A. W. Sampson, *The Revegetation of Over Grazed Range Areas,* USDA Forest Service Circular 158 (Washington, D.C.: USDA, 1908).

10. W. R. Chapline, "Range research in the United States," *Herbage Review* 5 (1937): 1–13.

11. J. S. Cotton, *The Improvements of Mountain Meadows,* USDA Bureau of Plant Industry Bulletin 127 (Washington, D.C.: USDA, 1908).

12. C. L. Forsling and W. A. Dayton, *Artificial Reseeding on Western Mountain Range Lands,* USDA Circular 178 (Washington, D.C.: USDA, 1931).

13. Young, "Range research in the far western United States: The first generation."

14. P. B. Kennedy, *The Depletion of the Ranges for Sheep and Cattle, with Suggestions for Range Improvement,* Bulletin 51, Nevada Agricultural Experiment Station, Reno, Nev., 1901.

15. A. C. Dillman, "History of crested wheatgrass in North America," *Agronomy Journal* 38 (1946): 237–250.

16. R. J. Lorenz, "Introduction and early use of crested wheatgrass in the Northern Great Plains," in *Crested Wheatgrass,* symposium proceedings, ed. K. L. Johnson (Logan: Utah State University, 1986), pp. 9–20.

17. A. C. Hull Jr. and G. J. Klomp, "Longevity of crested wheatgrass in the sagebrush-grass type in southern Idaho," *Journal of Range Management* 19 (1966): 5–11.

18. J. A. Young and R. A. Evans, "History of crested wheatgrass in the Intermountain Area," in *Crested Wheatgrass,* symposium proceedings, ed. K. L. Johnson (Logan: Utah State University, 1986), pp. 21–25.

19. F. H. Hillman, *Field Notes on Some Nevada Grasses,* Nevada Agricultural Experiment Station Bulletin 39, Reno, 1896.

20. Kennedy, *The Depletion of the Ranges for Sheep and Cattle.*

21. P. B. Kennedy and S. C. Dinsmore, *Digestion Experiments on the Range,* Nevada Agricultural Experiment Station Bulletin 71, Reno, 1909.

22. R. R. Blank, J. D. Trent, and J. A. Young, "Sagebrush communities on clayey soils of northeastern California: A fragile equilibrium," in *Proceedings of the Symposium on Ecology and Management of Riparian Shrub Communities,* USDA General Technical Report 289 (Ogden, Utah: USDA, 1992), pp. 199–203.

23. J. A. Young, C. D. Clements, and G. Nader, "Medusahead and clay: The rarity of perennial seedling establishment," *Rangelands* 21, no. 6 (1999): 19–23.

24. L. D. Humphrey and E. W. Schupp, "Seedling survival from locally and commercially obtained seeds on two semi-arid sites," *Restoration Ecology* 10 (2002): 88–95.

25. A. W. Sampson, "Collection and sowing of alfilaria seed (*Erodium cicutarium*), Sequoia National Forest," *Forest Service Investigations* 2 (1913): 14–17.

26. L. A. Stoddart, *Seeding Ranges to Grass,* Utah Agricultural Experiment Station Circular 122, Utah Agricultural College, Logan, 1943.

27. G. Stewart, *Reseeding Rangelands of the Intermountain West,* Farmer's Bulletin 1823 (Washington, D.C.: USDA, 1939).

28. J. A. Young and J. D. Budy, *Endless Tracks in the Woods* (Sarasota, Fla.: Cresting Publishing, 1989).

29. Young, "Range research in the far western United States: The first generation."

30. Pechanec et al., "Eradication of big sagebrush (*Artemisia tridentata*)."

31. Young and McKenzie, "Rangeland drill."

32. J. W. Robertson and A. P. Plummer, "Hints for the use of the wheat-land-type plows for brush eradication in connection with range reseeding," Research Paper 13, USDA Forest Service, Ogden, Utah, 1946.

33. Young and McKenzie, "Rangeland drill."

34. W. R. Chapline, "Early beginnings," 32d Annual Report, Vegetation Rehabilitation and Equipment Workshop, San Antonio, Tex., 1978, pp. 1–4.

35. Ibid., 2.

36. Ibid.

37. Ibid.

38. Young and McKenzie, "Rangeland drill."

39. J. F. Pechanec and A. C. Hull Jr., "Australian plow needs change," *Idaho Farmer-Stockman,* November 7, 1947, 17.

40. Young and McKenzie, "Rangeland drill."

41. J. A. Young, R. A. Evans, and D. McKenzie, *History of Brush Control on Western U.S. Rangelands* (Lubbock: Texas Tech University Press, 1985).

42. Young and McKenzie, "Rangeland drill."

43. Young, Martinelli, et al., *Halogeton.*

44. Ibid.

45. Ibid.

46. D. N. Hyder, "Controlling big sagebrush with growth regulators," *Journal of Range Management* 6 (1953): 109–116.

47. For example, J. H. Robertson and C. K. Pearse, "How to seed Nevada range lands," Research Paper 3, USDA Forest Service, Ogden, Utah, 1943.

48. J. A. Young, "Intermountain shrub steppe plant communities—pristine and grazed," in *Proceedings of the Western Raptor Management Symposium and Workshop*, ed. Beth Giron-Pendleton (Baltimore: National Wildlife Federation Scientific and Technical Series no. 12, 1999), pp. 33–44.

49. Ibid.

50. A. C. Hull, "The relation of grazing to establishment and vigor of crested wheatgrass," *Agronomy Journal* 36 (1944): 358–360.

51. J. A. Young, R. A. Evans, and R. E. Eckert Jr., "Effects of tillage operations on dispersal of downy brome caryopses in the soil," in *Proceedings of the Western Society of Weed Science*, Boise, Ida., 1968, 22.

11 : CONTROL AND SEEDING WITH HERBICIDES

1. F. L. Timmons, "History of weed control," in *Proceedings of the Washington State Weed Conference*, Yakima, 1963, pp. 1–11.

2. J. A. Young, P. C. Martinelli, R. E. Eckert Jr., and R. A. Evans, *Halogeton*, USDA Miscellaneous Publication 1553 (Washington, D.C.: USDA, 1999).

3. Timmons, "History of weed control."

4. B. W. Allred, "Distribution and control of several woody plants in Oklahoma and Texas," *Journal of Range Management* 2 (1949): 17–29.

5. J. A. Young, R. A. Evans, and D. W. McKenzie, *History of Brush Control on Western United States Rangelands* (Lubbock: Texas Tech University Press, 1985).

6. Anonymous, *A Survey of Extent and Cost of Weed Control and Specific Weed Problems*, ARS 34–23–1 (Washington, D.C.: USDA, Agricultural Research Service, 1965).

7. G. E. Peterson, "The discovery and development of 2,4-D," *Journal of Agricultural History* 41 (1967): 243–253.

8. J. W. Mitchell and C. L. Hammer, "Polyethylene glycol as carriers for growth regulating substances," *Botanical Gazette* 106 (1944): 482.

9. C. L. Hammer and H. B. Tukey, "Selective herbicidal actions of midsummer and fall applications of (2,4-dichlorophenoxy) acetic acid," *Botanical Gazette* 106 (1944): 232–233.

10. D. N. Hyder, "Controlling big sagebrush with growth regulators," *Journal of Range Management* 6 (1953): 109–116.

11. Timmons, "History of weed control."

12. H. D. Kerr and D. L. Klingman, "Rooting of grasses affected by siduron," Weed Science Society of America, Abstracts of Annual Meeting, 1972, p. 26.

13. J. A. Young, R. A. Evans, and R. E. Eckert Jr., "Siduron, a selective herbicide to aid in the establishment of perennial grasses," in *Proceedings of the Western Society of Weed Science*, Boise, Ida., 1968, p. 17.

14. R. E. Eckert Jr. and F. E. Kinsinger, "Effects of *Halogeton glomeratus* leachate on chemical and physical characteristics of soils," *Ecology* 41 (1960): 785–790; V. L. Kinsinger and R. E. Eckert Jr., "Emergence and growth of annual and perennial grasses and forbs in soil altered by halogeton leachate," *Journal of Range Management* 14 (1961): 194–197.

15. R. A. Evans, "Effect of different densities of downy brome (*Bromus tectorum*) on growth and survival of crested wheatgrass (*Agropyron desertorum*) in the greenhouse," *Weeds* 9 (1961): 216–223.

16. A. C. Hull Jr. and G. Stewart, "Replacing cheatgrass with perennial grasses on southern Idaho ranges," *Agronomy Journal* 40 (1948): 694–703.

17. C. A. Friedrich, "Seeding crested wheatgrass on cheatgrass land," Research Note 38, USDA Forest Service, Missoula, Mt., 1945.

18. R. A. Evans, R. E. Eckert Jr., B. L. Kay, and J. A. Young, "Downy brome control by soil-active herbicides for revegetation of rangelands," *Weed Science* 17 (1969): 166–169.

19. Ibid., p. 168.

20. Ibid.

21. R. E. Eckert Jr. and R. A. Evans, "A chemical-fallow technique for control of downy brome and establishment of perennial grasses on rangelands," *Journal of Range Management* 20 (1967): 35–41; H. P. Alley and E. Chamberlain, "Summary report of chemical fallow studies," Circular 166, University of Wyoming College of Agriculture, Laramie, 1962; R. W. Bovey and C. R. Fenster, "Aerial application of herbicides on fallow land," *Weeds* 12 (1964): 117–119.

22. S. J. Novak, R. N. Mack, and D. E. Soltis, "Genetic variation in *Bromus tectorum* (Poaceae) population differentiation in its North American range," *American Journal of Botany* 78 (1991): 1156–1161.

23. R. E. Eckert Jr., G. J. Klomp, R. A. Evans, and J. A. Young, "Establishment of perennial wheatgrass in relation to atrazine residue in the seedbed," *Journal of Range Management* 25 (1972): 219–224.

24. A. C. Hull Jr., *Regrassing Southern Idaho Rangelands*, Bulletin 146, Cooperative Agricultural Extension Service, University of Idaho, Moscow, 1944.

25. W. J. McGinnies, "The relationship of furrow depth to moisture content of soil and to seedling establishment on a range site," *Agronomy Journal* 51 (1959): 13–14.

26. A. P. Plummer, A. C. Hull Jr., G. Stewart, and J. H. Robertson, *Seeding Rangelands in Utah, Nevada, Southern Idaho, and Western Wyoming*, USDA Handbook 71 (Washington, D.C., USDA, 1949); A. L. Nelson, "Methods of tillage for winter wheat at Archer Field Station," Bulletin 300, Wyoming Agricultural Experiment Station, Laramie, 1950.

27. R. E. Eckert Jr., "Atrazine residue and seedling establishment in furrows," *Journal of Range Management* 27 (1974): 55–56.

28. J. A. Young, R. A. Evans, and R. E. Eckert Jr., "Population dynamics of downy brome," *Weed Science* 16 (1968): 20–26; R. A. Evans and J. A. Young, "Plant litter and establishment of alien annual species in rangeland communities," *Weed Science* 18 (1970): 697–703; R. A. Evans and J. A. Young, "Microsite requirements for establishment of annual rangeland weeds," *Weed Science* 20 (1972): 350–356.

29. R. A. Evans, H. R. Holbo, R. E. Eckert Jr., and J. A. Young, "Functional environment of downy brome communities in relation to weed control and revegetation," *Weed Science* 18 (1970): 154–162.

30. Young et al., "Population dynamics of downy brome."

31. Eckert and Evans, "A chemical-fallow technique for control of downy brome and establishment of perennial grasses on rangelands."

32. R. E. Eckert Jr., G. J. Klomp, J. A. Young, and R. A. Evans., "Nitrate-nitrogen status of fallowed rangeland soils," *Journal of Range Management* 23 (1970): 445–447.

33. J. A. Young and D. McKenzie, "Rangeland drill," *Rangelands* 4 (1982): 108–113.

34. McGinnies, "The relationship of furrow depth to moisture content of soil and to seedling establishment on a range site."

35. J. E. Asher and R. E. Eckert Jr., "Development, testing, and evaluation of the deep furrow drill arm assembly for the rangeland drill," *Journal of Range Management* 26 (1973): 377–379.

36. W. J. McGinnies, "Effect of seeding rate and row spacing on establishment and yield of crested wheatgrass," *Agronomy Journal* 62 (1970): 417–421.

37. J. S. Spencer, V. M. Rashelof, and J. A. Young, "Safety modifications for operation and transportation of the rangeland drill," *Journal of Range Management* 32 (1979): 406–407.

38. R. E. Eckert, J. E. Asher, M. Dale Christensen, and R. A. Evans, "Evaluation of the atrazine-fallow technique for weed control and seedling establishment," *Journal of Range Management* 27 (1974): 288–292.

39. J. A. Young and R. A. Evans, "Weed control in wheatgrass seedbeds with siduron and picloram," *Weed Science* 18 (1970): 546–549.

40. M. D. Christensen, J. A. Young, and R. A. Evans, "Control of annual grasses and revegetation in ponderosa pine woodlands," *Journal of Range Management* 27 (1974): 143–145.

41. R. A. Evans and J. A. Young, "Weed control-revegetation systems for big sagebrush/downy brome rangelands," *Journal of Range Management* 30 (1977): 331–336.

42. R. E. Eckert Jr., "Renovation of sparse stands of crested wheatgrass," *Journal of Range Management* 32 (1979): 332–336.

43. J. A. Young and R. A. Evans, "Etiolated growth of range grasses for an indication of tolerance to atrazine," *Weed Science* 26 (1987): 480–483.

44. R. A. Evans, R. E. Eckert Jr., and B. L. Kay, "Wheatgrass establishment with paraquat and tillage on downy brome ranges," *Weeds* 15 (1967): 50–55.

45. J. A. Young, R. A. Evans, and B. L. Kay, "Responses of medusahead to paraquat," *Journal of Range Management* 24 (1971): 4–43.

46. W. C. Robocker, D. H. Gates, and H. D. Kerr, "Effects of herbicides, burning, and seeding date in reseeding an arid range," *Journal of Range Management* 18 (1965): 114–118.

47. J. A. Young, R. A. Evans, and R. E. Eckert Jr., "Wheatgrass establishment with tillage and herbicides in a mesic medusahead community," *Journal of Range Management* 22 (1969): 151–155.

48. J. A. Young, B. A. Roundy, A. D. Brunner, and R. A. Evans, *Ground Sprayers for Sagebrush Rangelands*, Bulletin AAT-W-8 (Oakland, Calif.: USDA Agricultural Research Service, 1979).

49. J. A. Young and R. A. Evans, "Control of pinyon saplings with picloram or carbutilate," *Journal of Range Management* 29 (1976): 144–147.

50. B. L. Kay and J. E. Street, "Drilling wheatgrass into sprayed sagebrush in northeastern California," *Journal of Range Management* 14 (1961): 271–273.

51. R. Carson, *Silent Spring* (Boston: Houghton Miffin, 1962).

52. A. L. Hormay, "Rest-rotation grazing: a new management system for perennial bunchgrass ranges," Production Research Report No. 51, USDA Forest Service, Washington, D.C., 1961.

12 : REVEGETATION PLANT MATERIAL

1. D. R. Dewey, "The genomic system of classification as a guide to intergeneric hybridization with the perennial Triticeae," in *Gene Manipulation in Plant Improvement*, ed. J. P. Gustafson (New York: Plenum Press, 1984), pp. 209–279.

2. D. R. Dewey, "Taxonomy of the crested wheatgrasses," in *Crested Wheatgrass*, symposium proceedings, ed. K. L. Johnson (Logan: Utah State University, 1986), pp. 31–41, 32.

3. D. R. Dewey and K. H. Assay, "The crested wheatgrasses of Iran," *Crop Science* 15 (1975): 844–849; D. R. Dewey and K. H. Assay, "Cytogenetic and taxonomic relationships among three diploid crested wheatgrasses," *Crop Science* 22 (1982): 645–650.

4. Dewey, "Taxonomy of the crested wheatgrasses."

5. K. H. Assay and R. P. Knowles, "The wheatgrasses," in *Forages: The Science of Grassland Agriculture*, ed. M. E. Heath, R. F. Barnes, and D. S. Metcalfe (Ames: Iowa State University Press, 1985), pp. 166–176.

6. Dewey, "Taxonomy of the crested wheatgrasses."

7. J. R. Carlson and J. L. Schwendiman, "Plant materials for crested wheatgrass seedings in the Intermountain Area," in *Crested Wheatgrass,* symposium proceedings, ed. K. L. Johnson (Logan: Utah State University, 1986), pp. 45–52.

8. W. J. McGinnies, "Effects of seedbed firming on the establishment of crested wheatgrass seedlings," *Journal of Range Management* 15 (1962): 230–234.

9. L. E. Kirk, *Crested Wheat Grass,* Bulletin 54, University of Saskatchewan, Saskatoon, Canada, 1932.

10. Carlson and Schwendiman, "Plant materials for crested wheatgrass seedings in the Intermountain Area."

11. Ibid.

12. Ibid.

13. Ibid.

14. Ibid.

15. J. A. Young, R. R. Blank, W. S. Longland, and D. E. Palmquist, "Seeding Indian ricegrass in an arid environment in the Great Basin," *Journal of Range Management* 42 (1994): 2–7.

16. Carlson and Schwendiman, "Plant materials for crested wheatgrass seedings in the Intermountain Area."

17. G. A. Rogler, *Nordan Crested Wheatgrass,* Bulletin 16, North Dakota State University, 1954, 150–152.

18. K. H. Assay, "Breeding strategies in crested wheatgrass," in *Crested Wheatgrass,* symposium proceedings, ed. K. L. Johnson (Logan: Utah State University, 1986), pp. 53–57, 54.

19. K. H. Assay and D. R. Dewey, "Bridging ploidy differences in crested wheatgrass hexaploid × diploid crosses," *Crop Science* 25 (1985): 368–369; D. R. Dewey and P. C. Pendse, "Hybrids between *Agropyron desertorum* and *Agropyron cristatum,*" *Crop Science* 8 (1968): 607–611.

20. W. Tai and D. R. Dewey, "Morphology, cytology, and fertility of diploid and colchicine-induced tetraploid wheatgrass," *Crop Science* 6 (1966): 223–226.

21. Assay, "Breeding strategies in crested wheatgrass."

22. K. H. Assay, D. R. Dewey, F. B. Gomm, D. A. Johnson, and J. R. Carlson, "Registration of Hycrest crested wheatgrass," *Crop Science* 25 (1985): 368–369.

23. Carlson and Schwendiman, "Plant materials for crested wheatgrass seedings in the Intermountain Area."

24. Dewey, "The genomic system of classification as a guide to intergeneric hybridization with the perennial Triticeae."

25. L. A. Stoddart, *Seeding Ranges to Grass,* Circular 122, Agricultural Experiment Station, Utah Agricultural College, Logan, 1943; J. H. Robertson, "Seeding date and depth of planting crested wheatgrass and other grasses at lower eleva-

tions in northern Nevada," Research Paper 14, USDA Forest Service, Ogden, Utah, 1949; A. C. Hull Jr., "Depth, season, and row spacing for planting grasses on southern Idaho range lands," *Agronomy Journal* 40 (1948): 960–969; D. N. Hyder, F. A. Sneva, and W. A. Sawyer, "Soil firming may improve seeding operations," *Journal of Range Management* 8 (1955): 159–163; W. McGinnies, "Effect of seedbed firming on the establishment of crested wheatgrass seedlings," *Journal of Range Management* 15 (1962): 230–234.

26. J. Aldorson and W. C. Sharp, *Grass Varieties in the United States* (Boca Raton, Fla.: CRC, Lewis Publishers, 1995).

27. Ibid.

28. Ibid.

29. A. C. Hull Jr. and G. J. Klomp, "Thickening and spread of crested wheatgrass stands on southern Idaho ranges," *Journal of Range Management* 20 (1967): 222–227.

30. J. E. LaTourrette, J. A. Young, and R. A. Evans, "Seed dispersal in relation to rodent activities in several big sagebrush communities," *Journal of Range Management* 24 (1971): 118–120.

31. M. A. Hein, "Registration of varieties and strains of grasses. Wheatgrasses (*Agropyron* spp.) III. Whitmar beardless wheatgrass (reg. no. 4)," *Agronomy Journal* 59 (1958): 685–686; K. B. Jensen and K. H. Asay, "Cytology and morphology of *Elymus hoffmanii* (Poaceae:Triticeae): A new species from the Erzurum Province of Turkey," *International Journal of Plant Science* 157 (1996): 750–758.

32. S. R. Larson, T. A. Jones, Z.-M. Hu, A. J. Palazzo, and C. L. McCracken, "Genetic diversity of bluebunch wheatgrass cultivars and a multiple-origin polycross," *Crop Science* 40 (2000): 1142–1147.

33. J. L. Gibbs, G. Young, and J. R. Carlson, "Registration of 'Goldar' bluebunch wheatgrass," *Crop Science* 31 (1991): 1708.

34. T. A. Jones, S. R. Larson, D. C. Nielson, S. A. Young, N. J. Chatterton, and A. J. Palazzo, "Announcement of release of germplasm," USDA Agricultural Research Service, Logan, Utah, 2003.

35. R. A. Evans and J. A. Young. "Micro site requirements for downy brome (*Bromus tectorum*) infestation and control on sagebrush rangelands," *Weed Science* 332: 13–17.

36. J. A. Young and R. A. Evans, *Temperature Profiles for Germination of Cool Season Range Grasses*, Bulletin ARR-W-27, USDA Agricultural Research Service, Oakland, Calif., 1982.

37. J. A. Young, R. A. Evans, D. A. Johnson, and K. H. Asay, "Cold temperature germination of *Elytriga repens* × *Pseudorogenera spicata* hybrids," *Journal of Range Management* 39 (1986): 300–302.

38. J. A. Young, R. A. Evans, and D. E. Palmquist, "Soil surface characteristics and

emergence of big sagebrush seedlings," *Journal of Range Management* 43 (1990): 358–367.

39. N. C. Frischknecht and A. T. Bleak, "Encroachment of big sagebrush in seeded range in northeastern Nevada," *Journal of Range Management* 10 (1957): 165–170.

40. J. K. McAdoo, W. S. Longland, and R. A. Evans, "Nongame bird community response to sagebrush invasion of crested wheatgrass seedings," *Journal of Wildlife Management* 53 (1989): 494–502.

41. F. B. Gomm, "A comparison of two sweet clover strains and Ladak alfalfa alone and in mixtures with crested wheatgrass for range and dryland seeding," *Journal of Range Management* 17 (1964): 19–22; M. D. Rumbaugh and T. Thorn, "Initial stands of interseeded alfalfa," *Journal of Range Management* 18 (1965): 258–261; M. D. Rumbaugh and M. W. Pedersen, "Survival of alfalfa in five semiarid range seedings," *Journal of Range Management* 32 (1979): 48–51.

42. A. S. Hitchcock, *Manual of the Grasses of the United States,* Miscellaneous Publication 200 (Washington, D.C.: USDA, 1950).

43. R. R. Blank and J. A. Young, "Heated substrate and smoke: Influence on seed emergence and plant growth," *Journal of Range Management* 51 (1998): 577–583.

44. D. N. Hyder and F. A. Sneva, "A method for rating the success of range seeding," *Journal of Range Management* 7 (1954): 89–90.

45. J. L. Harper, *Population Biology of Plants* (London: Academic Press, 1977).

46. Hull and Stewart, "Replacing cheatgrass with perennial grasses on southern Idaho ranges."

47. John Vallentine, *U.S.-Canadian Range Management 1935–1977* (Phoenix: Oryx Press, 1978).

48. A. C. Hull Jr., R. C. Holmgren, W. H. Berry, and J. A. Wagner, "Pelleted seeding on the western range," USDA Forest Service Miscellaneous Publication 922, Ogden, Utah, 1963.

49. G. A. Baylan, "*Kochia prostrata* and its culture in Kirghhiza," Frunse, USSR, *Izdatel 'stov Kirghizo,* 1972 (authors' translation).

50. W. Keller and A. T. Bleak, "*Kochia prostrata:* a shrub for western ranges," *Utah Science* 35 (1974): 24–25.

51. C. D. Clements, K. Gray, and J. A. Young, "Forage kochia: To seed or not to seed." *Rangelands* 19 (1997): 29–31.

13 : CHEATGRASS AND NITROGEN

1. B. L. Kay and R. A. Evans, "Effects of fertilization on a mixed stand of cheatgrass and intermediate wheatgrass," *Journal of Range Management* 18 (1965): 7–11.

2. A. M. Wilson, G. A. Harris, and D. H. Harris, "Fertilization of mixed cheatgrass-bluebunch wheatgrass stands," *Journal of Range Management* 19 (1966): 134–137.

3. A. C. Hull Jr., "Growth periods and herbage production of cheatgrass and reseeded grasses in southwestern Idaho," *Journal of Range Management* 17 (1949): 261–264.

4. W. E. Martin, C. Pierce, and V. P. Osterli, "Differential nitrogen responses of annual and perennial grasses," *Journal of Range Management* 17 (1964): 67–68.

5. G. A. Harris, "Some competitive relationships between *Agropyron spicatum* and *Bromus tectorum*," *Ecological Monographs* 37 (1967): 89–111.

6. R. E. Eckert Jr. and R. A. Evans, "Response of downy brome and crested wheatgrass to nitrogen and phosphorus in nutrient solution," *Weeds* 11 (1963): 170–174.

7. Ibid.

8. B. L. Kay, "Fertilization of cheatgrass ranges in California," *Journal of Range Management* 19 (1966): 217–220.

9. N. E. West and J. Skujiš, "The nitrogen cycle in North American cold-winter semi-desert ecosystems," *Oecologia Plantarum* 12 (1977): 45–53.

10. J. A. Young and R. A. Evans, "Erosion and deposition of fine sediments from playas," *Journal of Arid Environments* 10 (1986): 103–115.

11. R. R. Blank, J. A. Young, and F. L. Allen, "Aeolian dust in a saline playa environment, Nevada, U.S.A.," *Journal of Arid Environments* 41 (1998): 365–381.

12. West and Skujiš, "The nitrogen cycle in North American cold-winter semi-desert ecosystems."

13. Young and Evans, "Erosion and deposition of fine sediments from playas."

14. N. E. West and J. Skujiš, "The nitrogen cycle in North American cold-winter semi-desert ecosystems."

15. R. E. Eckert Jr., G. J. Klomp, J. A. Young, and R. A. Evans, "Nitrate-nitrogen status of fallowed rangeland soils," *Journal of Range Management* 23: 445–447.

16. D. Tilman, "Nitrogen-limited growth in plants from different successional states," *Ecology* 67 (1986): 555–563; T. McLendon and E. F. Redente, "Effect of nitrogen limitation on species replacement dynamics during early succession on a sagebrush site," *Oecologia* 91 (1992): 312–317.

17. E. A. Paul and F. E. Clark, *Soil Microbiology and Biochemistry* (San Diego: Academic Press, 1988).

18. J. A. Young, J. D. Trent, R. R. Blank, and D. E. Palmquist, "Nitrogen interactions with medusahead (*Taeniatherum caput-medusae* ssp. *asperum*) seedbanks," *Weed Science* 46 (1998): 191–195.

19. R. R. Blank, J. D. Trent, and J. A. Young, "Sagebrush communities on clayey soils of northeastern California, a fragile equilibrium," in *Proceedings of the Symposium*

on Ecology and Management of Riparian Shrub Communities, USDA General Technical Report 289 (Ogden, Utah: USDA, Forest Service, 1992), pp. 198–200.

20. Ibid.

21. J. A. Young and R. A. Evans, "Invasion of medusahead into the Great Basin," *Weed Science* 18 (1970): 89–97.

22. J. A. Young, "Ecology and management of medusahead (*Taeniatherum caput-medusae* ssp. *asperum* [Simk.] Melderis)," *Great Basin Naturalist* 52 (1992): 245–252.

23. R. A. Evans and J. A. Young, "Enhancing germination of dormant seeds of downy brome," *Weed Science* 23: 354–357; C. M. Karssen and W. H. M. Hilorst, "Effect of chemical environment on seed germination," in *Seeds: The Ecology of Regeneration in Plant Communities,* ed. M. Fenner (Wallingford, U.K.: CAB International, 1992), pp. 327–347.

24. T. A. Monaco, D. A. Johnson, J. M. Norton, T. A. Jones, K. J. Conners, J. B. Norton, and M. B. Redinbaugh, "Contrasting responses of Intermountain West grasses to soil nitrogen," *Journal of Range Management* 56 (2003): 282–290.

25. J. A. Young and R. A. Evans, "Conversion of medusahead to downy brome communities with diuron," *Journal of Range Management* 25 (1972): 40–43.

26. J. A. Young, R. A. Evans, and R. E. Eckert Jr., "Population dynamics of downy brome," *Weed Science* 17 (1968): 20–26; J. A. Young and R. A. Evans, "Germinability of seed reserves in a big sagebrush community," *Weed Science* 23 (1975): 358–364.

27. J. A. Young and R. A. Evans, "Conversion of medusahead to downy brome communities with diuron," *Journal of Range Management* 25 (1972): 40–43.

28. J. A. Young, C. D. Clements, and R. R. Blank, "Influence of nitrogen on antelope bitterbrush seedling establishment," *Journal of Range Management* 50 (1997): 536–540.

29. J. A. Young and C. D. Clements, *Purshia: The Wild and Bitter Roses* (Reno: University of Nevada Press, 2002).

30. A. L. Hormay, "Bitterbrush in California," Research Note 34, USDA Forest Service, Berkeley, Calif., 1943; E. C. Nord, "Autecology of bitterbrush in California," *Ecological Monographs* 35 (1965): 307–334.

31. Young et al., "Influence of nitrogen on antelope bitterbrush seedling establishment."

32. J. E. LaTourrette, J. A. Young, and R. A. Evans, "Seed dispersal in relation to rodent activities in seral big sagebrush communities," *Journal of Range Management* 24 (1971): 451–454.

33. J. A. Young, R. R. Blank, and C. D. Clements, "Nitrogen enrichment and immobilization influences on the dynamics of an annual grass community," in *People and Rangelands: Building the Future,* ed. D. Eldridge and D. Freudenberger, Pro-

ceedings of the Sixth International Rangeland Congress, Townsville, Queensland, Australia, 1999, pp. 279–281.

34. R. R. Blank and J. A. Young, "Influence of matric potential and substrate characteristics on germination of Nezpar Indian ricegrass," *Journal of Range Management* 45 (1992): 205–209.

35. Ibid.

14 : GRAZING MANAGEMENT

1. Glen Fulcher and William Mathews, Discussion, in *Proceedings of the Cheatgrass Symposium* (Portland, Ore.: USDI, Bureau of Land Management, 1965), pp. 81–83.

2. J. A. Young and B. A. Sparks, *Cattle in the Cold Desert* (Reno: University of Nevada Press, 2002).

3. Ibid.

4. B. Z. Lang and J. A. Young, "Red water," *Rangeman's Journal* 5 (1978): 107–109.

5. M. Clawson, *The Western Range Livestock Industry* (New York: McGraw-Hill, 1950); A. W. Sampson, *Range and Pasture Management* (New York: McGraw-Hill, 1923); L. A. Stoddart and A. D. Smith, *Range Management* (New York: McGraw-Hill, 1943).

6. Ibid.

7. Young and Sparks, *Cattle in the Cold Desert.*

8. Fulcher and Mathews, Discussion, in *Proceedings of the Cheatgrass Symposium,* pp. 81–83.

9. K. Platt and E. R. Jackman, *The Cheatgrass Problem in Oregon,* Bulletin 668, Agricultural Extension Service, Oregon State University, Corvallis, 1946, p. 9.

10. For example: M. A. Shipley, C. E. Fleming, and B. S. Martineau, *Estimating the Value of Forage for Grazing Use by Means of an Animal-Unit-Month Factor Table,* Bulletin 160, Agricultural Experiment Station, University of Nevada, Reno, 1942.

11. M. M. Caldwell, J. H. Johnson, R. S. Nowak, and R. S. Dzurec, "Coping with herbivory: Photosynthetic capacity and resource allocation in two semiarid *Agropyron* bunchgrasses," *Oecologia* 50 (1992): 14–24.

12. C. W. Cook and L. A. Stoddard, "Some growth responses of crested wheatgrass following herbage removal," *Journal of Range Management* 6 (1953): 267–270; J. R. Gray and H. W. Springfield, *Lambing on Crested Wheatgrass,* Bulletin 461, Agricultural Experiment Station, New Mexico State University, Las Cruces, 1962.

13. D. W. Hedrick, "Managing crested wheatgrass for early spring use," *Journal of Range Management* 20 (1967): 53–54.

14. J. A. Young, R. A. Evans, and J. Major, "Alien plants in the Great Basin," *Journal of Range Management* 25 (1972): 194–201.

15. B. L. Kay, "Fertilization of cheatgrass ranges in California," *Journal of Range Management* 19 (1966): 217–222.

16. D. N. Hyder and F. A. Sneva, "Fertilization on sagebrush-bunchgrass range: A progress report," Miscellaneous Paper 115, Oregon Agricultural Experiment Station, Corvallis, 1961; F. A. Sneva and D. N. Hyder, "Yield, yield-trend, and response to nitrogen of introduced grasses on the Oregon high desert," Special Report, Oregon Agricultural Experiment Station, Corvallis, 1965; F. A. Sneva, "Crested wheatgrass response to nitrogen and clipping," *Journal of Range Management* 26 (1973): 47–50.

17. P. W. McCormick and J. P. Workman, "Early range readiness with nitrogen fertilizer: An economic analysis," *Journal of Range Management* 28 (1975): 181–184.

18. H. F. Heady and J. Barttolome, *Vale Rangeland Rehabilitation Program: Desert Repaired in Southeastern Oregon,* Research Bulletin 70 (Portland, Ore.: USDA Forest Service, 1977).

19. Caldwell et al., "Coping with herbivory: Photosynthetic capacity and resource allocation in two semiarid *Agropyron* bunchgrasses."

20. J. R. Bentley and M. W. Talbot, "Annual-plant vegetation of the California foothills as related to range management," *Ecology* 29 (1948): 72–79.

21. T. L. DeFlon, "The case for cheatgrass," *Rangelands* 8: 14–17.

22. Platt and Jackman, *The Cheatgrass Problem in Oregon,* p. 10.

23. Ibid.

24. J. O. Klemmedson and R. B. Murray, "Research on cheatgrass ranges," in *Proceedings of the Cheatgrass Symposium* (Vale, Ore.: USDI Bureau of Land Management, 1965), pp. 38–50.

25. R. B. Murray, H. F. Mayland, and P. J. Van Soest, "Growth and nutritional value to cattle of grasses on cheatgrass range in southern Idaho," Research Paper 199, USDA Forest Service, Ogden, Utah, 1978.

26. C. E. Fleming, M. A. Shipley, and M. R. Miller, *Bronco Grass* (Bromus tectorum) *on Nevada Ranges,* Bulletin 159, Nevada Agricultural Experiment Station, Reno, 1942.

27. R. McCall, R. T. Clark, and A. R. Patton, *The Apparent Digestibility and Nutritive Value of Several Native and Introduced Grasses,* Bulletin 418, Montana Agricultural Experiment Station, Montana State College, Bozeman, 1943.

28. Fleming et al., *Bronco Grass* (Bromus tectorum) *on Nevada Ranges.*

29. Ibid.

30. Murray et al., "Growth and nutritional value to cattle of grasses on cheatgrass range in southern Idaho."

31. R. B. Murray and J. O. Klemmedson, "Cheatgrass range in southern Idaho: seasonal cattle and grazing capacities," *Journal of Range Management* 21 (1968): 308–313.

32. Platt and Jackman, *The Cheatgrass Problem in Oregon.*

33. Ibid.

34. A. Nelson, *The Brome Grasses of Wyoming*, Bulletin 46, Wyoming Agricultural Experiment Station, Laramie, 1901.

35. J. A. Young, "Range research in the far western United States: The first generation," *Journal of Range Management* 53 (2000): 2–11.

36. Young and Sparks, *Cattle in the Cold Desert*.

37. Ibid.

38. J. A. Young and C. D. Clements, "Rangeland monitoring," *Arid Land Research and Management* 17 (2003): 439–447.

39. F. L. Emmerich, J. A. Young, and J. W. Burkhardt, "A Nevada ranch family: Their success through four generations," *Rangelands* 14 (1992): 66–70.

40. J. A. Young and F. Tipton, "Invasion of cheatgrass into arid environments of the Lahontan basin," in *Symposium on Cheatgrass Invasion, Shrub Die-off and Other Aspects of Shrub Biology and Management*, General Technical Report 276 (Ogden, Utah: USDA Forest Service, 1990), pp. 37–40; F. H. Tipton, "Cheatgrass, livestock, and rangeland," in *Symposium on Ecology and Management of Annual Rangelands*, General Technical Report 313 (Ogden, Utah: USDA Forest Service, 1994), pp. 414–416.

41. Tipton, "Cheatgrass, livestock, and rangeland."

42. G. K. Miller, J. A. Young, and R. A. Evans, "Germination of seeds of perennial pepperweed," *Weed Science* 34 (1986): 252–255.

15 : CHEATGRASS AND WILDLIFE

1. J. Peterson, "Gone with the sage," Range Improvement Notes, USDA Forest Service, Ogden, Utah, 16 (1971): 1–5.

2. F. E. Clements and E. S. Clements, "The sagebrush desclimax," *Carnegie Institute Yearbook* 38 (1939): 139–140; O. Julander, "Range management in relation to mule deer habitat and herd productivity," *Journal of Range Management* 15 (1962): 278–281; O. Julander and J. B. Low, "A historic account and present status of mule deer in the west," in *Mule Deer Decline in the West*, symposium proceedings (Logan: Utah State University, 1976), 3–20; J. A. Young, R. E. Eckert, and R. A. Evans, "Historical perspective regarding the sagebrush ecosystem," in *The Sagebrush Ecosystem*, symposium proceedings, Utah State University, Logan, 1978, pp. 1–13; G. E. Gruell, *"Post-1900 Mule Deer Irruptions in the Intermountain West: Principal Causes and Influences*, General Technical Report INT 206 (Logan, Utah: USDA Forest Service, 1986), pp. 1–37.

3. Robert McQuivey, personal communications, 2000; T. R. Vale, "Presettlement vegetation in the sagebrush-grass area of the Intermountain West," *Journal of Range Management* 28 (1975): 32–36.

4. F. A. Sneva, "Grazing return following sagebrush control in eastern Oregon," *Journal of Range Management* 25 (1972): 174–178.

5. J. Work, *The Journal of John Work. Fur Brigade to the Boneventura California Expedition 1832–1833 for the Hudson Bay Company*, ed. Alice Maloney and Herbert Eugene Bolten (San Francisco: California Historical Society, 1833); A. S. Leopold, "Deer in relation to plant succession," *Transactions of the North American Wildlife Conference* 15 (1950): 571–580; D. Helfrich, E. Helfrich, and T. Hunt, *Emigrant Trails West* (Reno: Trails West, 1984).

6. Work, *Journal*.

7. Helfrich et al., *Emigrant Trails West*.

8. F. M. Phillips, *Desert People and Mountain Men: Exploration of the Great Basin, 1824–1865* (Bishop, Calif.: Chalpant Press, 1977).

9. R. R. Kindschy, C. Sundstrom, and J. Yoakum, "Pronghorns," in *Wildlife Habitats in Managed Rangelands: The Great Basin of Southeastern Oregon*, ed. J. W. Thomas and C. Maser, General Technical Report PNW-160, USDA Forest Service; USDI, Bureau of Land Management, Portland, Ore., 1976.

10. D. Mathis, *Following the Nevada Wildlife Trail: A History of Nevada Wildlife and Wildlife Management* (Reno: Nevada Agricultural Foundation, 1997).

11. Nevada Division of Wildlife, 2000–2001 Big Game Status, compiled by Mike Cox, Big Game Staff biologist, Reno, Nev.

12. C. D. Clements and J. A. Young, "A viewpoint: Rangeland health and mule deer habitat," *Journal of Range Management* 50 (1998): 129–138.

13. Ibid.

14. D. R. Dietz and J. G. Nagy, "Mule deer nutrition and plant utilization," in *Mule Deer Decline in the West* (Logan: Utah State University, 1976), pp. 71–78; A. S. Leopold, "Deer in relation to plant succession," *Journal of Forestry* 48, no. 10 (1950): 675–678.

15. Clements and Young, "A viewpoint: Rangeland health and mule deer habitat."

16. Ken Gray, Nevada Division of Wildlife, personal communications, 2003.

17. Jim Jeffress, Nevada Division of Wildlife, personal communications, 2003.

18. Ken Gray, Nevada Division of Wildlife, personal communications, 2003.

19. Ibid.

20. M. W. Call and C. Maser, "Sage grouse," in *Wildlife Habitats in Managed Rangelands: The Great Basin of Southeastern Oregon*, ed. J. W. Thomas and C. Maser, General Technical Report 187 (Ogden, Utah: USDA Forest Service, 1985).

21. K. H. Johnson and C. E. Braun, "Viability and conservation of an exploited sage grouse population," *Conservation Biology* 13 (1999): 77–84.

22. D. A. Klebenow, "Sage grouse nesting and brood habitat in Idaho," *Journal of Wildlife Management* 33 (1969): 649–662.

23. Jim Jeffress, personal communications, 2003.

24. K. H. Johnson and C. E. Braun, "Viability and conservation of an exploited sage grouse population," *Conservation Biology* 13 (1999): 77–84.

25. Anon., *To Research California History, Noble's Trail* (Susanville, Calif.: Fukawee Tribe, 1993).

26. L. W. Higby, "A summary of Long Creek sagebrush control," *Proceedings of the Western States Sage Grouse Workshop* 6 (1969): 164–168.

27. D. A. Klebenow, "The habitat requirements of sage grouse and the role of fire in management," *Proceedings of the Annual Tall Timbers Fire Ecology Conference*, June 8–9, 1972.

28. Anon., "6th Biennial Western States Sage Grouse Workshop," Rock Springs, Wyo., 1969.

29. G. E. Rogers, *Sage Grouse Investigations in Colorado*, Technical Publication 16 (Denver: Colorado Game, Fish and Parks Department, 1964), p. 132.

30. B. L. Welch, "Add three more to the list of big sagebrush eaters," in *Proceedings: Shrubland Ecotones*, ed. D. E. McArthur, W. K. Ostler, and C. L. Wambolt, General Technical Report 11, RMRS-P-11, 171–174 (Ogden, Utah: USDA Forest Service, 1999).

31. K. J. McAdoo, D. A. Klebenow, and R. A. Evans, "Nongame bird community responses to sagebrush invasion of crested wheatgrass seedings," *Journal of Wildlife Management* 53 (1989): 494–502.

32. C. D. Clements, "Influence of rodent predation on antelope bitterbrush seeds and seedlings" master's thesis, University of Nevada, Reno, 1994.

33. C. D. Clements and J. A. Young, "Influence of rodent predation on antelope bitterbrush seedlings," *Journal of Range Management* 49 (1996): 31–34.

34. J. E. LaTourrete, J. A. Young, and R. A. Evans, "Seed dispersal in relation to rodent actives in seral big sagebrush communities," *Journal of Range Management* 24 (add missing year}: 451–454; M. H. McMurray, S. H. Jenkins, and W. S. Longland, "Effects of seed density on germination and establishment of a native and an introduced grass species dispersed by granivorous rodents," *American Midland Naturalist* 138 (1997): 322–330.

35. J. Yoakum, *Habitat Management Guides for the American Pronghorn Antelope*, Technical Note 347 Washington, D.C.: USDI Bureau of Land Management, 1980).

36. Mathis, *Following the Nevada Wildlife Trail.*

37. Robin Tausch, Rocky Mountain Research Station, USDA Forest Service, Reno, Nev., personal communication.

38. G. C. Christensen, *The Chukar Partridge in Nevada*, Bulletin 1 (Reno: Nevada Department of Fish and Game, 1954).

39. J. A. Young and R. A. Evans, "Invasion of medusahead into the Great Basin," *Weed Science* 18 (1970): 89–97.

40. D. A. Savage, J. A. Young, and R. A. Evans, "Utilization of medusahead and downy brome caryopses by chukar partridge," *Journal of Wildlife Management* 33 (1969): 975–977.

1. J. A. Young, R. A. Evans, and J. Major, "Alien plants in the Great Basin," *Journal of Range Management* 25 (1972): 194–201.

2. P. E. Wigand, M. Hemphill, S. Sharpe, and S. Patra, "Great Basin semi-arid woodlands dynamics during the late Quaternary," in *Climate Change in the Four Corners and Adjacent Regions: Implications for Environmental Restorations and Land-Use Planning*, ed. W. J. Waugh, Conf-9409325, U.S. Department of Energy, National Technical Information Service, Springfield, Va., 1995, pp. 51–70.

3. R. F. Miller and R. J. Tausch, "The role of fire in pinyon juniper and pinyon woodlands: a descriptive analysis," in *The Role of Fire in the Control and Spread of Invasive Species*, Proceedings of the Invasive Species Workshop, ed. K. E. M. Galley and T. P. Wilson, Publication 11, Tall Timbers Research Station, Tallahassee, Fla., 2001, pp. 15–30.

4. P. K. Wells, "Paleobiogeography of montane islands in the Great Basin since the last glacial pluvial," *Ecological Monographs* 53 (1983): 341–382.

5. P. E. Wigand, "Diamond Valley Pond, Harney County, Oregon: Vegetation history and water table in the eastern Oregon desert," *Great Basin Naturalist* 47 (1987): 427–445.

6. P. J. Mehringer, "Prehistoric environments," in *Handbook of North American Indians*, vol. 11: *Great Basin*, ed. W. L. D'Azevedo (Washington, D.C.: Smithsonian Institution Press, 1968), pp. 31–50.

7. J. O. Davis, "Bits and pieces: The last 35,000 years in the Lahontan area," in *Man and the Environment in the Great Basin*, ed. D. B. Madson and J. F. O'Connell (Washington, D.C.: Society American Archeology Paper 2, 1981), pp. 53–75.

8. Miller and Tausch, "The role of fire in pinyon juniper and pinyon woodlands."

9. C. I. Miller, D. Germanoski, K. Waltman, R. Tausch, and J. Chambers, "Influences of late Holocene hill slope processes and landforms on modern channel dynamics in upland watersheds in central Nevada," *Geomorphology* 38 (2001): 373–391.

10. P. E. Wigand, M. L. Hempill, S. Sharpe, and S. Patra, "Great Basin semi-arid woodland dynamics during the late Quaternary."

11. T. W. Swetman and C. H. Baisan, "Historical fire regime patterns in the southwestern United States since AD 1700," in *Fire Effects on Southwestern Forests*, ed. C. D. Allen, General Technical Report 286 (Fort Collins, Colo.: USDA Forest Service, 1995), pp. 11–32.

12. Miller and Tausch, "The role of fire in pinyon juniper and pinyon woodlands."

13. S. F. Arno and K. M. Sneck, *A Method for Determining Fire History in Coniferous Forest of the Mountain West*, General Technical Report 42 (Ogden, Utah: USDA Forest Service, 1977).

14. J. W. Burkhardt and E. W. Tisdale, "Natural and successional of western juniper vegetation in Idaho," *Journal of Range Management* 22 (1969): 264–270.

15. J. A. Young and R. A. Evans, "Demography and fire history of a western juniper stand," *Journal of Range Management* 34 (1981): 501–506.

16. P. J. Mehringer Jr., S. R. Arno, and K. L. Peterson, "Postglacial history of Lost Trail Pass Bog, Bitterroot Mountains, Montana," *Arctic Alpine Research* 9 (1977): 345–368.

17. J. A. Young and R. A. Evans, "Erosion and deposition of fine sediments from playas," *Journal of Arid Environments* 10 (1986): 103–115; R. R. Blank, J. A. Young, and F. L. Allen, "Aeolian dust in a saline playa environment, Nevada, U.S.A," *Journal of Arid Environments* 41 (1999): 365–381.

18. R. R. Blank, T. J. Svejcar, and G. M. Riegel, "Soil genesis and morphology of a montane meadow in the northern Sierra Nevada Range," *Soil Science* 169 (1999): 136–152.

19. J. A. Young and R. A. Evans, "Population dynamics of green rabbitbrush in disturbed big sagebrush communities," *Journal of Range Management* 27 (1974): 127–132.

20. S. G. Whisenant, "Changing fire frequencies on Idaho's Snake River Plains: Ecological and management implications," in *Cheatgrass Invasion, Shrub Die-off, and Other Aspects of Shrub Biology and Management,* symposium proceedings, ed. E. D. McArthur, E. M. Romney, S. D. Smith and P. T. Tueller, General Technical Report INT 256, USDA Intermountain Research Station (Ogden, Utah: U.S. Forest Service, 1989), pp. 4–14.

21. C. D. Clements and J. A. Young, "A viewpoint: rangeland health and mule deer habitat," *Journal of Range Management* 50 (1997): 129–139.

22. Young, Evans, and Major, "Alien plants in the Great Basin."

23. J. A. Young and J. K. McAdoo, "Jackrabbits," *Rangelands* 2 (1980): 135–138.

24. D. H. Gates, "Sagebrush infested by leaf defoliating moth," *Journal of Range Management* 17 (1964): 209–210.

25. J. A. Young and C. D. Clements, *Purshia: The Wild and Bitter Roses* (Reno: University of Nevada Press, 2002).

26. D. K. Griffiths, *Forage Conditions on the Northern Border of the Great Basin,* Bulletin 15 (Washington, D.C.: USDA Bureau of Plant Industry, 1902).

27. G. D. Pickford, "The influence of continued heavy grazing and promiscuous burning on spring-fall ranges in Utah," *Ecology* 13 (1932): 159–171.

28. J. F. Pechanec and A. C. Hull Jr., "Spring forage lost through cheatgrass fires," *National Wool Grower* 35, no. 4 (1945): 13; A. C. Hull Jr. and J. F. Pechanec," Cheatgrass—a challenge to range research," *Journal of Forestry* 45 (1947): 555–564.

29. J. A. Young, J. D. Budy, and R. A. Evans, "Use of shrubs for fuel," in *The*

Biology and Utilization of Shrubs, ed. C. M. McKell, 479–491 (San Diego: Academic Press, 1989).

30. J. A. Young, "Range research in the far western United States: The first generation," *Journal of Range Management* 53 (2000): 2–11.

31. J. A. Young and B. A. Sparks, *Cattle in the Cold Desert* (Reno: University of Nevada Press, 2002).

32. J. A. Young and J. D. Budy, *Endless Tracks in the Woods* (Osceola, Wisc.: Crestline, 1989).

33. R. J. Tausch, "Historic woodland development," in *Ecology and Management of Pinyon-Juniper Communities within the Interior West,* ed. S. B. Monsen, R. Stevens, R. J. Tausch, R. Miller, and S. Goodrich (Ogden, Utah: USDA Forest Service, 1999), pp. 12–19.

34. H. Weaver, "Fire as an ecological factor in southwestern ponderosa pine," *Journal of Forestry* 49 (1951): 93–98; H. H. Biswell, "The use of fire in wildland management in California," *Proceedings of the Tall Timbers Fire Ecology Conference* 2 (1963): 62–97; R. R. Humphrey, "The role of fire in the desert and desert grassland areas of Arizona," *Proceedings of the Tall Timbers Fire Ecology Conference* 2 (1963): 44–61; R. Daubenmire, "Ecology of fire in grasslands," *Advances in Ecological Research* 5 (1968): 209–266.

35. For example, S. J. Pyne, *Year of the Fire* (New York: Viking Press, Penguin Putnam, 2001).

36. J. A. Young and D. McKenzie, "Rangeland drill," *Rangelands* 4 (1982): 108–113.

37. J. Y. Young, R. A. Evans, and R. E. Eckert Jr., "Successional patterns and productivity potentials of sagebrush and salt desert ecosystems," in *Developing Strategies for Rangeland Management,* ed. Committee on Developing Strategies for Rangeland Management (Boulder: Westview Press, 1984), pp. 1259–1298.

38. J. W. Connelly, S. T. Knick, M. T. Schroeder, and S. J. Stiver, *Conservation Assessment of Greater Sage Grouse and Sagebrush Habitats* (Cheyenne, Wyo.: Western Assoc. Fish and Game Agencies, 2004).

39. J. F. Pechanec, G. Stewart, and J. P. Blaisdell, *Sagebrush Burning—Good and Bad,* Farmer's Bulletin (Washington, D.C.: USDA, 1954).

40. K. Platt and E. R. Jackman, *The Cheatgrass Problem in Oregon,* Bulletin 668, Oregon Agricultural Experiment Station, Corvallis, 1946.

41. J. A. Young, R. A. Evans, and R. E. Eckert Jr., "Population dynamics of downy brome," *Weed Science* 17 (1969): 20–26.

42. R. A. Evans and J. A. Young, "Effectiveness of rehabilitation practices following wildfire in a degraded big sagebrush–downy brome community," *Journal of Range Management* 31 (1978): 185–188.

43. D. A. Bronstein, D. Baer, H. Bryan, J. F. C. DiMento, and S. Narayan, "National Environmental Policy Act at 35," *Science* 307 (2005): 674–675.

44. J. A. Young, "Ecology and management of medusahead," *Great Basin Naturalist* 52 (1992): 245–252.

45. H. A. Wright and J. O. Klemmedson, "Effects of fire on bunchgrasses of the sagebrush-grass region in southern Idaho," *Ecology* 46 (1965): 680–688.

46. R. R. Blank, F. L. Allen, and J. A. Young, "Influence of simulated burning on soil-litter from low sagebrush, squirreltail, cheatgrass and medusahead on water-soluble anions and cations," *Journal of Wildland Fire* 6 (1996): 137–143.

Index

Italic page numbers refer to figures; page numbers with the letter *t* refer to the tables.

antelope bitterbrush: as animal food source, 251, 261; Indian ricegrass compared to, 217; nitrogen fertilizer impact on, 212–13; rodent interaction with, 259–60, *260;* seeding of, 194, 213–14

anthers, extrusion of, 131

arsenical herbicides, 160

Asher, Jerry, 170

Asia, Central: cheatgrass evolution in, 35; Great Basin compared to, 36–37, 38; weeds originating in, 36, 71

Assay, Kay, 188

assemblages, plant: climate change effect on, 54; replacing, 56; seedbed characteristics modified by, 66

atrazine: discovery of, 163; evaluation of, 171, 177–78; fallow technique with, 171–72, 173, 178; karbutilate compared to, 177; nitrate nitrogen treatment compared to, 206–7; patent expiration for, 180; residues of, 167, 168; sulfometuron methyl compared to, 181; weed control with, 166–67

Badger Mountain fire, 254

Bailey, Art, 271

Baker, Herbert G., 127, 133

Balkan Peninsula, cheatgrass in, 40

barbwire Russian thistle, 71–72, 73, 237

Barker, Frank W., 18–19

barley, 82, 133

basin wild rye, 82, 235

Beatley, Janice, 72

Beckwith, E. G., 42

Bedell Flat, cheatgrass control experimental site at, 96

beet leafhopper, as virus transmitting vector, 88

Belle Fourche Experiment Station (Newell, S.Dak.), 144

Bennion, Glenn: burning advocated by, 141; free-range policy criticized by, 49, 140

bighorn sheep, 272

Billings, W. D.: on plant distribution in salt deserts, 26; salt desert described by, 136

Billy Meadows, range seeding attempts in, 142

bioassay sampling: from herbicide-treated plots, 102; seed banks without need for, 99; seed viability through, 98

biological control of cheatgrass, 289

biological diversity, 196

birds: food requirements of, 9; Russian thistle seeds eaten by, 72; species diversity, 258

Biswell, Harold, 276

Black Rock Desert, 14–15

black sagebrush/desert needle-grass, 274

Blank, Robert, 208, 209, 215, 217–18, 286

bluebunch wheatgrass: in abandoned fields, 90; cheatgrass competition with, 123–25, 202; cheatgrass suppression by, *106;* dominance, former of, 122; domination by, 196; grazing of, 123; leaves of, 124; nitrogren fertilizer impact on, 211; nomenclature for, 183–84; range of, 146; reestablishment of, 194; and restoration efforts of sagebrush/bunchgrass rangelands, 143; root depth of, 118–19; roots of, 123, 124–25; seed production of, 123; variants of, 193–94

—seedlings: *v.* cheatgrass seedlings, 110; competitive, developing, 125; survival of, 146

blue-green algae, 27

Bonneville, Lake, 25

borate-clorate mixtures, vegetation control through, 160

Braun, Clait, 255

Brewer's sparrows, 72

broadleaf species, 195

brome grasses: in California, 44, 131; in Great Basin, 57; phylogenetic relations among, 134–35; species of, 38–39, 130–31; studies of, 117–18

Bromus (genus), 38

Bromus mango, 38
Bromus rubens (cheatgrass scientific name), 11, 44
Bromus tectorum (cheatgrass scientific name), 11, 38, 44
bronco grass (term), 4, 11, 39, 63, 94
browse species, seeding of, 145
Brunner, Alan, Great Basin described by, 28
Buckley, Samuel Betsford, 40–41
buffalo grass, revegetation with, 52
Buffalo Meadows Ranch, 215
bunchgrass: cheatgrass in communities of, 90; cheatgrass suppression, role in, 157; death and succession of, 91; decline of, 42, 59; drought, recovery from, 116; grazing of, 15, 114; as horse food supply, 42; as livestock food supply, 31; overgrazing of, 4, 107, 108, 255; pre-European arrival, 136; restoration of, 64; watershed protection provided by, 113; wildfire effect on, 32. See also sagebrush/bunchgrass
Bureau of Land Management: cheatgrass conferences convened by, 220; Grazing Service merged into, 154; grazing studies sponsored by, 235; research plot location chosen by, 20; wild horse removal by, 243
Burkhardt, Wayne, 93, 270, 282
burned cheatgrass ranges, herbicide application to, 181
burning
—prescribed: Forest Service recommendations concerning, 141–42; as stand renewal process type, 267, 283
—promiscuous: attitudes concerning, 280; cheatgrass, impact on, 46, 116–17, 157; cheatgrass in spread of, 11; controversy over, 49–50, 140; grazing effects with, 58; obstacles to, 141; as range management tool, 46–47, 48; as stand renewal process type, 267, 275; writings on, 57

burros (wild), protection of, 33

calcium nitrate, 210
Caldwell, Marten, 232
California: adobe bricks, plant material in, 41, 77; cheatgrass introduction in, 41–42; wheat production in, 5–6
calving, assistance with, 225–26
Canada: cheatgrass in, 41; wheatgrasses in, 185–86
Canadian thistle, 110
carbon enrichment, nitrogen immobilization influenced by, 211
Carson, Rachel, 178, 179
Carson Desert, plant communities of, 26, 27
cattle
—food requirements for, 1, 2, 3, 30, 148, 223, 224–25, 226–27, 234; grazing by, 4, 15, 31–32, 42, 63, 63, 72, 274
—industry: cheatgrass control advocated by, 11; fires (promiscuous) use in, 47; introduction of, 42; seeding and, 153; sheep industry competition with, 44–45, 47, 48, 276; World War II, 150
—introduction into desert, 34
—land requirements for, 2
—production system for, 223–26
—seasonal movement of, 30, 37–38
—sheep compared to, 223
cereal grain crops: in Asiatic Russia, 67; cheatgrass as competitor with, 20; evolution of, 35; in Great Plains, 69; herbicide application to, 161; pests of, 76; production, 34, 45; salt tolerance of, 82; weeds associated with, 128; yield, increase in, 201
Changes in Weed Plant Cover on Cleared Sagebrush Land and Their Probable Causes (Piemeisel), 112
Chapline, W. R.: grazing, concerns over, 150; on Hurtt's seeding work, 151; range management techniques by, 148; seeding attempts described by, 142

Charles Sheldon National Pronghorn
Refuge, 254, *255*

cheatgrass, 5, 6, 39, 57, 44

—control: cattle industry as advocate of, 11;
experimental sites for, 96, 102; herbicide
use in, 95–96; research on, 94

—ecology of, 89: misconceptions
concerning, 118–19; seedbed, 102–3;
studies of, 109–10, 122

—flowers: bud formation, 122; exerted
anthers of, 118, 131

—leaves *v.* bluebunch wheatgrass leaves, 124

—production: grazing exceeded by, *236;*
studies and records of, 114, 115 (*see also*
forage production: cheatgrass *v.* native
perennial grasses)

—red color in, 112

—roots: *v.* bluebunch wheatgrass roots,
123–24, 125; growth, rapid of, 157

—seedling competition: *v.* annual and
established plant competition, *106;*
v. bluebunch wheatgrass seedling
competition, 110; *v.* perennial seedlings,
157–58; writings on, 105, 107–9

—seedling emergence: cold-weather
germination role in, 194; conditions
favoring, 192

—seeds: as animal food source, 9, 33;
banks, 98–99, 100, 122, 137, 157, 168,
181, 187, 280–81; burning effect on, 135,
284; collecting, 250; description of, 6,
237–38; in dormancy state, 100–101,
109–10, 173; experiments with, 109–10;
germination of, 85, 101–2, 103–4, 112,
113, 122, 123, 135, 173–74, 194, 211,
216, 236, 288; hazards of, 57, 237–38;
location of, 99–100, *100,* 102; longevity
of, 122; production of, 113, 123, 136;
smut infection effect on, 61–62; viability,
testing, 96–99, *97;* wildfire effect on, 59

—variations, 134–35

chemical fallow (term), 167

Chepil, W. S., 121–22

chloracetic acids, 163

chlorobenzoic acids, 163

chloro-phenylurea compounds, 163

chromosome numbers for brome grasses,
118

chukar partridge: cheatgrass, association
with, 263, *264;* food source of, 9, 20;
hunting of, 7–9

Churchill Canyon, 274

Civilian Conservation Corps (ccc):
firefighting under, 17–20, 144;
manpower of, *14;* overview of, 12–13

Classbook of Botany (Wood), 40

clay-textured seedbeds, 83

Clements, Frederic E.: on plant ecology, 68;
on succession stages, 79; teachings and
influence of, 51, 54, 56, 65, 66, 107

climate: cheatgrass, effect on, 59, 63; plant
communities impacted by, 53, 54

climax community: cheatgrass as member
of, 89, 90, 111; definition and overview
of, 53–54; invasion of, 107

climax dominants, requirements for, 53

closed communities: cheatgrass and, 109;
invasion of, 107; natural tendency to
remain as, 107; of sagebrush, 109

clover, seeding of, 172, 174

collecting of cheatgrass, 45

colonizing species: genetics of, 132; habitat
variation of, 133; self-fertilization in,
129

combines, 5

competition: growth and reproduction
capacity in, 107, 116; as plant succession
stage, 53; and seeding experiment results,
109

competitive native annuals, lack, reasons
for, 91, 93

competitiveness of cheatgrass: factors
contributing to, 125, 126; *v.* seedlings,
167; succession truncation linked to,
104. *See also* native grasses, perennial:
cheatgrass as competitor with

Comstock Lode, role in Nevada nickname, 25

conifers, fire effect on, 269

conservation movement: birth of, 51; following droughts, 144; wildfire policy of, 9, 10, 141, 276

control of cheatgrass: antelope bitterbrush role in, 213–14; attitudes concerning, 180, 220–21; competition as tool in, 198; cooperative efforts, 159; and curable problem, 89; grazing management approach to, 90, 110, 220; herbicide use in, 163–64, 166, 167, 173–74, 180–82; innovations in, 158; myths concerning, 157; and perennial establishment, 198; plant material role in, *199;* requirements for, 289; seeding and reseeding approach to, 125, 158; through weed control, 156; tillage as means of, 158

Cook, C. W., 232

cottonwood trees, 30

cow-and-calf operations, 223, 224, 226, 237, 288

Cowles, H. C., 52

Craddock, George, watershed protection comparisons by, 113

crested wheatgrass: attitudes concerning, 252–53; breeding programs for, 188–89; cheatgrass competition with, 120–21, 147, 156, 172, 203–4; cheatgrass occurring with, *106,* 109, 233; cheatgrass suppression through, 197, 203, 251, 253, 289; demise of, 191; fire prevention, role in, 251, *253;* as forage source, 224, 231, *231,* 232, 288–89; grasses, other seeded with, 194; halogeton suppressed by, 75–76, 154, 155, *189;* intermediate wheatgrass compared to, 190; introduction of, 147; longevity of, 192; "monocultures" of, 194; as non-self-invasive plant, 191, 192; origin of, 144, 145; phenology of, 232; rangeland restoration, role in, 147–48, 150; seeding

of, 144–45, 151, *152,* 153, 155, 156, 157, 158, 169, 185–86, 189, *189,* 191, 192, 195, 197, 232, 234, 243, *252,* 255, 256, 258, *259,* 264, 278, 281, 282; taxonomy of, 183, 184; varieties and nomenclature for, 184–85, 186–88; yields of, 113, 114

crested wheatgrass seedlings: cheatgrass domination of, 156, 157; establishing, 121, 158, 170–71

crop production, land expanded for, 58

crop cultivation, cheatgrass with, 5–6, 22, 35

crop species, Triticeae tribe members of, 183

cross-vertilizing populations, genetic variability in, 130

cultivars (term), 184–85

cultivated land, 48, 110

curly top virus, 88

dalapon, 163, 174

Daubenmire, Rexford F.: cheatgrass studied by, 90, 111, 116; on old-field succession, 89; on wildfire policy, 276

DDT, 179

deep furrow drills, 171

densities of cheatgrass: burning impact on, 116, 135; studies of, 113

desert: federal management of, 139; nitrate deposition in, 205; plant communities, nutrient dynamics of, 28

desert needlegrass, 145

desert ranges: grazing in, 42, 48; winter access to, 37

desert seedings v. meadow seedings, 142

Dewey, Douglas, 183: on genomic classification system, 183–84, 185

Dewey, Lyster Moxie, on Russian thistle, 67–68, 71

Diamond Craters (Ore.), 270

diploid-hexaploid hybrids, 188

diploids, 184, 185, 186, 188

disk plow, 149, 151, 158

disturbed environment, cheatgrass introduction into, 126

diuron, 163, 211

dominant plants: annuals *v.* perennials, 52, 53; cheatgrass, 57, 58, 66, 115, 117, 140, 156–57; environmental modification by, 89; tumble mustard, 87. *See also* plant dominance

dormancy: breakdown, factors contributing to, 100, 101; cheatgrass contribution to, 121

downy brome (term), 39, 94

downy chess (term), 59–60

droughts: cheatgrass adaptability to, 11, 21, 60, 115–16, 116, 137; crested wheatgrass adaptability to, 147; fields abandoned following, 89, 144; forage affected by, 59; grazing following, 185; restoration incentive of, 143–44; role of closed communities in, 107; seedings during, 185–86; wildfires caused by, 269

Dutton, Walt, 150–51

dyer's woad, 46

"early spring grazing," 228, 232, 233

Eckert, Richard: and cheatgrass-crested wheatgrass program, 164, 172, 188, 189, 190, 191, 192, 288; cheatgrass research by, 94–95; control and revegetation studies by, 134; W. B. Ensis on, 162; and Raymond Evans, 159; funding received by, 162–63; halogeton germination demonstrated by, 75; herbicidal fallowing practiced by, 165, 169, 171, 172; herbicide research by, 161; herbicide testing by, 166, 167, 168; nitrogen fertilizer experiments by, 203–4; seeding equipment favored by, 170; seeding project participation of, 158

ecological sites, creation of new, 126

ecosystem, plant assemblage modification of, 79

ecotones (defined), 53

ecotypes of cheatgrass: evolution of, 118; herbicide resistance of, 135

elk, 249–50, 261, 272

Emigrant Pass, 134, 167

Ensis, W. B., 162

environmental impact studies, 281, 282–83

environmentalism, pesticides opposed in name of, 179

environmentalists: on crested wheatgrass "monocultures," 194; herbicides, attitudes concerning, 178; native grasses promoted by, 193

environmental review regulations, 178–79

environmental safety, herbicides and, 182

environmental stress, on plant breeding, 188

environments, and cheatgrass adaptability to, 118, 126, 132–33, 138

eolian soils, 15

'Ephraim' crested wheatgrass, 184, 186–87

Evans, Raymond: and crop dusting, 175, 176, 177; and cheatgrass-crested wheatgrass program, 164, 174, 188, 189, 190, 191, 192; cheatgrass research by, 94–95, 120–21; control and revegetation studies by, 134; W. B. Ensis, observations concerning, 162; fertilizer trials researched by, 201; funding received by, 162–63; herbicidal fallowing practiced by, 165, 169, 171, 172; herbicide research by, 161; herbicide testing by, 166, 167, 168, 177; moisture interactions researched by, 125, 126; on nitrogen fertilizer experiments, 203, 204; on prehistoric fires, 270, 271; weed control and seeding efforts by, 158–59

evolutionary advantage, self-pollinated line development as, 130

evolutionary success: defined, 127; measure of, 133

exotic species: brome grasses as, 118; cheatgrass as, 220; introduction of, 66–67, 122–23; invasive plants as, 129, 264; native meadow replacement with,

142; perennial grass failure to replace, 203; plant communities dominated by, 66; rangeland restoration, role in, 145

fairway crested wheatgrass (*Agropyron spicata*): colchicine treatment for, 188; as diploid form, 185; nomenclature for, 185; overview of, 186
fallowing: herbicide treatment after, 165–66, 167, 171–72; nitrate nitrogen treatment compared to, 206–7; weed control during, 166
farmers, and cheatgrass, 6
fertilization experiments on cheatgrass. *See* nitrogen fertilizers
field capacity (defined), 85
filaree: introduction of, 147; as invasive plant, 41, 48, 76–78
fire as cheatgrass ally, 287
firefighters: in action, *14;* casualties among, 16–20, 179; composition of, 10; equipment and training of, 13; recruitment and hiring of, 144
firefighting, tactical errors in, 17
fires. *See* wildfires
fire scar technique, 269–70
Fisher, G. W., 122
five-needle pine species, 267–68
Flanigan (Nev.): burn site, experiments at, 214, 215–19, *216;* wildfire at, 214–15
Flathead Valley (Mont.), cheatgrass-based grazing in, 110
Fleming, C. E.: cheatgrass grazing overseen by, 230; forage production researched by, 235; forages evaluated by, 236; writings of, 62–64
foliar herbicides, cheatgrass control attempts with, 174
food chain, wildfire impact on, 260–61
foothills ranges, federal managing of, 140
forage cycle, annual, 221, 226–36, *230,* 238–40
forage period, green, 61, 231

forage production: assessing, difficulties of, 240; cheatgrass *v.* native perennial grasses, 62, 114, 115, 116, 235; lack in early spring, 241
forage species: cheatgrass as, 20, 64–65, 94, 111–12, 114, 125, 220, 222, 225, 228, *229,* 229–30, 234, 237, 240, 243, 245 (*see also* grazing: cheatgrass-based); as cheatgrass competitors, 76; crested wheatgrass as, 147, 155, 252; phenological comparison of, 235; production of, 111–12, 225; Triticeae tribe members of, 183; for wildlife, 251
forb: as animal food source, 251; native perennial, 195; seeding of, 145; term, 53
forests, national: advocacy of, 44–45; establishment of, 140, 276; experiment stations on, 55–56; grazing, use for, 45, 55, 150, 238
Forest Service: burning recommendations concerning, 141–42; creation and purpose of, 54–55, 139, 276; grazing regulated by, 58, 238; manpower for, 13; mountain ranges managed by, 140; natural resource conservation by, 12; plant collecting overseen by, 153; publications of, 155; range management in, 149, 150; range research conducted by, 111, 142; seeding policy and efforts, early of, 139; wildfires combated by, 10, 19; in World War I, 150; in World War II, 150, 151, 152
Forsling, C. L., 49, 142
fossil pollen record, 270–71
four-wing saltbush, 194
foxtail (term), 39
free-range policy. *See* open-range policy
Frémont, John C., 41, 77
frost effect on cheatgrass, 112–13
furrowing drill, 170–71
furrow seeding, 168, 169
furrows *v.* flat seedbeds, 168

galinsoga, as invasive weed, 128
garden bean, as self-pollinated plant, 130
garlic promotion, for cheatgrass control, 195
Gates, Dillard, 174, 202
Gem County (Idaho), cheatgrass in, 62
"genetic pollution," fear of, 145–46
genetic segregation following wildfire, 136
Genetics of Colonizing Species, 1964
    symposium on, 128–29
genome (term), 183
genotype, ideal: for cheatgrass, 134, 137; in
    uniform environment, 133
genotypic variability: of cheatgrass, 126,
    137; in opportunistic species, 126; in
    self-fertilizing *v.* sexually reproducing
    plants, 129–30
geranium family (Geraniaceae), 76–77
germination: ecological principles of, 86–
    87; factors promoting, 103–4; moisture
    required for, 84–85; osmotic potential
    in, 85–86; temperature requirements for
    germination, 70
Germ-N-8, 196
gibberellin, cheatgrass emergence by, 101,
    210
glyphosate, 180, 181
goats, domestication of, 35
'Goldar' bluebunch wheatgrass, 193
golden eagles, 260–61
goosefoot family (Chenopodiaceae): below
    sagebrush zone, 136; domination by, 78;
    geranium family (Geraniaceae) members
    compared to, 77; members of, 73, 76;
    succession, role in, 80
government funding for range
    improvement, 155
government-led restoration efforts: agencies
    involved in, 144 (*see also under specific
    agency, e.g.:* Forest Service); attitudes
    concerning, 143
grain drill, 153
grains: evolution of, 35. *See also* cereal grain
    crops

grama grass, revegetation with, 52
granivores, cheatgrass seed as food source
    for, 33
Grass and Forage Plant Investigations Unit,
    Bureau of Plant Industry, 44
grasses, annual: brome grasses, 131; fire
    effect on, 49–50, 267, 283; fires burning
    in, 283–86; nitrogen immobilization
    effect on, 211; perennial grass
    replacement by, 202–3; suppression of,
    172
grasses, classification of, 38. *See also* native
    grasses
grasses, perennial: lack of adapted, 142–43;
    as animal food source, 261; antelope
    bitterbrush seeded with, 213–14;
    cheatgrass communities closed to, 218;
    cheatgrass stand conversion into, 178;
    cheatgrass suppression by, 105, 264;
    and crested wheatgrass seeding, 155;
    depletion of, 146; diminishing of, 49;
    dominance following wildfire, 261;
    exotic annuals not replaced by, 203; fire
    effect on, 46; forage period, green for,
    61; grazing of, 146, 228, 230–31, 274;
    growth renewal of, 232; herbicide effect
    on, 161, 279; medusahead, control of,
    284; native shrubs mixed with, 195;
    nitrogen fertilizer application to, 201–2,
    203, 211, 212–13, 233; phenology of,
    202; protection of, 58; regrowth of,
    236; reproduction of, 93–94; restoration
    of, 230–31; in sagebrush/bunchgrass
    communities, 196, *248;* sagebrush
    competition with, 156–57; seeding
    and reseeding, 109, 180; seedlings,
    establishment of, 180, 281; seedlings *v.*
    cheatgrass seedlings, 157–58, 161, 174;
    stand renewal in communities of, 91;
    as wildfire fuel, 269. *See also under grass
    type, e.g.:* crested wheatgrass; native
    grasses, perennial
grass genetics, 183–84

Gray, J. R., 232
Gray, Ken, 251, 253
grazing: bluebunch wheatgrass, impact on,
194; cheatgrass ability to withstand, 111;
cheatgrass-based, 58, 59, 60, 63, *63*,
110, 157, 221, 230, 231, 233, 234, 235,
236, 237, 240, 244, 245, 278; cheatgrass
dominance following, 156; cheatgrass
invasion following, 146; cheatgrass
presence for reasons other than, 89,
233–34; cheatgrass suppression, possible
due to, 233–34; crested wheatgrass
impacted by, 147, 186, 191, 192; criticism
of, 93; effects of, 58; opposition to,
254; environment impact from, 241;
plant replacement caused by, 56–57;
promiscuous burning during, 47; public
land use for, 288; reseeding results
on, 142; seasonal patterns of, 227–31;
seeding protection from, 157, 167; by
wild animals, 33. *See also* overgrazing
grazing management: advocacy of, 151;
benefits of, 42–43; cheatgrass-based,
245; cheatgrass control through, 90,
110, 220; cheatgrass increase in spite of,
156, 279; environmental impact of, 241;
feedlot and factory raising of animals
combined with, 222; individual attempts
at, 140; perennials favored through, 126;
range improvement replaced with, 179;
rise of, 155–56; and shrub-perennial
grass, 48; stand renewal process, role in,
267, 281–83
Grazing Service: duties of, 140;
establishment of, 12, 50; land
administered by, 151; manpower
for, 13; merger into Bureau of Land
Management, 154; wildfire investigation
testimony from officials of, 19
greasewood: appearance of, *27*; salt
adaptability of, 26; woody chenopod
haloxylon compared to, 36
Great Basin: Central Asia rangelands

compared to, 36–37; cheatgrass in,
136–37; climate of, 23–24, 26, 132;
description and characteristics of,
22–23, *23*; and evolving environmental
conditions, 32–33; growing conditions
on, 137–38; mountains of, 26; nitrogen
cycling in, 205–6; population of, 23;
prehistoric conditions, 23–24, 25;
rangelands, fenced on, 140; sheep
industry in, 154; shrub-grass balance in,
47–48; soil conditions of, 25–26, 26–27,
79; soil formation in, 205
Great Basin Experiment Station, 55–56
Great Basin Forest and Range Experiment
Station, 105
Great Plains: drought in, 107; grasslands,
ecology of, 52; Russian thistle in, 67, 69
Griffiths, David: and promiscuous fires,
47, 275; grazing and range conditions
observed by, 43, 44
ground sprayer, herbicide administration
with, 164–65
groundwater contamination, herbicides
blamed for, 180
growth regulator herbicides, 161, 163
*Guidelines for Habitat Protection in the
Sage Grouse Range* (Western States Sage
Grouse Workshop, Sixth Biennial), 256

Hallelujah Junction wildfire, 1973, 170,
171, 274
halogeton: control of, 154, 155, 156, 158,
160; identification of, 153–54; overview
of, 73–76, *74;* spread of, 154–55
Hammer, Charles L., 161
Hanson, H. C., 119
Harding grass, 203
Harper, J. L., 81, 83–84, 197
Harris, Grant A.: on cheatgrass ecology
and competition, 122–25; on moisture
interactions, 126; on nitrogen fertilizer
experiments, 202, 203; rooting habits
studied by, 125

harvester ants, 73

Hassan, N. A., 123

hay meadows, grazing, 239–40, 243, 244

hay production: for brood cows, 225; challenges of, 2–3, 10–11, 45; and conserving, 37; crested wheatgrass use in, 186; cycle, traditional, drawbacks of, 241; drop, compensation for, 244; fertilizer use in, 240; importance of, 30; irrigated land use in, 43, 153; land fenced for, 140; negative consequences of, 45; poor-quality, 227; in Quinn River area, 15, 16; requirements of, 238–39. *See also under specific type of hay, e.g.:* smooth brome grass

hay supply, 224–25

herbaceous vegetation: depletion of, 258; domination, pre-European, 247, *248;* grazing of, 257; as wildfire fuel, 269

herbaceous/woody plant ratio, 195, 251

herbage density, 113–14

herbage production, 113–14, 117

herbicides: administration of, 164–65, 174–77; attitudes concerning, 178, 289; bioassay sampling and, 102; cheatgrass control through, 95–96; cheatgrass resistance to, 135, 156–57; cheatgrass seed plots and, 102–3; classification of, 163; crested wheatgrass resistance to, 190; halogeton control through, 154; injuries due to, 168, 181; innovations in, 160–62, 163–65, 177, 180–82; residues of, 171; stand renewal, role in, 267, 279–80; treatment of, 166, 173–74. *See also* weed control: herbicide use in

herbivores: cheatgrass as food source for, 40; domestication of, 34–35, 36; food supply of, 34–35; lack in Intermountain Area, 91–92; pre-European, 92–93, 272

herdsmen: in Central Asian, 37, 38; in the West, 38

Hereford cattle: diet of, 63; smut-infected grasses grazed by, 61

hexaploids, 184, 188

Hillman, F. H., 145

Hinds, W. Ted, 116, 133

Hitchcock, A. S., 40

Holland, Richard: alfalfa introduced by, 195; dismissal of, 179; seeding achievements of, 170, 171

Homer (ghost), 214–15, 219

homesteading: lands, unmanaged open to, 50; public lands closed to, 12, 139

Hormay, A. L., 212, 233, 282

horses
—domestic, grazing by, 42, 45, 61, 110, 274
—wild: food supply of, 3, 221; protection of, 33; roundup of, 243

How to Reseed Range Lands (Forest Service), 155

Hudson's Bay Company, 7, 23

Hulbert, Lloyd C.: brome grasses studied by, 134–35; on cheatgrass genetic variability, 123; cheatgrass physiology study by, 117–18; cheatgrass root depth study by, 119; plant flowering study by, 122

Hull, A. C.: on cheatgrass as invasive plant, 117; on cheatgrass control methods, 158; cheatgrass study by, 111–14, 202; on crested wheatgrass establishment, 198; and crested wheatgrass seeding, 168; Raymond Evans's experiments repeated by, 121; grazing management advocated by, 90; herbicide use advocated by, 279; on sagebrush eradication, 148–49; tillage methods reviewed by, 166; on wheatgrass seeding, 190, 191

human activity: cheatgrass adaptability to, 21–22; and food supply, 34

human health safety, herbicides and, 182

Humboldt County, geography and terrain of, 13–15

Humboldt River, hay meadows along, 239, 244

Humboldt River valley, 13

hunters: animal species introduction by, 7; cheatgrass, experience with, 246
hunting as Western tradition, 8–9
Hurtt, Leon C., 110
hybridization of cheatgrass, 57
hybrid vigor, 136, 138
Hycrest crested wheatgrass, 189, *189*, 196
hydraulic conductivity, of soil water, 81, 83–84, 86

ice, as cheatgrass ally, 287
Idaho, cheatgrass in, 46, 111, 112, 117, 181
imazethapyr, 181–82
'immigrant' forage kochia, *199*, 199–200, 251, *252*
inbred lines, crossing of, 136
inbreeding depression, 130, 136
Indian ricegrass: cheatgrass in former areas of, 215; dominance by, 196; dormancy of, 217, 218; grazing on, 72, 228–29, 244; rodent interaction with, 259–60; seed banks of, 123; seeding, obstacles to, 187, 196; seeding of, 214, 217–18; seedling survival, 146
insect herbivory, 273–74
Interagency Range Seeding Equipment Committee, 149
Interior Department, 19
intermediate wheatgrass: ammonium sulfate applied to, 201; cheatgrass production effect on, 202; seeding of, 192, 196
Intermountain Area: burning in, 47; Central Asia environment compared to, 36; cheatgrass collecting in, 45; cheatgrass introduction in, 57; cheatgrass spread in, 60; climate variations in, 49; climatic change in, 92; environmental degradation in, 22; European culture impact on, 122–23; exotic species in, 66–67; native plant degradation in, 107; perennial grasses in, 60; perennial *v.* cheatgrass variability in, 115; range

research in, 90; range sheep operations in, 37, 154
Intermountain Forest and Range Experiment Station: data from, 113; employees of, 105; grazing study sponsored by, 235; publications of, 117; range seeding overseen by, 148
invasive plants: cheatgrass as, 90, 117, 247, 250, 256, 261, 284–85, 288; colonizing, 128–29; competition among, 73; succession, role in, 107; wildlife habitats affected by, *257*
invasive (definition), 129
Irano-Turanian floristic region, weeds originating in, 36
irrigation: forage base increased by, 43; in Idaho, 46; land, rarity of, 45; mountain stream runoff used in, 15, 30; rancher use of, 2; sagebrush cleared away for, 88; soil erosion and, 18
Iverson, Floyd, 153

jackrabbits: food supply of, 3, 221, 222; loss, impact of, 260–61; pioneer encounters with, 255; pre-European, 272
James, Walter, 18–19
Jeffress, Jim: crested wheatgrass, attitude concerning, 252–53, 258; experiences of, 265; on sage grouse and sagebrush loss, 254
juniper, 261, 275

karbutilate, 177
Kay, Burgess L.: on cattle grazing, 229; cheatgrass control and seeding endeavors by, 159; and cheatgrass production records, 114; fertilizer trials researched by, 201, 204–5; herbicide preferences of, 177; nitrogen fertilizer experiments of, 233; paraquat study, participation in, 172, 173, 174
Keller, Wesley, 162
Kennedy, George J., 18–19, 20

mountain ranges, federal management of, 140

mountains, water runoff from, 29–30

mucilaginous seeds. *See* seed mucilage

mule deer: cheatgrass impact on, 246; food supply of, 3, 212, 251, 252, 261, 274; habitat, changing of, 251; population expansion of, 33, 248, 250, 255; pre-European, 272

mules, food supply for, *253*

mullein, common, 110

Murray, Robert B., 235, 237

mustard family (Brassicaceae): germination of, 87; seedlings of, 87; seeds of, 80–81, 84, 85–86; species, increase in, 87; succession, role in, 78, 80, 104

mustard species, nitrate content of, 205

Narregang, S. W., Russian thistle threat noted by, 67

National Environmental Policy Act (NEPA), 281, 283

Native Americans: in Great Basin, 22–23; hunting by, 248; plants named by, 41, 77

native grasses: cultivars of, 193; micro-nutrient products tested on, 195–96; wildfire burning of, 267

—perennial: cheatgrass as competitor with, 48, 58, 60, 91, 116, 120–21, 123–25, 133–34, 213, 228, 235–36; drought effect on, 115–16; fire effect on, 285; grazing effect on, 147, 234, 278; growth start *v.* cheatgrass, 116; and lack of adapted perennials, 142–43; livestock depletion of, 247; nitrogen fertilizer impact on, 213; protein content of, 237; reestablishing, 63–64, 109, 140, 152; scientist perceptions concerning, 118; seed collecting, difficulty of, 143; seeding of, 234; on slopes, 151; yields of, 113, 114

native species, cheatgrass exclusion of seedlings of, 287

native vegetation: cheatgrass competition with, 247; destruction of, 5, 140; fire effect on, 46; forage kochia as invader of, 200; low-water adaptability of, 26; plants competitive with, 73; rangeland restoration attempts with, 145–47; and rodent burrows, 260; salt effect on, 76; seed industry, 143; seeding, difficulties of, 144, 185

native wildlife: cheatgrass as food source for, 20, 22, 40; species population changes in, 33; vegetation supporting diverse, 195

natural resource management: plant ecology application to, 51; public impact on, 178

natural selection, 54, 130

Nebraska Agricultural Experiment Station, 186

needle-and-thread grass, 145, 235

needlegrass, 145, 196

Nei's genetic diversity values, for phenogram construction, 134

New Deal, 144, 150

'Newhy' (quackgrass-bluebunch wheatgrass hybrid), 193, 194, 199

Newlands project, initiation of, 46

Newman, E. I., germination studied by, 100–101

nitrapyrin, 208, 213, 218

nitrate accumulation, fallowing as cause of, 169

nitrate nitrogen, 206–7

nitrification, inhibition of, 208, 210, 211, 213, 216, *216,* 218–19

nitrogen: deficiency in, 216; fire release of, 32; immobilization of, 208, 210, 211, 216, *216,* 218–19; mineralization of, 126; plant competition interaction of, 218–19

nitrogen cycling, 205–6

nitrogen deficiencies in cheatgrass, 203

nitrogen fertilizers: and cheatgrass impact, 204–5; cheatgrass production increased by, 201–2, 203, 211, 212, *216,* 233; and Harding grass, 203; in hay production,

240; overview of, 201; as obstacles to
wheatgrass, 207
nitrogen-phosphorus-sulfur testing method,
202, 203
nitrogen stress, 112, 215, 216
noncultivated colonizing species, 127–28,
128t
nonnative species: attitudes on use of,
264–65; seeding of, 252; as wildlife aid
measure, 252
Nord, E. C., 212
'Nordan' crested wheatgrass: classification
of, 184; overview of, 186, 187–88;
seeding of, 189, 190
North America, cheatgrass introduction in,
40–41
Northern Great Plains Experiment Station
(Mandan, N.Dak., 144
Novak, Stephan J., 134, 167
nuclear weapon test sites, invasive plants
at, 69
nutrient cycling, impact of sagebrush
clearing on, 278
nutrient deficiencies: in cheatgrass, 203;
plants with, experiments on, 202
nutrients: burning process release of, 135;
increase following wildfire, 286
nutritive content of cheatgrass, 236

Ogden, Peter Skene, 7, 14, 23, 247, 248
open-range policy: consequences of, 50;
criticism of, 49, 140; end of, resistance
to, 31
origin and discovery of cheatgrass, 35
Orvada Mercantile, 15–16
osmotic potential, of soil water, 81, 82, 83,
86, 87
Otley Brothers' ranch (Diamond, Ore.),
232
overgrazing: challenges of, 37; of cheatgrass,
116; as cheatgrass proliferation factor, 58,
110, 111, 116, 140; cheatgrass suppressed
by, 155; criticism of, 49; evidence of,

56–57; fields abandoned following,
88–89; as halogeton proliferation
factor, 154–55; halogeton susceptibility
following, 73, 75; by horses, 45; impact
on perennial grasses, 204, 209; livestock
losses following, 30; native vegetation
depleted through, 4, 5; observations
of, 42–44, 44; pattern of, 48; plant
growth following, 140; rationale for,
24; restoration following, 139; seedling
protection from, 142
overseeding syndrome, 197
oxygen tension, partial, seed sensitivity to, 86

Pacific bluestem (wheat variety), 5
paraquat, 172–77, 178, 180
Parkway (crested wheatgrass cultivar), 184,
186
Pearce, C. Kenneth: on cheatgrass
competitive nature, 120; on cheatgrass
seedling competition, 105, 107–9;
and Grant A. Harris, 123; watershed
protection comparisons by, 113
Pechanec, Joseph: cheatgrass studied by,
111–14; on grazing management, 90; on
sagebrush eradication, 148–49
pelleted seeds, cheatgrass control through,
198
perceptions, of cheatgrass, 1, 111. See also
under groups of people, e.g.: ranchers: and
cheatgrass
perennial plants: brome grasses, 130–31;
conditions favorable to, 63; dominance,
factors determining, 133; establishment,
cheatgrass control required in, 198;
plants competitive with, 73; plants
suppressed by, 69, 73; salt effect on, 76;
seed banks of, 123; seedlings, herbicide
effect on, 182. See also grasses, perennial
permanent wilting point (PWP), 85, 120–21
pesticides: attitudes concerning, 178;
environmental impact of, 179; wildlife
habitat enhancement with, 213

pests, plants resistant to, 188
Peterson, Joel, 246–47
Phaseolus genus, 130
phenology of cheatgrass, 228
phenotype of cheatgrass: *v.* perennial
   grasses, 202; plasticity in, 136, 138;
   production and, 112–13; variability in,
   115, 126
phenotypic variability (definition), 21
phenyl compounds, 163
phosphorus enrichment, and cheatgrass
   production, 203, 204
physiology of cheatgrass: *v.* bluebunch
   wheatgrass, 123–25; studies and writings
   on, 109–10, 117–18
Pickford, G. D., 48, 49, 57, 58, 275
picloram, 171–72
Piemeisel, R. L., 88, 112, 115, 116, 289
pigmy rabbits, 272–73
Pinchot, Gifford, 55
Pine Nut Mountains, vegetation of, *29*
pinyon, 261, 275
pinyon-juniper woodlands: basin wild-rye
   seeds, germination in, 82; boundary,
   northwestern of, 32; encroachment of,
   262, 268; mortality of, 269
plant-animal interactions, 258–61
plant classification system, 38, 183–84
plant communities: cheatgrass impact on,
   66; climate effect on, 5, 53, 54; closed,
   107, 109; destruction of, 79; formation
   of, 52; invasion of, 38; vacant space in,
   70–71
plant dominance: requirements for, 53
plant ecology, origin and development of,
   51–54
Plant Ecology (Weaver and Clements), 51,
   52, 54
Planter Jr. Vegetable Seeders, 165
plant families, subdivision of, 183
plant invasions, Western: of cheatgrass
   communities, 88; before European
   colonization, 41, 77; fire role in, 50;

immigrant role in, 46, 67, 71. *See also*
   invasive plants; *under plant type, e.g.:*
   Russian thistle
plant nomenclature, modern, 128, 128t
plant succession. *See* succession
playa: basin wild-rye seeds, germination in,
   82; description and characteristics of, *28;*
   erosion from, 79; surroundings as seen
   from, 28
Plummer, A. Perry, on sagebrush
   eradication, 148–49
Podosporiella verticillata (seed-pathologic
   fungus), 97
Point Springs (*later* Lee Sharp) Experimental Area (Idaho), 114–15
polyethylene glycol, 86
polypoidy, induced in plant breeding, 188
population growth, and food production, 37
population structures, evolution of, 130
potassium nitrate: and bioassay sample
   germination, 211; cheatgrass emergence
   promoted by, 101; cheatgrass germination
   promoted by, 101
pot test technique. *See* nitrogen-
   phosphorus-sulfur testing method
precipitation, scarce, conditions of, 24
preharvest desiccant, paraquat use as, 175
prickly lettuce, 128
pristine environment, perceptions
   concerning, 248–49, 265
*Proceedings of the National Academy of
   Science,* 1863, 40–41
pronghorn antelope: pioneer encounters
   with, 7, 255; population expansion of,
   248, 249, *250,* 261; pre-European, 272
propionic acids, 163
protein content of cheatgrass, 236, 237,
   244
Pseudoroegneria spicata (term), 184
p-7 bluebunch wheatgrass, 193
public land management, 178–79, 221
Pyne, Stephen J., wildfire, writings on,
   276–77

Pyramid Lake Indian Reservation, Russian
thistle on, 72

Quinn River, 14
Quinn River valley, 14–15

radiocarbon dating, 271
railing (defined), 50
railroad role in cheatgrass introduction, 4
rainfall effect on cheatgrass, 112
ranchers: accountability of, 241; challenges
faced by, 1–4, 30–31; and cheatgrass, 4–
5, 11, 116; environmental consciousness
among, 178; grazing land stewardship
by, 140; national forest establishment
advocated by, 44–45; overgrazing, views
on, 43–44; scientific reports funded by,
62
range improvement: golden age, end of,
281–82; government funding for, 155;
grazing management as replacement
for, 179; seeding role in, *252;* wildlife,
impact on, 263–65
rangeland drill, 153, 169, *169,* 170, 178
rangeland plow, 152, *152,* 153, 155
rangelands: cheatgrass impact on, 59;
irrigated land proportion to, 43
rangeland seeders, 170
range management, Forest Service role in,
149, 150. *See also* grazing management
range restoration. *See* restoration
rattlesnake brome, 210
red brome grasses, 39, 122
red stem filaree, 77, 127–28
red water vaccine, 224
replacement of cheatgrass, attempts at, 64
reseeding: cheatgrass control through, 125;
grazing effect on, 142; obstacles to, 143;
of sagebrush/bunchgrass, 105, 107
resources, overutilization of, 37
restoration: attitudes concerning, 139,
143–44, 248–49; experiments in, 108–9;
importance of, 123; methods advocated

for, 140 (*see also under specific method,*
e.g.: burning, promiscuous); personnel
for, 144
rest-rotation grazing, cheatgrass dominance
following, 155–56, 233, 282
rhizomatous C_4 grasses, 92–93
Richfield Oil Company, 160
roadside plants, invasive: factors limiting
type of, 79; halogeton, 73, *74;* Russian
thistle, 70
Robbins, W. W., 41
Robertson, Joseph H.: on cheatgrass
competitive nature, 120; on cheatgrass
seedling competition, 105, 107–9; files
of, 60; and Grant A. Harris, 123; plant
ecology theories challenged by, 54; range
management challenges of, 149–50; on
sagebrush control, 148, 149; in seeding
projects, 151; on wheatgrass seeding, 190
Robinson, Joe, crested wheatgrass seeded
by, 192
Rocky Mountain elk, 249, 272
rodents: cheatgrass seeds harvested by, 215,
216, 260; Russian thistle seeds eaten by,
72; seeds harvested by, 259–60, *260;*
wheatgrass seeds harvested by, 192;
wildfire impact on, 260
Romaine lettuce, nitrogen fertilizer tested
on, 202
Roosevelt, Franklin, 12, 144
root depth of cheatgrass, misconceptions
concerning, 118–19
root growth, 164, 203
RS-H hybrid wheatgrass, 193
Ruby Mountains (Nev.), seeding in and
near, 151, 153
Ruff crested wheatgrass, 184, 186
ruminants, nutrition for, 221–22
Rural Resettlement Administration,
144–45, 185
Russell, Israel C., 27, 42
Russia, grasses of, 144, 145, 186
Russian thistle: abandoned fields colonized

by, 89, 110; as animal food, 72; cereal grain association of, 69; cheatgrass competition with, 60, 80; cheatgrass stand conversion into, 180; description of, 67–68; environment, ideal for, 76, 79; germination of, 69–70, 72, 77; grazing of, 237; halogeton compared to, 73, *74*, 75, 153; introduction of, 46, 67, 71; near rodent burrows, 260; reproduction of, 68–69; spread of, 68, 70–71; as successful weed, 128; succession, role in, 104; as virus reservoir, 88; wheat production destroyed by, 67. *See also* barbwire Russian thistle

safe site parameters (term), 81
sagebrush: alluvial fan support of, *29;* cheatgrass as competitor with, 48, 246; cheatgrass in biome of, 45, *108,* 136, 193; cheatgrass replacement of, 214, 215; clearing equipment for, 149, 151–52, 153, 277–79; clearing of, 88, 141, 150, 255–56, 263; controlling, 141–42, 148, 149; cover provided by, 261; crested wheatgrass seedings invaded by, 195; decline of, 257–58, 268; dominance, rise in, 140, 251; drought impact on, 16; ecosystems, evolution of, 34; environmental potential for, 156; fire effect on, 32, 46, 47, 187, *278;* as fuel, 275; grazing followed by increase in, 255; herbicide treatment of, 279; invasions by, 155, *257, 259;* as livestock food supply, 3, 4, 15, 31–32, 247, *249;* nitrogen in areas of, 205, 207; odor of, 8; overabundance of, 141; and perennial grasses competition, 156–57; rangelands, changes in, 49; reproduction of, 93–94, *95;* reproductive capacity of, 107; restoration of, 141, 252, 264; secondary compounds of, 274; seeds and seedbed ecology of, 194; as shrub, 31; soft chess growing with, 57; stand renewal

process for, 266; succession impact on, 52; tillage role in spread of, 158; weed control-revegetation systems for, 172; as wild animal food source, 7, 254, 258, 261, *262,* 274; wildfire spread through, 10, 108, 187; zone, surface soils of, 99
sagebrush, degraded: cheatgrass in areas of, 156; communities of, 87; seeding in areas of, 109, 150, 153
sagebrush/bluebunch wheatgrass: cheatgrass dominance stages in communities of, *108,* 204
sagebrush/bunchgrass: burning of, 267; cheatgrass in areas of, 59, 89, 117, 135, 228, 238, 253; crested wheatgrass in areas of, 155; cultivars adapted to, 196; grazing of, 42, 93, 153, 223, 257–58; modification of, 274; and nutrients increase after wildfire, 286; perennial grass conversion of, *248;* perennial grass portion of, 196; restoration, obstacles to, 141, 142–43, 233; scarcity of, 90; seeding of, 105, 107, 196, 198; soft chess in areas of, 57; and other species, 92t; tillage role in spread of, 158; wildfire in areas of, *248;* wildfire intervals, pre-European, 267, 269, 271, 272
sagebrush/bunchgrass, degraded: cheatgrass in areas of, 126, 171, 178, 231, *273;* crested wheatgrass seeding in areas of, 151, 171, 232; drought and overgrazing compared, 107; forage, insufficient in areas of, 224
sagebrush defoliator, 273–74
sage grouse: cheatgrass impact on, 246; as endangered species, 272–73; food supply of, 254, 255, 258, *262;* habitat of, *259;* hunting of, 7; population expansion of, 248, 255; protection of, 256, 257; sagebrush clearing impact on, 255–56
salt desert vegetation: cheatgrass, 136–37, 193, 214, 220, 223, 238, 288; shrubs, 268; succession impact on, 52

salt (NaCl), weed suppression through, 160

salts: dry lake beds, accumulation in, *28;* plant adaptability to, 25–26; soil water concentration of, 82

Sampson, Arthur William: on filaree, 147; as Great Basin Experiment Station director, 55–56; influences and career of, 54; on meadow seeding, 150; and range management techniques, 148; seeding trials conducted by, 54, 142

Sandberg bluegrass: grazing of, 228; in sagebrush/bunchgrass communities, 196

sand-textured seedbeds, 83

Santa Rosa Mountains: description of, 13–14; soil of, 15; wildfire in, 1939, 16–20

Sawyer, Art, 220–21

Saylor Creek, grazing study at, 235–36, 237

scarlet runner bean, as self-pollinated plant, 130

Schwendiman, L. R., 122

Scotch thistle, control of, 172

'Secar' Snake River wheatgrass, seeding of, 197

seed banks, 75. *See also* cheatgrass seeds: banks

seedbed: alterations in, 104; benefits of, 83; competition in, 70; conditions in, 101, 102; microdifferences in, 138; moisture relations of, 81–87; quality of successional change, 56; and surface seeds, 69. *See also under type of seedbed, e.g.:* sand-textured seedbeds

seeding, range: attempts and experiments, 139; attitudes concerning, 143, 152, 179; cheatgrasss suppression through, 197, 281; equipment for, 153, 154, 168, 169, *169,* 170–71; failed experiments, 142; funding for, 151; rules of, 142, 157; in 1930s, 148–49; in sagebrush communities, degraded, 109; in World War II, 152

seeding competition, 194

seedings, restoration, 139, 144

seeding technology, obstacles to, 139

seedlings, establishment of, 168

seedling stage, as phenological competition stage, 105

seed longevity, 121–22

seed mucilage: benefits of, 80–81; cheatgrass lacking, 103; domination linked to, 87; formation of, 86; hydraulic conductivity with, 83–84

seed production, 133

seeds, laboratory testing of, 84–87

seed self-burial, 77–78, 103

selective herbicides, 161, 163–64, 165

self-fertilization: of barley plants, 133; of cheatgrass, 134, 138; as derived characteristic, 131–32, 136; studies on, 129–30

self-pollinated species, genetics of, 135

Shadscale/Bailey greasewood, *27*

Shantz, H. L., 52

Shantz, H. T., 48

Shaw, W. C., war record and career of, 162

sheep: cheatgrass spread by, 5, 40; domestication of, 35; filaree consumed by, 77; food requirements for, 148; food supply of, 223; grazing by, 44, *44,* 55, 58, 60, 274

Sheep Experiment Station (Dubois, Idaho), 144

sheep industry, range: cattle industry, competition with, 44–45, 47, 48, 278; expansion of, 33; fires, promiscuous, use in, 47; forage requirements for, 116–17; halogeton plant impact on, 154; in Intermountain Area, 37; past *v.* present, 223; seeding and, 153

shield cress: domination by, 87; germination of, 85, 86, 87; overview of, 80; salt tolerance of, 82; seedlings of, 87; seeds of, 84, 86; in transition zone, *82*

Shippley, M. A., 62–64

shrub-grass balance, livestock impact on, 47–48

shrubs: mounds, nutrient cycling of, 27–28; protection of, 253; wildfire impact on, 251, 254, 261

Shrub Sciences Laboratory, 187

Siberia, grasses of, 145, 186, 188

Siberian wheatgrass (*Agropyron fragile*), 185

siduron, 163–64

Sierra-Cascade Mountains, 26

Silent Spring (Carson), 178, 179

Simpson, J. H., 249

six-weeks fescue (*Vulpia octoflora*), 91

Smith, Jedediah, 23, 247

smooth brome grass, 38, 130

smut (*Ustilago*) infections, 61–62, 112

Snake River Plains: grazing in, 58; native perennial grass yields in, 113; plant research in, 88; wildfire intervals, prehistoric in, 271–72

Snake River Valley, cheatgrass in, 60

Snyder, James A., on Sierra Nevada plants, 42

Society for Range Management, 111

sodium chlorate as soil sterilant, 160

soft chess: in Intermountain Region, 57; as invasive species, 129

soil active herbicides: action mechanism of, 163; application techniques for, 181; pros and cons of, 165; testing and evaluating, 166, 167

soil-building process, 79

Soil Conservation Service, 144, 200

soil environments: in dry lake beds, *28;* exotic invasive weed adaptability to, 25–26

soil erosion: of abandoned fields, 144; acceleration of, 126; cheatgrass affected by, 187; herbicide hazards following, 181; as obstacle to native plant reestablishment, 146; nitrate nitrogen affected by, 205; in Pleistocene era, 29; ravines created by, 18–19; seedlings, effect on, 171. *See also* eolian soils

soil moisture: at bottom of furrow, 168; as

cheatgrass *v.* perennial grass competition, 121, 125, 218–19; and nitrogen fertilizer in cheatgrass, 205; wildfire effect on availability of, 135

soil protection, cheatgrass role, possible in, 111

soils, rangeland, mapping of, 202

soil water solution, 81–84

Spain, cheatgrass in, 40

Spaller, T. F., 191

Spanish settlers, 41, 77

Spence, L. E., 118–19

Spence's Bridge (bc), cheatgrass collecting in, 45

spread of cheatgrass: attitudes concerning, 63; and environmental change, 126; grazing management and, 233; in Great Basin, 136–37; history of, 40–42, 45–46; human activity role in, 123; second wave of, 59–60; in 21st century, 220

spring as relative term, 227–31

Springfield, H. W., 232

squirreltail: grazing of, 228; nitrogren fertilizer impact on, 211; in sagebrush/bunchgrass communities, 196

standard crested wheatgrass (*Agropyron desertorum*): fairway crested wheatgrass compared to, 186; name origin of, 184; nomenclature for, 185; seedings of, 185

stand renewal: cheatgrass role in, 266; factors, nongrazing contributing to, 90–91; partial, 80; as succession stage factor, 79; types of, 266–67

Stebbins, G. Ledyard, on self-pollination and fertilization, 129–30, 131–32, 135

Steins Mountains, overgrazing in, *44*

Stevens, O. A., Russian thistle sighted by, 71

Stewart, George: on cheatgrass studies and writings, 58–62, 114, 117; tillage methods reviewed by, 166

stocker cattle, 223–25

stock-raising homestead act, 2

Stoddart, L. A., 232

Wallowa Mountains, sheep range seeding in, 54

Warg, S. A.: on bluebunch wheatgrass, 90; on cheatgrass seeds, 99, 109–10; and W. S. Chepil, 122

War Production Board, 150, 151

water, potable in Great Basin, 23, 30

waterfowl, pioneer encounters with, 255

water-logged soils, seedbed moisture relations in, 86

watershed protection, cheatgrass stand role in providing, 113

Weaver, John E.: on plant ecology, 68; students of, 105; succession stages outlined by, 79; teachings and influence of, 51, 54, 56, 66, 107; theories of, 53, 65

weed control: aerial applications in, 174–77; cheatgrass control through, 156, 281; equipment used in, 142, 149, 164–65, 175; herbicide use in, 154, 159, 160, 161, 166, 167–68, 181; integrated, 289; nitrogen immobilization and nitrification inhibition in, 218–19; range seeding and, 142, 158–59; rooting growth impact on, 125

weeds: Asia-to-North America migration of, 36, 40; cereal grains, occurrence with, 35, 67; colonies, transitory of, 132; evolution of, 35; grazing animals, association with, 35–36; import of, 46; invasive classification of, 128, 129; reproduction of, 131–32; seed longevity of, 121–22; study of, 127; successful, 128–29

Weed Science Society, 39

Welch, Bruce, 258

West, Neil: on competitive native annual species, 93; nitrogen cycling research by, 205; sagebrush, commentary on, 31

western juniper, 257, 270

Western landscape, prior to Europeans, 247–49

Western States Sage Grouse Workshop, Sixth Biennial, 256

Westover-Enslow Expedition, 1934, 187

"wet freeze" seeds, dormancy in, 109–10

wheat: cheatgrass as weed in, 5–6, 22; evolution of, 35; production, expansion in, 45, 143; Russian thistle effect on, 67

wheatgrass: control of, 163; genetically engineered, 182; germination of, 194; introduction of, 64; plants, other mixed with, 195; seeding, fallowing prior to, 166; variations of, 193. See also crested wheatgrass

wheatgrass seedlings: establishment, requirements for, 164; herbicide effect on, 163–64, 165, 166, 167, 174, 182

Whisenant, Steven, 271–72

'Whitmar' bluebunch wheatgrass, 193

wildfire, 1939, 16–20

wildfire elimination policy: burning, prescribed v., 141; consequences of, 9–10; defense of, 49; as stand renewal process type, 267, 276–77

wildfires: cheatgrass, effect on, 135, 138; cheatgrass-fueled, 11, 20, 21, 59, 108, 112, 137, 146, 156, 157, 179, 209, 213, 234, 242–43, 251, 254, 258, 266, 267, 273, 274, 280–81, 284; cheatgrass spread, role in, 48, 111, 136, 251, 253, 261; experiments at sites of, 214, 215–19; grazing management and, 282; hazard of, 112, 242–43, 254; and herbaceous/woody plant maintenance, 195; increased, pattern of, 48; interval, shortened between, 126; plant renewal through, 32; plant succession affected by, 53; pre-European, intervals between, 267–72; prevention of, 251, 253, 264; reintroduction, advocacy of, 49; Russian thistle role in, 68; sagebrush, 10, 108, 187; sagebrush defoliator interaction with, 274; seedings following, 171, 197, 243, 281, 283; sites of, 255; suppression of, 175, 180. See also firefighters